Cool Buildings for Hot Weather

by John Krigger ©2024
Saturn Resource Management Inc.
srmi.biz

Cool-Building Science

Shading and Reflectivity

Air Sealing and Insulation

Air Conditioning and Evaporative Cooling

Ventilation and Dehumidification

Photovoltaic Electricity to Power Cooling

Disclaimer

Please work safely when you attempt the suggestions in this book. Attempt only those tasks for which you are qualified and willing to assume responsibility. We recommend that you hire a professionals for specialized tasks such as air sealing and HVAC work. Mistakes may cause building damage, injury, or even death.

Cool Buildings for Hot Weather
Published by Saturn Resource Management, Inc.
805 North Last Chance Gulch, Helena MT 59601

Order books from **Saturn Resource Management** at *www.srmi.biz*.

Book-Publication Acknowledgments
Cover design by John Krigger and Bjorn Smalbro
Cover artwork by Neil Pangilinan; Cover layout by Bjorn Smalbro
Uncredited art by John Krigger, Bob Starkey, and Wayne Harny
Uncredited photos by John Krigger
Borrowed graphics by organizations, as credited on the graphics
Publishing contributions by Mary Coster and Bjorn Smalbro
Printing by: InterPress Global

Publisher's Cataloging-in-Publication
Krigger, John 1950–
Cool Buildings for Hot Weather John Krigger. 1st ed.
p. cm.
Includes index, glossary, and online resources
ISBN 978-1-880120-00-2
1. Residential buildings—Energy efficiency. 2. Building enclosure—HVAC systems.
TJ163.5.D86K752023 696
QB112-600018

BISAC Subject Codes
ARCHITECTURE / Buildings / Residential
ARCHITECTURE / Methods & Materials
ARCHITECTURE / Sustainability Green Design
HOUSE & HOME / Energy
HOUSE & HOME / Smart Home
SCIENCE / Energy
SCIENCE / Global Warming & Climate Change

Print and Kindle Publishing Codes
Energy efficiency
Building science
Energy conservation
Energy retrofit
Weatherization
Building insulation
Home energy
Residential energy
Renovation

Preface

This Book's Content and Strategy

Many existing residential buildings are poorly air-sealed, insulated, and shaded. Retrofitting existing buildings and constructing efficient new ones is one of the two major topics of this book. The other main topic is HVAC.

Heating, ventilation, and air conditioning (HVAC) includes air conditioning (AC), evaporative cooling, ventilation, and dehumidification. HVAC is a difficult technology to understand — one reason that many HVAC systems function poorly. Another reason for HVAC malfunction and inefficiency is neglected maintenance. Typical HVAC systems provide less comfort, consume more energy, and cost more money than well-designed, well-installed, and well-maintained systems. You'll learn much about HVAC from this book.

Chapter 1 explains the science of cost-effective cooling. You'll understand the remaining chapters better if you learn the science behind my climate-based cooling recommendations.

Many practical ideas, discussed in this book, improve the building enclosure to reduce electricity costs and to increase comfort. In Chapters 2 through 5, I discuss how to shade, air seal, and insulate the building enclosure. These chapters will help you to provide comfort and minimize energy cost, while preserving or improving health and safety.

The biggest learning challenge, posed by this book, is to understand humidity. Controlling indoor relative humidity in hot humid climates is a requirement for modern expectations of comfort and energy efficiency. I've divided the first two chapters equally between understanding heat gain and understanding why humidity is so important to indoor comfort in humid climates.

In Chapter 6, I cover the installation, performance, and maintenance of HVAC systems. In dry climates, most of the AC energy and evaporative-cooling energy goes to reducing indoor temperatures. Strangely enough, evaporative coolers increase relative humidity to provide dry-climate comfort. In humid climates, much of the AC energy and all dehumidifier energy goes to reducing indoor relative humidity.

In Chapter 7, we discuss solar photovoltaic (PV) systems, which generate electricity for cooling. This chapter discusses the different types of PV equipment and PV systems available. Solar photovoltaic systems are logical way to reduce air-conditioning costs because HVAC systems use electricity during a similar time schedule as PV systems produce electricity.

If you follow the suggestions in this book, you'll increase comfort, save money, and promote health, safety, and durability for your building. Thank you for reading.

A Short History of Cooling Devices and Systems

Four thousand years ago, servants fanned wealthy Egyptians with palm leaves. In 180 A.D. the Chinese developed human-powered fans. In the 1880s, American companies began selling fans powered by electric motors.

Ancient cultures cooled buildings with earth tubes, wind towers, and evaporative cooling. The first electric evaporative cooler appeared at the Adams Hotel in downtown Phoenix Arizona in 1916. Before air conditioning, evaporative coolers were popular in the dry climates of the Southwest USA, and many buildings still use them.

American inventor Willis Carrier built the first dehumidifier in 1902 to dehumidify a printing plant in Brooklyn, New York. In 1922, Carrier Engineering Corporation installed the first refrigerated air-conditioning system at the Metropolitan Theater in Los Angeles.

In the early 1940s, easy-to-install room air conditioners appeared in stores. By the 1960s, contractors were installing central air conditioners in millions of buildings. Now almost 90% of all US homes, apartments, and commercial buildings cool with air conditioners.

Dedication

I dedicate this book to my readers who desire comfort during hot weather and to workers who provide summer comfort to their customers.

I recognize the contributions of scientists and engineers; energy auditors and retrofitters; energy raters and home inspectors; and HVAC service techs and installers. Also contractors, manufacturers, and distributors of products that provide comfort with energy efficiency during hot weather. Your work saves electricity and mitigates climate change.Thank you for your contributions to comfort and energy efficiency.

Acknowledgments

I'm grateful to all the businesses and organizations who contributed illustrations for this book. To maintain my credibility, I've never accepted money from manufacturers or suppliers of energy-saving products. Borrowed illustrations in this book promote generic products that save energy and improve comfort. I've credited these images with the names of organizations and/or their Internet addresses. These illustrations and credits don't indicate my endorsement of specific products, services, or organizations. **For more information about the contributing companies and their products, visit their websites.**

Many thanks to my wife Mary, my employees, customers, collaborators, teachers, and friends for their support throughout my 45-year career in energy-conservation. Please accept my appreciation and gratitude.

To everyone who advanced building science and contributed to the substance of *Cool Buildings for Hot Weather*, my sincere thanks!

Best regards — John Krigger; Helena Montana, USA; March 2024

Table of Contents

1 The Building Science of Cooling

This building-science chapter presents a mixture of information about the building enclosure and mechanical cooling systems. This mixture of topics helps you to develop an comprehensive and cost-effective cooling strategy based on building science.

Cooling Strategies

A smart cooling strategy recognizes that there is an ideal sequence of building improvements that produces the best overall outcome. This smart-cooling-improvement strategy prioritizes cost-effective improvements to the building enclosure, before making investments in a new air conditioning system. Here are the two reasons why.

1. An airtight, well insulated, and well shaded building will allow you to buy a smaller air conditioner that costs less than a larger one and is also more efficient.
2. If you buy a new air conditioner before improving the building enclosure, the air conditioner will be oversized, inefficient, and probably incapable of removing enough moisture for comfort in humid weather.

Don't neglect improvements to the building enclosure. If you depend entirely on air conditioning for summer comfort, you will spend more money and enjoy less comfort in the years to come, unless you follow a smart cooling strategy.

Cooling is Different from Heating

Cooling is the most variable type of energy use in buildings. Two similar buildings in the same climate could differ by a factor of 50 for cooling-electricity use. For example, an inefficient home with air conditioning could use 1000 kWh of electricity per summer month, while a neighbor in a well-designed home — with no air conditioning — might use only 20 kWh per month to operate fans. The efficiency of the building enclosure and the expectations of the residents create this large variation.

How to minimize cooling-electricity use — Some procedures to minimize cooling electricity are the same as procedures to minimize heating energy and some are different.

These procedures have similar value for cooling and heating.

- Insulate the attic.

- Air-seal the building enclosure.
- Retrofit existing windows or choose multi-pane low-e windows when you replace them.
- Dehumidify when indoor air is above 60% relative humidity.

These procedures work for cooling only.

- Shade windows with trees and window-shading devices.
- Install a reflective roof.
- Ventilate with cool night air if the outdoor temperature is less than indoors and the relative humidity is moderate.
- Circulate indoor air with fans to create a cooling "wind chill effect."

Table 1-1: *Electricity Consumption for Air Conditioning*

Region & AC Type	Single Family	Multi-family per apartment	Mobile Home
North	**kWh/yr**	**kWh/yr**	**kWh/yr**
Room	200–500	100–300	300–600
Central	900–1400	400–600	1200–1800
South	**kWh/yr**	**kWh/yr**	**kWh/yr**
Room	1100–1500	300–600	1000–1400
Central	3000–4600	1000–1600	2600–3400

From: Lawrence Berkeley National Laboratory, Energy Information Administration, and utility sources. For U.S, households.

Figure 1-1: *Components of Summer Heat Gain*

Component	Low	High
Solar Gains –Windows	15%	35%
Solar Gains – Roof	10%	30%
Solar Gains –Walls	3%	8%
Internal Gains – Heat	10%	25%
Internal Gains – Humidity	5%	15%
Air Leakage – Heat	10%	20%
Air Leakage – Humidity	5%	25%
Delta-T Indoor-Outdoor	4%	12%

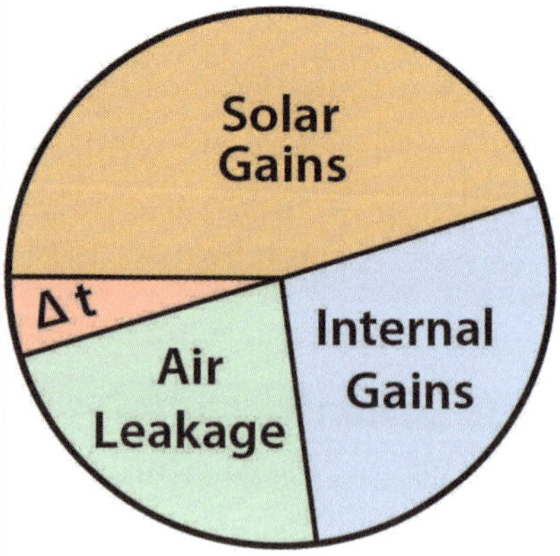

Summer Comfort Principles

The combination of air temperature, radiant temperature, relative humidity, and air movement determine summer comfort. To be comfortable, the human body prefers to cool itself by convection and radiation. Sweating is the body's last resort when radiation and convection aren't sufficient.

Three factors determine how the human body cools itself.

1. Air temperature and airflow determine the rate that a human body loses heat by convection.
2. Radiant temperature determines the rate that a human body loses or gains heat by radiation.
3. In hot weather, your body sheds most of its metabolic heat by evaporating sweat. Relative humidity determines the rate that sweat evaporates. Higher relative humidity slows sweat evaporation, which reduces comfort.

Air temperature and air movement — Air temperature and moving air combine to cool your body. Moving air carries heat away from the skin and increases the evaporation of sweat. Rapid airflow aids air conditioners, evaporative coolers, and whole-house fans to cool residents.

Whenever the outdoor night air is relatively cool and dry, ventilate with outdoor air to remove accumulated heat from the building.

Radiant temperature and comfort — Radiant temperature is almost as important as air temperature for comfort. The radiant temperature of walls and ceilings around you affects your comfort by either absorbing your body's radiant heat or instead by heating your body with radiant heat from your surroundings. Absorbed solar radiation increases wall and ceiling temperatures, sending radiant heat from these surfaces to the building's residents.

Relative humidity and comfort — Relative humidity can increase or decrease comfort, depending on whether your climate is dry or humid. High relative humidity slows down the evaporation of sweat, which reduces comfort in humid climates. In dry climates with low relative humidity, evaporative coolers add humidity, which decreases the temperature of the outdoor air, that the cooler uses to cool the building. See "Evaporative-cooler Operation" on page 186.

Relative humidity is the air's percent of water-vapor saturation at any temperature. Air at 100% relative humidity is saturated and can hold no more water vapor. Warm air can hold more water vapor than cooler air. See "Relative Humidity (Rh), Absolute Humidity, and Comfort" on page 51. See "Dehumidification Strategies" on page 241.

Dew point is the temperature at which condensation begins. At 100% relative humidity, the dew point is the same as the air temperature. Below 100% relative humidity, the dew point is less than the air temperature.

Humidity affects the choice of a building's cooling strategy during hot weather. At low relative humidity and low dew point, evaporative cooling and ventilation are effective cooling methods. Nighttime ventilation performs well below 55% outdoor relative humidity. Many first-world people expect air conditioning to dehumidify, during hot weather when the dew point is above 68°F (20°C) or when the relative humidity outdoors is above 70%. See “How air conditioners fail to dehumidify” on page 198.

At 70% relative humidity or above, the air feels either hot and sticky, or cold and clammy. (Clammy means having wet skin and feeling temporarily cold). Most first-world people aren't comfortable in these changing conditions. Air conditioners must remove moisture from indoor air to achieve acceptable comfort.

Figure 1-2: *Visualizing Summer Heat Gain*

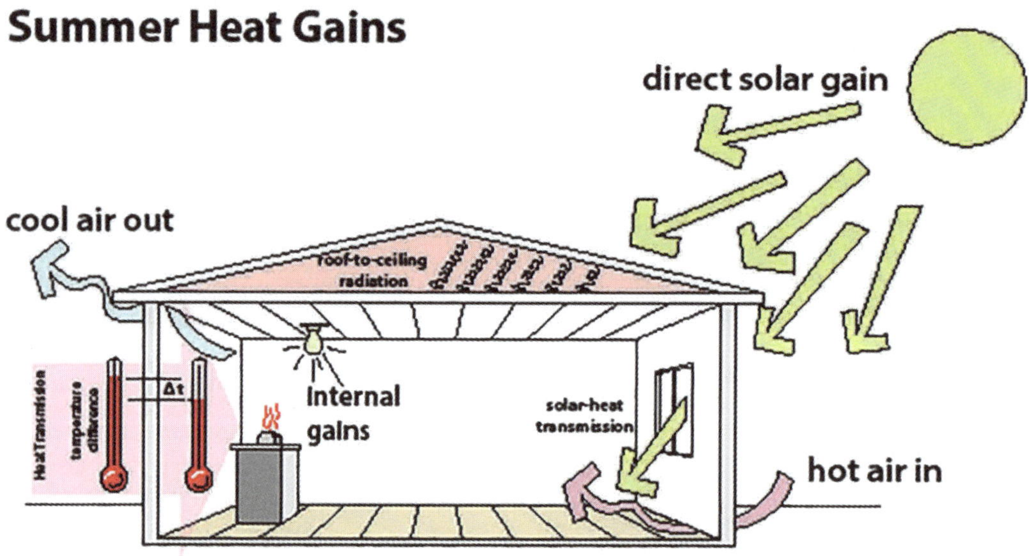

Temperature and Heat

Temperature indicates the average kinetic energy or motion of molecules. Molecules in a solid are stationary, but they vibrate faster as they absorb heat and their temperature increases. A fluid's temperature indicates how fast the molecules in a fluid move around.

Heat flows because of a difference in temperature between two places. *Joules* or *British thermal units (BTU)* measures the heat. A BTU is the amount of heat required to increase a pound of water's temperature 1°F. A BTU or 1055 joules is approximately the amount of heat released by burning one wooden kitchen match.

The number of BTUs of heat that a pound of any material absorbs or releases for each degree of temperature change is called its *specific heat*. The Imperial system of measurement expresses specific heat in *BTUs per pound per degree Fahrenheit* (BTU/lb./°F). The metric (SI) system expresses specific heat in *joules per kilogram per degree Celsius.*

Water has a specific heat of 1 BTU/lb./°F. Aluminum has a specific heat of 0.2 BTU/lb./°F. Thus, takes only 0.2 BTU to raise a pound of aluminum 1°F. If we add one BTU to a pound of aluminum, it becomes 5°F warmer.

Figure 1-3: *Temperature Scales and Reference Temperatures*

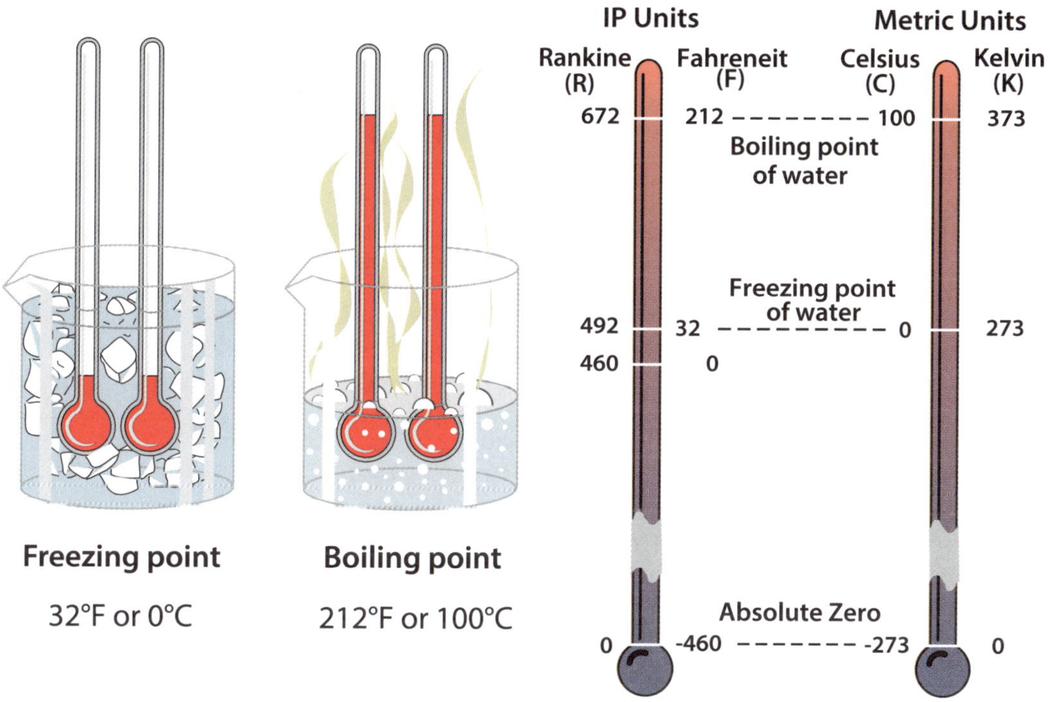

The temperature of a given weight or volume of material represents the amount of energy that the material contains, which we call *enthalpy*. We measure enthalpy in BTUs per pound or joules per kilogram. We also use volumetric heat capacity, with the measurement units explained in this table.

Sensible Heat and Latent Heat

The relationship between water's temperature and its heat content is predictable — add heat to water and the water temperature increases. A pound of water (0.455 kg) becomes one degree Fahrenheit degree warmer with each BTU you add. Add 150 BTUs to a pound of 50°F water, and its temperature increases 150°F to the temperature of 200°F. This *sensible* relationship ends at 212°F or 100°C — water's boiling point. With additional heat, the pound (0.455 kg) of water remains at 212°F or

100°C, while it absorbs 970 BTUs or about 1 megajoule during its complete evaporation into steam. This heat absorbed in evaporation is more than six times the heat that the water absorbed going from 50°F to 212°F (10°C to 100°C).

Figure 1-4: *Water Changing States: Specific Heat and Volumetric Heat Capacity:*

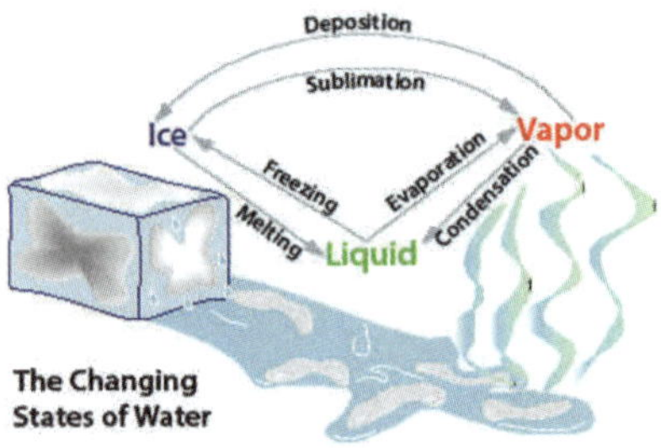

Latent heat is the heat absorbed or released when a material changes phase between a solid and a liquid or a liquid and a gas.

Imperial	Specific Heat	British thermal unit (BTU) / pound (lb) • degree Fahrenheit (F°)
	Volumetric Heat Capacity	British thermal unit (BTU) / cubic feet • degree Fahrenheit (F°)
Metric	Specific Heat	joule (J) / kilogram (kg) • degree Celcius (C°)
	Volumetric Heat Capacity	joule (J) / cubic meter • degree Celcius (C°)

Specific heat and volumetric heat capacity are two ways of expressing the embodied heat of a material.

We call this unexpected or hidden heat that a material releases or absorbs as a changes form — *latent heat*. Our pound (0.455 kg) of liquid water vaporized when we added 970 BTUs, which is water's *latent heat of evaporation*. If that vapor or steam condenses back to a liquid in a radiator, the steam releases 970 BTUs or 1,023,350 joules or 1023 kilojoules (1023 kj) or 1.23 megajoules (1.23 mj). Steam heating uses latent heat to heat buildings.

Our one pound of water (0.455 kg) goes through similar changes if we cool it. The water loses 1°F for every BTU we remove, until the water reaches its freezing point: 32°F or 0°C. We must then remove 144 BTUs — water's *latent heat of fusion* — to turn the pound of water into ice at 32°F or 0°C. If we reverse the process, it requires 144 BTUs or 151,920 joules (152 kj) of heat to melt this pound of ice into liquid water again at the same temperature: 32°F or 0°C.

How we use latent heat — Steam-heating systems, air conditioners, refrigerators, and human bodies use latent heat to move heat energy from one place to another. In steam heating systems, water vaporizes at a boiler and condenses back to a liquid in radiators because the radiators are cooler than the steam.

Air conditioners use a fluid called a *refrigerant* that vaporizes at the *evaporator*, and cools off dramatically during that vaporization from ~40°C or 104°F to ~7°C or 45°F. (~ means "approximately") The cool evaporator coil absorbs heat from building's air flowing through it. Then the compressor compresses the cool refrigerant gas into a pressurized hot gas. The hot gas (~45°C or 113°F) moves outdoors to a *condenser*, where it condenses back to a liquid, releasing its latent heat into the outdoor air that a fan moves through the condenser coil.

Heat flows from a high temperature to a lower temperature. Thus the outdoor air, moving through the condenser, must be cooler than the condenser for this heat-transfer to succeed. Heat only flows with a temperature difference, which we abbreviate ΔT. If the outdoor air is and the condenser is 40°C or 104°F, then the ΔT is 5°C or 9°F. Thus, heat flows from the condenser to the outdoor air.

When the temperature around you increases, your brain signals your heart and blood vessels to move more blood to your skin. This increased circulation increases your skin temperature, which expels sensible heat and evaporates sweat from your body. Your hot skin (35°C or 95°F) expels heat, if the surrounding air is cooler than your skin. Sweat evaporates and takes body heat with it — specifically 970 BTUs per pound or 1023 kJ per kilogram for each pound or kilogram of sweat that evaporates.

Figure 1-5: *Quantifying Sensible Heat and Latent Heat*

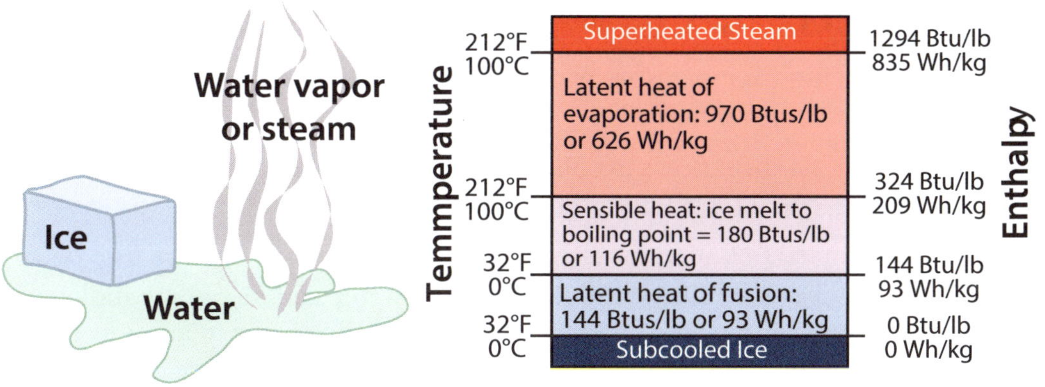

Summer Heat Gain

Heat gain is collected indoor heat, during the summer cooling season. The capacity of our cooling equipment, needed to counteract the heat gain, is called the *cooling load*. There are four types of heat gains in a building: solar heat, internal heat, air leakage, and temperature-driven heat transmission. For information about the intensity of solar radiation near you, see "Solar-Irradiation Map" on page 293.

Solar gain — In most climates, solar heat is the largest heat gain, contributing more than 50% of the unwanted heat that accumulates indoors. Solar energy falling on the roof and coming through the windows accounts for most of this heat gain. Walls are less important as a source of solar gains. Thus, shade windows and roofs, and use reflectivity whenever possible to minimize solar gain. See "Shade and Reflectivity" on page 35.

Internal gains — Internal gains usually contribute approximately 20% of the total summer heat gain. Besides cooking, internal gains include waste heat from lights, refrigerators, water heaters, and other appliances. People's body heat also contributes. Efficient appliances and lights produce less waste heat than less-efficient appliances and lights.

Air leakage — Air sealing helps reduce summer heat gain. Air leakage contributes approximately 20% to summer heat gain. Air leakage allows hot outdoor air to leak into the home, and cooler indoor air to leave. Seal leaks to minimize the heat gain from air leakage. See "Air-Sealing Methods and Materials" on page 113.

Heat transmission — Heat transmission represents approximately 10% or less of the total cooling load, except in very hot weather where heat transmission could rise to 20%. Insulate roofs, attics, and walls to minimize summer heat transmission.

Types of Heat Flow

Heat travels from high-temperature areas to low-temperature areas in three ways: *conduction, convection,* and *radiation.*

Conduction is the way heat flows in solids. Heat flows through a solid when its stationary molecules vibrate, spreading heat throughout the solid material by vibration.

Convection is the way heat flows in fluids, such as water and air, where molecules move around. Winds and ocean currents transmit heat from warm areas to cooler areas around the globe by convection.

Radiation is the way heat flows in a line of sight between bodies with different temperatures. Heat radiates between all objects that can "see" each other through space or through a gas, like air.

Conduction

Conduction is the most familiar and predictable type of heat flow. Heat conducts through solid objects and between solid objects touching one another. When you grab a hot frying pan, your hand feels hot because the pan's heat conducts into your hand. As a solid object becomes warmer, the molecules vibrate, bump, and rub against each other more vigorously, passing heat through the solid material. This heat flow always moves from higher temperature to lower temperature. When a squirrel lays down on a shaded slab of concrete, the squirrel is conducting excess heat from its body to the concrete.

Convection

Convection is heat transferred by a moving fluid like air or water. Convection moves the fluid because temperature and density differences in the fluid makes one part of the fluid heavier than the other. *Density* measures how many pounds per cubic foot or kilograms per cubic meter that a fluid weighs. Warmer fluid-segments rise, and cooler denser fluid-segments fall.

As you add heat to one part of a fluid, the molecules there race around faster, colliding more often and driving each other farther apart. The greater spacing caused by the collisions reduces the density of the heated fluid. The hotter fluid rises, and the cooler, denser fluid falls. The old saying, "heat rises," is actually incorrect because heat moves in all directions. The truth is, a fluid's hottest molecules rise to the top.

Most convection, which is relevant to residential energy, occurs between a surface and a fluid — between your skin and a cool wind, for example. Hot combustion gases convect against the metal surfaces of a heat exchanger, transferring heat to the metal. Blow on your coffee to cool it, and you are using convection.

See "Night Ventilation with and without a Fan" on page 178, for information about convective cooling.

Radiation

Radiant heat flies through space from one object to another. The sun's radiant heat on your face or a cold window pane sucking radiant heat from the back of your neck are two examples of how you feel radiant heat. Objects, within a line of sight of one another, exchange heat radiation continuously. In this exchange, there is a net heat flow from the high-temperature object to the low-temperature object. The high temperature object gets cooler and the low-temperature object gets warmer as a result of this radiant-heat exchange.

Figure 1-6: *Ideal Orientation and Ideal Shading*

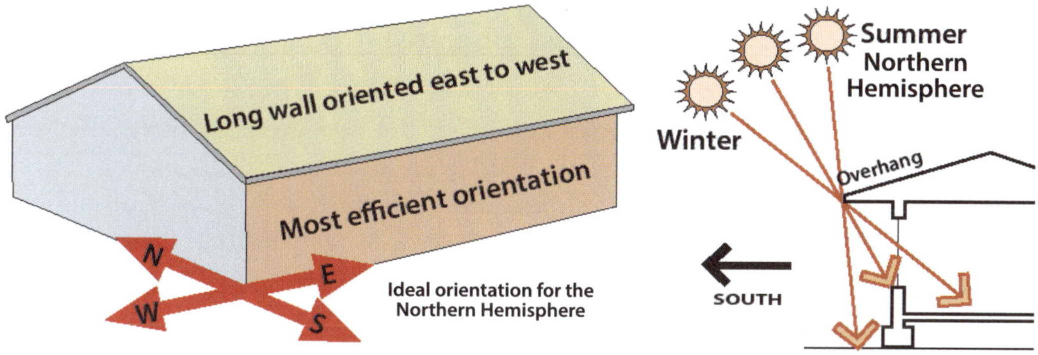

Solar radiation — About 49% of the sun's energy comes to earth as solar heat, 46% comes as visible light, and the remaining 5% is ultraviolet radiation. The earth's ozone layer filters most of the ultraviolet (UV) radiation, which is fortunate, because life on earth can't survive constant bombardment by unfiltered UV rays. The atmosphere absorbs 10% to 20% of incoming solar radiation. The earth reflects 35% to 40% of the solar radiation. The earth absorbs the remaining 40% to 55%.

The sun's rays impose their maximum power on surfaces at a 90° angle from their path (90° incidence angle). As the incidence angle varies from 90°, solar radiation's density and power decrease because the rays spread farther apart. Also, when the sun is directly overhead, solar rays are relatively more intense because they travel a shorter distance through earth's heat-absorbing atmosphere. When the sun is lower in the sky, the rays are less intense because they travel relatively farther through the earth's atmosphere. Part of the ray's power is depleted, striped off by earth's atmosphere.

Solar movement and direction in earth's two hemispheres — In both the Northern and Southern Hemispheres, the sun rises in the East and sets in the West. In both hemispheres, the east wall absorbs solar heat in the morning, and the west wall absorbs heat in the afternoon.

However most of the sun's solar heat falls on the south side of a building in the Northern Hemisphere. In the Southern Hemisphere, most of the sun's solar heat falls on the north wall. These facts require that you shade the correct wall surfaces — North or South — depending on which hemisphere you live in. See "Reducing Heat Gain" on page 35.

Earthly infrared radiation — The amount of infrared radiation emitted by an object depends on its temperature and its surface characteristics. Warmer objects emit radiant heat more rapidly than cooler ones. Most common objects emit infrared radiation readily. The exceptions are objects with shiny metalized surfaces. The numerical rating of *emittance* is based on a theoretically perfect emitter having an emittance of 1, or 100%. Most common objects have emittance of 85% to 95%. Metalized surfaces like aluminum foil and galvanized steel have emittance of 5% to 20%.

When solar or infrared heat rays strike an object, the rays are absorbed, reflected, and transmitted, depending on surface properties of the stricken object. These properties are called *absorbance*, *reflectance*, and *transmittance*. Because all radiant heat striking the object is either absorbed, reflected, or transmitted, the values for the object's properties, added together, equal 1 or 100%.

Figure 1-7: *Behaviors of Radiation / Concentration of Solar Radiation*

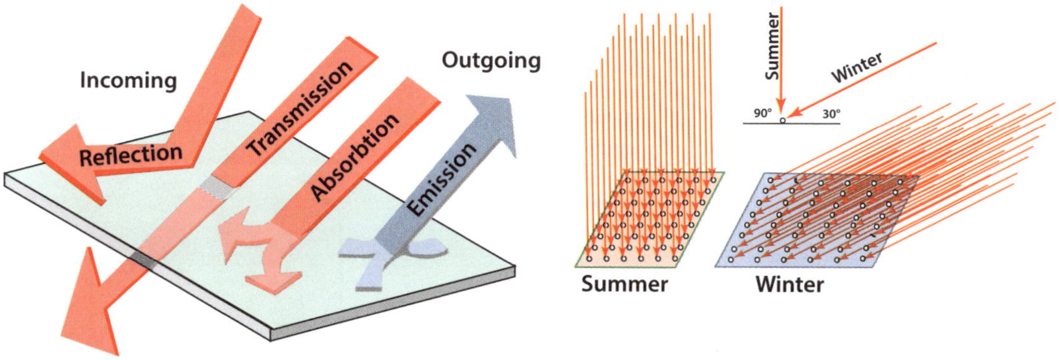

Radiation characteristics of surfaces — Bright white and polished metal objects **reflect** 80% to 98% of incoming solar energy. Other objects **absorb** 40% to 95% of incoming solar radiation depending on their color — darker colors absorb a greater percentage than lighter colors. Glass has special qualities, absorbing or reflecting only 10% to 20% of incident solar radiation and transmitting 80% to 90% of incident solar radiation.

Most common materials, even glass, absorb almost all incoming infrared heat from earthly objects. The exceptions are opaque polished metal surfaces that reflect most infrared radiation. Polished aluminum, steel, and certain metal alloys of tin, silver, and nickel are the only common substances that reflect both solar and infrared radiation.

Glass' solar transmittance and infrared absorptance explain how glass traps heat inside a greenhouse, causing the *greenhouse effect*. Solar heat passes through a greenhouse window and strikes objects inside, warming them. The warm objects in the greenhouse emit infrared heat, which the glass absorbs. Part of that absorbed heat escapes outside, and part radiates and convects back inside, reheating the greenhouse.

Metalized glass coatings called low-e (low emissivity) decrease the glass's heat emittance. Low-e glass has a greater thermal resistance than un-coated glass because of its low emittance. Low-e coatings also reduce the glass's solar transmittance by reflecting part of the entering solar radiation.

See "Thermal Transmittance (U-factor or U-value)" on page 59.

Figure 1-8: *Solar Position in the Sky / The Greenhouse Effect*

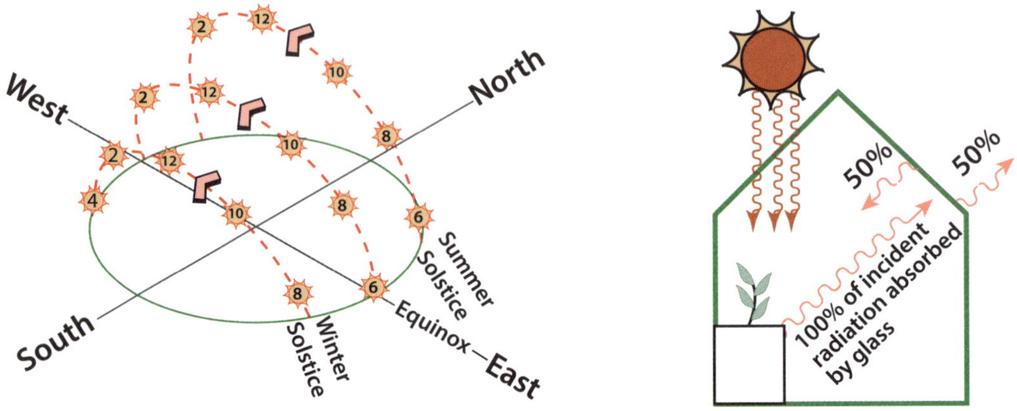

Heat Transmission

The collective action of conduction, convection, and radiation in a building assembly is called *heat transmission*. The driving force for heat transmission is the temperature difference (ΔT) between the indoors and outdoors. The greater the temperature difference between indoors and outdoors, the faster that the heat transmits. We quantify heat transmission with the *U-factor* or *U-value*, which measures heat flow in units that include surface area, temperature difference, and time.

Figure 1-9: *Comfort Factors and Cooling the Body*

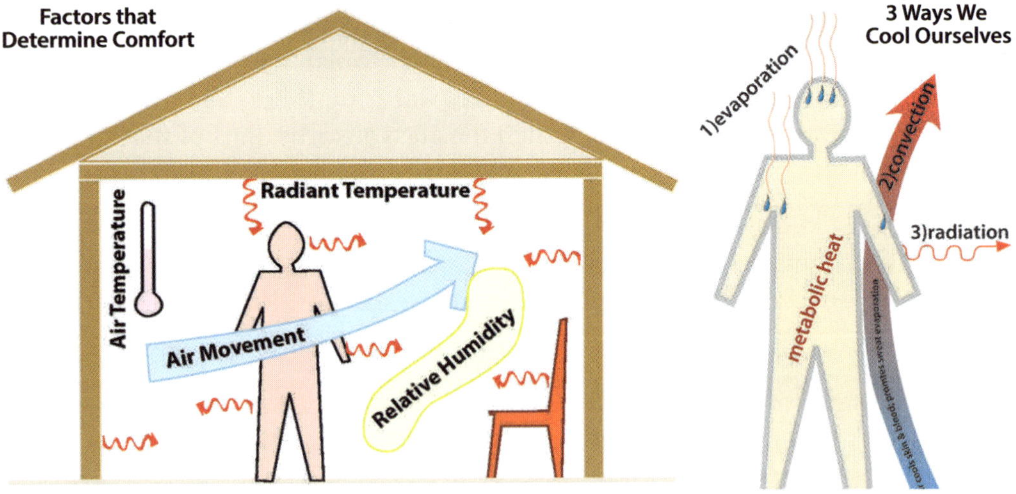

Energy, Comfort, and Climate

The outdoor climate has the most influence on human comfort of any common factor. The temperature, relative humidity, solar radiation, precipitation, and wind affect the immediate comfort of people outdoors. The conditions outdoors determine how much *space-conditioning* a building must provide for acceptable indoor comfort.

We feel comfortable when we are in a state of *thermal equilibrium or homeostasis* with our environment without having to sweat excessively. In thermal equilibrium or homeostasis, the human body loses as much heat to its surroundings as it gains from metabolism.

Air temperature is usually the primary factor determining comfort, but the temperature of walls, ceilings, floors, and furnishings, called *radiant temperature,* is also important. Together these two temperatures create a composite effect that determines comfort, along with Rh, during hot weather.

Air Temperature

Outdoor air temperature is the most important climate factor in determining heating energy consumption but a less important factor in determining cooling energy consumption, compared to solar heat gain. Outdoor temperature is always changing according to the season, the weather, and the time of day. For information about he average outdoor temperature near you, see "Annual-Average-Temperature Map" on page 291.

Summer comfort inside buildings depends on these factors.

- The thermostat setpoint for the air conditioner or evaporative cooler
- Energy cost: What the building owner and residents can afford
- Outdoor temperature and relative humidity
- Thermal resistance and airtightness of the building-enclosure
- Heat-tolerance of the residents

Radiant Temperature

The temperature of walls, ceilings, floors, and furnishings, called *radiant temperature,* is also a very important comfort factor. Together, the air temperature and radiant temperature create a combined effect that determines summer comfort. A low summer radiant temperature can counteract the comfort effects of high air temperature.

Air Movement

Moving air is important for summer comfort — rapidly moving air increases bodily heat losses through convection of air against the skin. Air currents cool the skin and its surface capillaries, and also promotes sweat evaporation. Air circulation also helps avoid air stagnation, condensation, and room-temperature variations.

Relative Humidity (Rh)

Relative humidity (Rh) measures how saturated the air is with a percentage of water vapor. Completely saturated air has 100% Rh. *Relative humidity* is a very important summer comfort factor, because it determines how rapidly sweat can evaporate from your skin.

Warmer air can hold more moisture than cooler air. For example, if outdoor air at 91°Fand 50% Rh cools to 70°F (21°C), the relative humidity rises to 100%. This cooled air contains exactly the same amount of water vapor, but at 91°F (33°C), the air is only 50% saturated, while at 70°F (21°C), it is 100% saturated. Saturated air (100% Rh) at 91°F (33°C) holds twice as much water vapor as 70°F (21°C) saturated air. For information about the average Rh near you, see "Yearly-Average-Relative-Humidity Map" on page 292. See also "Cooling Strategy for Humid Climates" on page 195.

Rh and climate — The outdoor relative humidity depends on rainfall, nearness to bodies of water, cloudiness, windiness, and other environmental factors. Indoor humidity depends on these factors.

- Amount of moisture generated within the building
- Ability of an air conditioner or dehumidifier to control Rh
- Temperature of indoor air
- Temperature and humidity of outdoor air
- Rate at which outdoor air exchanges with indoor air.

Rh and dew point — The *dew point* is the temperature where a particular indoor or outdoor air mass condenses water vapor to liquid water. If the temperature of an outdoor air mass falls below its dew point, it rains, or the air mass deposits its condensation (dew) on the coolest outdoor surfaces.

When humid air touches a cool indoor surface, tiny bead condensation form on its surface if the surface temperature is below the dew point. Condensation is undesirable because it damages building materials and may allow microbes and insects to colonize those materials. Indoor humidity should be less than 60% during hot weather to prevent condensation on surfaces cooled by an air conditioner.

Equilibrium moisture content (EMC) and building durability **—** The moisture content of building materials, called equilibrium moisture content (EMC) is directly related to the relative humidity of the air surrounding the building materials. At a particular EMC, the material may suffer damage from rot, insect infestation, freezing, or *efflorescence*.

Rh and comfort **—** The above heat-index chart provides guidance about the danger of high temperatures and high relative humidity. High relative humidity suppresses sweat's evaporation from human skin, which inhibits the body's evaporative-cooling function. Keeping indoor relative humidity at less than 60% during the summer promotes comfort and prevents condensation on the cool surfaces of an air-conditioned building.

The temperature and relative humidity of the air surrounding a person determines not only comfort, but also a person's health and safety. There is a wide variety of human tolerance to high temperature and high relative humidity. This variation depends on an individual's heat-and-humidity tolerance, based on their long-term exposure to heat and humidity.

Figure 1-10: *Relative Humidity, Material EMC, and Durability*

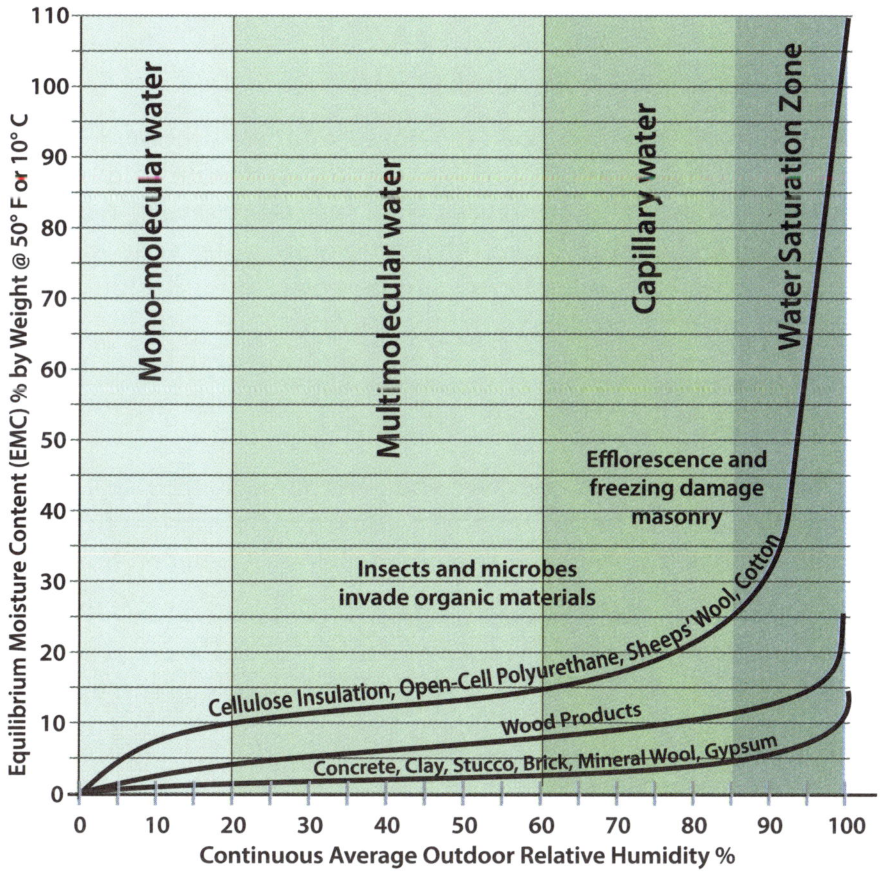

The Physiology of Heat Tolerance — Acclimatization

Acclimatization is the human body's physiological adaptation that results from repeated exposure to heat, humidity, and solar radiation, especially during physical exertion. These physiological changes result from blood vessel expansion, called *vasodilation*. Vasodilation increases blood flow, skin temperature, and blood cooling. Sweating also increases and begins earlier as a person acclimatizes to hot, humid weather. These are the beneficial adaptations of acclimatization.

- Increased skin temperature sheds more heat by radiation and convection
- Sweating begins sooner
- Sweat production increases
- The body loses less electrolytes as the sweat evaporates

Acclimatization's increased sweating efficiency promotes more stable blood circulation, increased blood flow to the skin, reduced core temperature, and reduced heart rate, compared to a non-acclimatized person. Increasing blood flow at the skin's surface cools the blood and evaporates sweat efficiently. This removes the blood's metabolic heat effectively, before hot blood circulates back to the heart.

Manual laborers and athletes should learn how temperature, humidity, and solar radiation affect their bodies. If you exert yourself in a hotter, more humid, or sunnier climate than you are acclimatized to, you may suffer heat ailments if you don't rest more often and drink more water than usual.

Workers and athletes, who developed the physical fitness necessary to exert themselves in a cooler climate, can acclimatize to a hotter climate within 7 to 14 days. During that time, they should increase their exertions slowly, take more rests, drink more water, and consume electrolytes. If you increase your water consumption without consuming essential electrolytes, a variety of health problems may occur. Some commercial drinks contain electrolytes. You can also ingest electrolytes as pills and salt candy, or add electrolyte supplements to your drinking water.

Measuring Heat's Effect on the Body

There are two accepted ways to measure and communicate the effect of heat on the human body. The *Heat Index* is the most common measurement because weather reporters use it to communicate the combined effects of heat and humidity.

The second heat measurement is the *Wet Bulb Globe Temperature (WBGT). WBGT* accounts for a greater number of weather factors plus the average acclimatization of people living in a particular region. The WBGT is more useful to workers, athletes, managers, coaches, and the military, compared to the Heat Index. When a large number of people exert themselves in hot weather, some group members may become sick or even die from heat stroke. The U.S. military created WBGT to reduce the risk of heat-related illnesses in 1950.

Figure 1-11: *Heat Index: the Combined Effect of Temperature and Relative Humidity*

Temperature / Relative humidity	80 °F (27 °C)	82 °F (28 °C)	84 °F (29 °C)	86 °F (30 °C)	88 °F (31 °C)	90 °F (32 °C)	92 °F (33 °C)	94 °F (34 °C)	96 °F (36 °C)	98 °F (37 °C)	100 °F (38 °C)	102 °F (39 °C)	104 °F (40 °C)	106 °F (41 °C)	108 °F (42 °C)	110 °F (43 °C)
40%	80 °F (27 °C)	81 °F (27 °C)	83 °F (28 °C)	85 °F (29 °C)	88 °F (31 °C)	91 °F (33 °C)	94 °F (34 °C)	97 °F (36 °C)	101 °F (38 °C)	105 °F (41 °C)	109 °F (43 °C)	114 °F (46 °C)	119 °F (48 °C)	124 °F (51 °C)	130 °F (54 °C)	136 °F (58 °C)
45%	80 °F (27 °C)	82 °F (28 °C)	84 °F (29 °C)	87 °F (31 °C)	89 °F (32 °C)	93 °F (34 °C)	96 °F (36 °C)	100 °F (38 °C)	104 °F (40 °C)	109 °F (43 °C)	114 °F (46 °C)	119 °F (48 °C)	124 °F (51 °C)	130 °F (54 °C)	137 °F (58 °C)	
50%	81 °F (27 °C)	83 °F (28 °C)	85 °F (29 °C)	88 °F (31 °C)	91 °F (33 °C)	95 °F (35 °C)	99 °F (37 °C)	103 °F (39 °C)	108 °F (42 °C)	113 °F (45 °C)	118 °F (48 °C)	124 °F (51 °C)	131 °F (55 °C)	137 °F (58 °C)		
55%	81 °F (27 °C)	84 °F (29 °C)	86 °F (30 °C)	89 °F (32 °C)	93 °F (34 °C)	97 °F (36 °C)	101 °F (38 °C)	106 °F (41 °C)	112 °F (44 °C)	117 °F (47 °C)	124 °F (51 °C)	130 °F (54 °C)	137 °F (58 °C)			
60%	82 °F (28 °C)	84 °F (29 °C)	88 °F (31 °C)	91 °F (33 °C)	95 °F (35 °C)	100 °F (38 °C)	105 °F (41 °C)	110 °F (43 °C)	116 °F (47 °C)	123 °F (51 °C)	129 °F (54 °C)	137 °F (58 °C)				
65%	82 °F (28 °C)	85 °F (29 °C)	89 °F (32 °C)	93 °F (34 °C)	98 °F (37 °C)	103 °F (39 °C)	108 °F (42 °C)	114 °F (46 °C)	121 °F (49 °C)	128 °F (53 °C)	136 °F (58 °C)					
70%	83 °F (28 °C)	86 °F (30 °C)	90 °F (32 °C)	95 °F (35 °C)	100 °F (38 °C)	105 °F (41 °C)	112 °F (44 °C)	119 °F (48 °C)	126 °F (52 °C)	134 °F (57 °C)						
75%	84 °F (29 °C)	88 °F (31 °C)	92 °F (33 °C)	97 °F (36 °C)	103 °F (39 °C)	109 °F (43 °C)	116 °F (47 °C)	124 °F (51 °C)	132 °F (56 °C)							
80%	84 °F (29 °C)	89 °F (32 °C)	94 °F (34 °C)	100 °F (38 °C)	106 °F (41 °C)	113 °F (45 °C)	121 °F (49 °C)	129 °F (54 °C)								
85%	85 °F (29 °C)	90 °F (32 °C)	96 °F (36 °C)	102 °F (39 °C)	110 °F (43 °C)	117 °F (47 °C)	126 °F (52 °C)	135 °F (57 °C)								
90%	86 °F (30 °C)	91 °F (33 °C)	98 °F (37 °C)	105 °F (41 °C)	113 °F (45 °C)	122 °F (50 °C)	131 °F (55 °C)									
95%	86 °F (30 °C)	93 °F (34 °C)	100 °F (38 °C)	108 °F (42 °C)	117 °F (47 °C)	127 °F (53 °C)										
100%	87 °F (31 °C)	95 °F (35 °C)	103 °F (39 °C)	112 °F (44 °C)	121 °F (49 °C)	132 °F (56 °C)										

USA National Oceanic and Atmospheric Administration (NOAA) Heat-Index Table

Heat index — Heat Index is a good indication of how hot a person feels during hot weather. The higher the relative humidity, the hotter a person feels at a specific dry-bulb temperature. The Heat Index derives from a variety of climatic factors and assumptions. The assumed local weather conditions include shade and a breeze. The person, assumed to experience the weather conditions, is of average height, healthy weight, and walks at the rate of 3 mph or 5 km/h. The most important factors are dry-bulb temperature and relative humidity. Heat Index is much easier to understand than WBGT, but it's inadequate to evaluate the risk of exertion in hot and humid conditions.

Figure 1-12: *Instruments to Measure Wet-bulb, Dry-bulb, and Globe Temperatures*

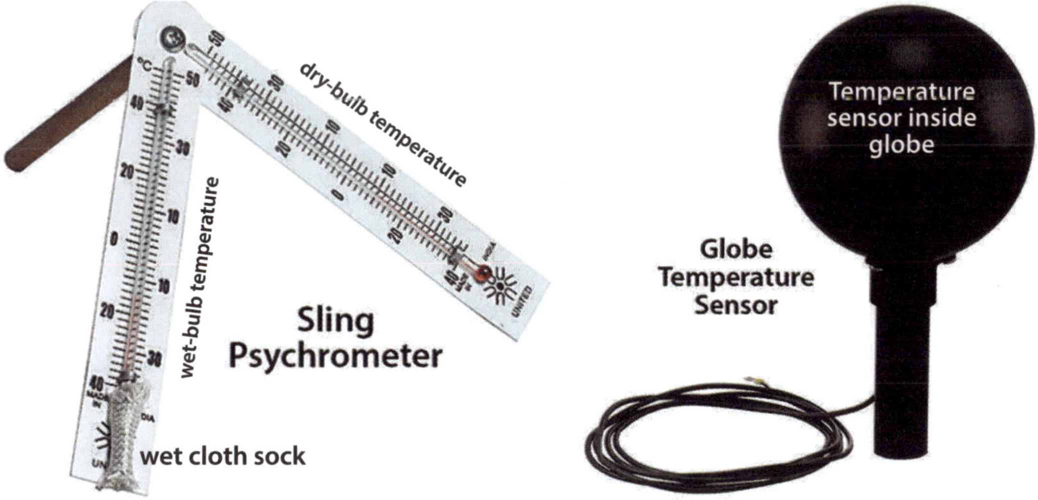

Wet-bulb globe temperature (WBGT) — Weather experts calculate WBGT using four factors: dry-bulb temperature, wet-bulb temperature, climate category, and globe temperature. The three WBGT climate categories create a scale from 1 to 3 of increasingly hot-humid-sunny climates. The categories consider wet-bulb temperature, dry-bulb temperature, solar radiation, and also the typical acclimatization of the region's population. WBGT is a better risk-analysis tool than Heat Index for athletes, manual laborers, solders, supervisors, and their organizations.

A 6-inch or 75 mm diameter black copper globe with its heat sensor inside measures a combination of dry-bulb temperature, wet-bulb temperature, and radiant temperature. The globe-temperature device is expensive, which is one reason why the WBGT is less-often reported by the media. The WBGT equation is a simple weighted average of these 3 temperatures.

WBGT Temperature-Weighted Average

WBGT = 0.7 x wet-bulb + 0.2 x globe + 0.1 x dry-bulb

See "Estimating Wet-Bulb Globe Temperature" on page 295, "Advice for Coping with WBGT Conditions" on page 296, and"World Wet-Bulb-Globe-Temperature Map" on page 294 for more information.

Heat-related illnesses — People often become ill when the temperature and relative humidity around them are in the **high-risk** or **extreme-risk** zones on the "World Wet-Bulb-Globe-Temperature Map" on page 294. By order of increasing severity, the heat illnesses are these.

1. *Heat rash* — Red clusters of small blisters that look like pimples on the skin, usually on the neck, chest, groin, or in elbow creases
2. *Heat cramps* — heavy sweating during intense exercise; muscle cramps, pain, or spasms
3. *Heat exhaustion* — heavy sweating; cold and clammy skin; fast, weak pulse; muscle cramps; nausea or vomiting; dizziness or headache; fainting
4. *Heat stroke* — Red, dry, or damp skin; fast pulse; dizziness, confusion; nausea, sweating cessation; loss of consciousness

For more information on heat illness, see "Heat Illnesses: Symptoms & Solutions" on page 290 and "Advice for Coping with WBGT Conditions" on page 296.

Figure 1-13: *Water and Earth's Environment*

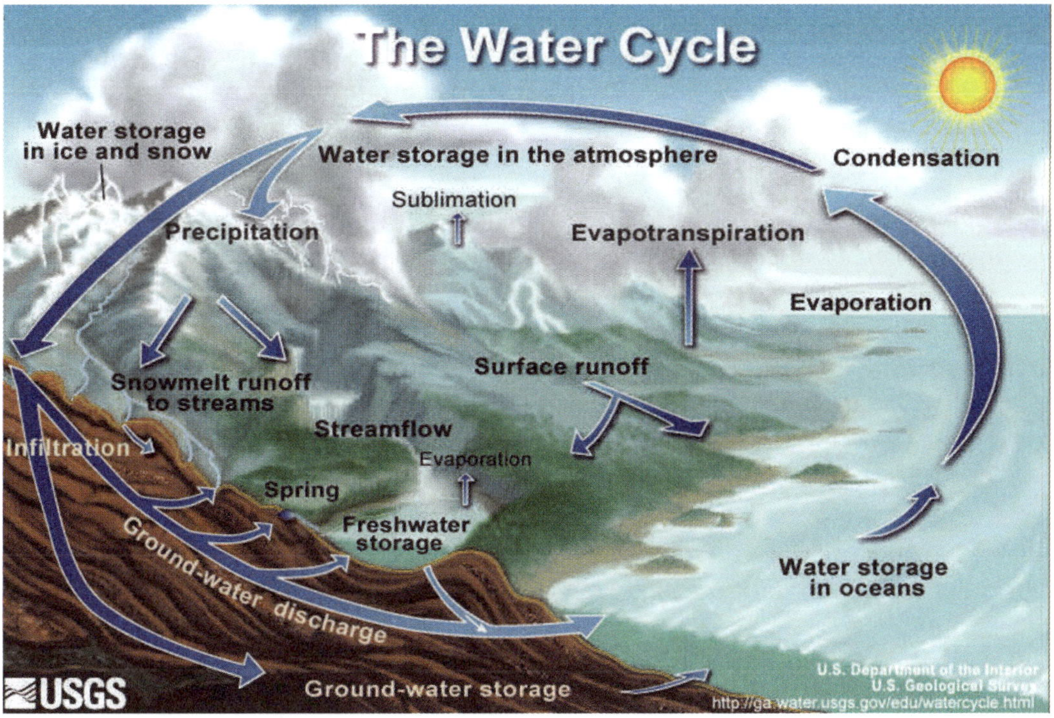

Moisture's Behavior

Moisture together with fire are the two most potent building destroyers. You should understand moisture's behavior in and around buildings and how it affects comfort, energy costs, and your home's durability. This section explains how water and water vapor affect buildings. See also "Moisture Management" on page 47, and "Dehumidification" on page 239

Moisture Movement through Buildings

Moisture enters buildings and moves through them as both liquid water and water vapor. Moisture movement happens in four ways.

1. *Liquid flow.* Driven by gravity or pressure differences, water flows into a building's holes and cracks. Roof leaks and plumbing leaks can deposit large amounts of water into a building. See "Foundation Water Leakage: Prevention and Treatment" on page 48.

2. *Capillary seepage.* Liquid water creates a suction of its own as it moves through tiny spaces within and between building materials. This capillary suction draws water up from the ground. Seepage also redistributes water from leaks, spills, and condensation. See "Solutions for foundation water leakage" on page 49.

3. *Air movement.* Air movement carries water vapor into and out of buildings. Air-pressure differences drive this air movement, and holes in the building enclosure are the leaks. If the air temperature falls below the dew point, condensation occurs on surfaces or within building materials, which is a common cause of damage to buildings. See "Air Leakage and Duct Leakage" on page 46.

4. *Vapor diffusion.* Water vapor moves through building assemblies depending on their *permeance* and the vapor-pressure difference across the assembly. See "Controlling Water Vapor" on page 51.

Water-vapor diffusion — When water vapor moves through a solid material, we call this *vapor diffusion*. Materials vary in their permeability to water vapor. Porous materials like brick and insulation have a high permeability and transmit water vapor rapidly. House wrap is a specially designed material that repels liquid water while letting water vapor through because of house wrap's high vapor permeability. Plywood and drywall have a medium permeability. Metals and plastic films, often called *vapor barriers*, restrict vapor diffusion to a negligible amount.

A force called *vapor pressure* drives vapor diffusion. A difference in the amount of water vapor between two bodies of air, separated by some barrier, creates vapor pressure. The greater the vapor pressure, the faster that water vapor diffuses through the porous building materials that separate the moister air mass from the drier air mass. See "Preventing Moisture Problems" on page 52.

Figure 1-14: *Refrigeration Cycle / Split-System Air Conditioner*

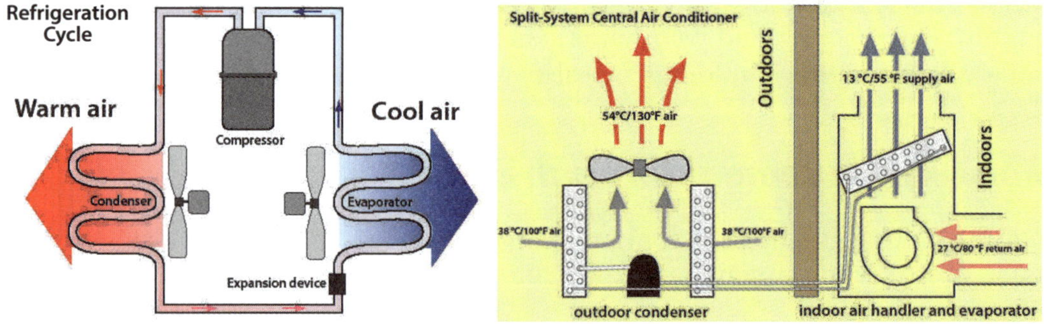

The Refrigeration Cycle for Air Conditioning

Refrigerators, air conditioners, heat pumps and dehumidifiers move heat from one location to another using latent heat. The refrigerated air-conditioning system cools indoor air and heats outdoor air.

Liquid *refrigerant* vaporizes in the *evaporator* of a refrigeration system, sucking heat out of the evaporator's metal. The evaporator coil becomes cold and removes heat from the warm indoor air moving through it. The vaporized refrigerant carries the heat it collected from the air to the *compressor* outdoors. The compressor heats the gaseous refrigerant by compressing it. The compressor then forces the now-hot refrigerant into the *condenser*. There the refrigerant condenses back to a liquid, releasing its latent heat and heating the condenser coil. Then the condenser coil has a higher temperature than the outdoor air moving through it, so heat flows from the condenser to the cooler outdoor air circulating through its coil.

The liquid refrigerant collects in the condenser and flows back toward the evaporator indoors, pushed by the compressor's pressure. The *expansion device*, which is like a spray nozzle, sprays liquid refrigerant into the evaporator, where it evaporates again. The evaporating refrigerant removes heat from the evaporator coil, and the refrigeration cycle repeats itself. See "Air Conditioners" on page 189.

2 Reducing Heat Gain

Reduce solar-heat gain — Solar radiation is almost always the most potent force that heats a building in hot weather. Thus, shade and reflectivity are the most important solutions to provide comfort and minimize electricity costs. See "Solar-Irradiation Map" on page 293 for information about the intensity of solar radiation in your climate.

Control relative humidity — In humid climates, minimizing relative humidity may be as important as shading and reflectivity. If an air conditioner or dehumidifier must dry the air to create comfort, this may be difficult and expensive. Passive moisture control, such as drainage and humidity-control, is essential to provide comfort and minimize electricity costs in humid climates. See "Yearly-Average-Relative-Humidity Map" on page 292 for information about average Rh in your climate. If the indoors is uncomfortably dry, excessive air leakage from outdoors is usually the cause. See "Finding Air Leaks" on page 99.

Shade and Reflectivity

When solar radiation falls on the roof and enters through windows, these two building components account for most of the solar heat that accumulates indoors. Walls are less important as a source of solar gain.

Thus the most important places to use shading and reflectivity are on the roof and windows. Energy conservation measures that block the sun, before it strikes the roof or windows, are the most effective. Trees and other plants that provide shade are the best long-term investment for minimizing cooling costs.

Shading with Landscape

A well-planned landscape can reduce an unshaded building's summer air-conditioning costs by 15% to 50%.

Notice the coolness of parks and forests when compared to unshaded city streets. Cool microclimates use trees and shrubs to minimize the temperature and the absorbed solar heat.

Trees provide shade and *evapotranspiration.* Trees use evapotranspiration to evaporate water like we evaporate sweat. Shade and evapotranspiration can reduce outdoor air temperatures as much as 9°F or 5°C compared to treeless areas. Since cool air falls toward the ground, temperatures directly under trees can be up to 25°F or 14°C cooler than air temperatures at nearby unshaded streets.

Planting trees is ten times more cost-effective than building new power plants for summer cooling, according to the Lawrence Berkeley National Laboratory (LBNL).

Figure 2-1: *Trees Provide Shade*

Studies by the LBNL found summer daytime air temperatures 3°-to-6°F (1.5°-to-3°C) less in neighborhoods with mature trees than in neighborhoods with no trees. Large urban parks are up to 7°F (4°C) cooler than nearby neighborhoods. A 25% increase in tree cover decreases a city's average mid-afternoon July temperature by 6°-to-10°F (3°-to-6°C) according to U.S. Department of Energy simulations for Sacramento, California.

Figure 2-2: *Plot Plans for Cool Buildings*

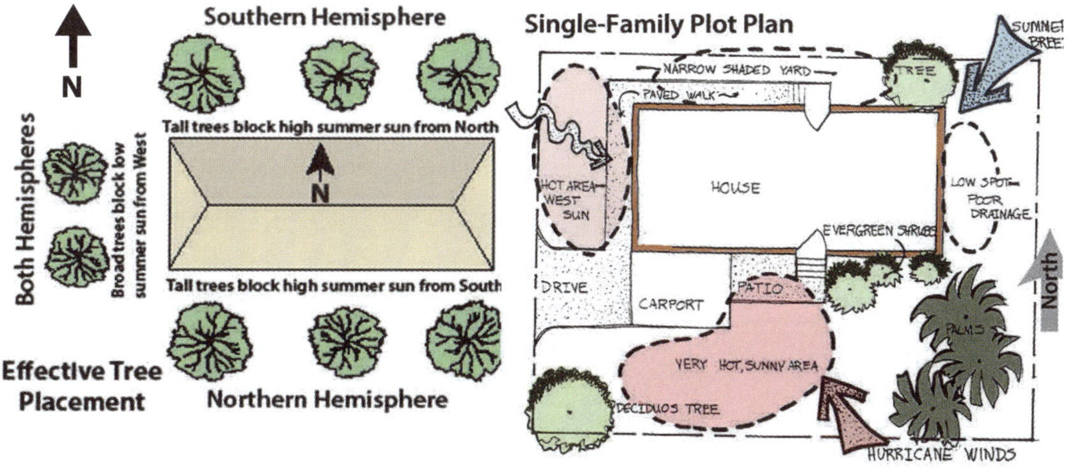

Shading Windows

Single-pane, unshaded windows transmit about 85% of the solar heat striking them. Solar gain through windows produces up to 40% of a building's unwanted summer heat. Many property owners have strong feelings and opinions about their windows. The number of window openings, the frequency in which residents operate their windows, and the windows' prominence as architectural features also affect decisions. For all these reasons, energy specialists must balance window energy efficiency with many other factors. See "Window Treatments" on page 71.

Consider these factors when deciding which windows to shade.

- **Direction the windows face.** South windows transmit the most solar heat. West windows contribute solar heat in the afternoon, when the heat has the most negative effect on comfort. East windows begin heating the home early in the morning.
- **Location of natural shade from trees, overhangs, and other objects.** If windows aren't already shaded by trees, nearby buildings, or large overhangs, additional shading is very beneficial.
- **Surface area of your windows.** Shading devices for larger windows are more cost-effective than for smaller windows.

After considering these factors, use the shade options discussed here to block 60% to 90% of the solar heat through your windows.

Figure 2-3: *Solar Screens: Fixed and Rolling*

Solar screens — Solar screens are often the least expensive window-shading option that retains a view through the window. A solar screen is fabric stretched over an aluminum frame, like an insect screen. The fabric absorbs or reflects 65% to 70% of the solar heat before it enters the building.

Install solar screens on the exterior of a window, to be effective. Thus, they aren't practical for outwardly opening windows, such as awning or casement windows, unless they attach to the movable window sash.

Figure 2-4: *Window Films and Solar Screens*

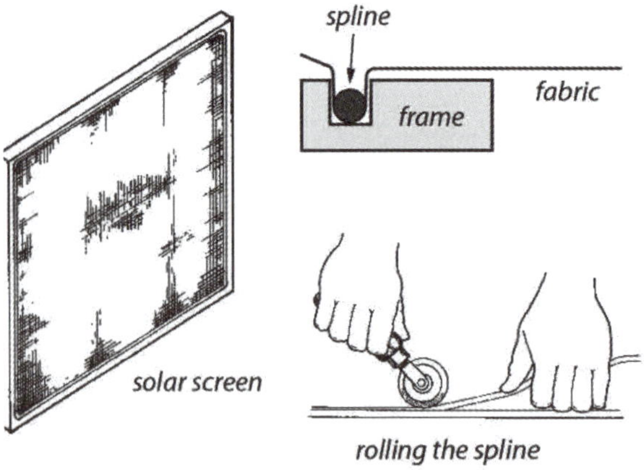

Solar screens: Intercept and absorb solar heat before it enters the building. Building a solar screen is much like building an insect screen.

Courtesy:lustalux.co.uk

Reflective or Low-e window films Manufacturers provide step-by-step instructions for installing their products.

Reflective films — Metalized plastic window films (similar to those applied to automotive windows) can block 50% to 75% of the solar heat from single-pane glass. A microscopic layer of metal on the film reflects solar radiation.

Installed on the interior side of single-pane glass, reflective window films reflect solar heat, reduce glare, and prevent fabric fading. Tinted films that merely color the glass aren't as effective at blocking solar heat as metalized films. Modern window films transmit sufficient light, while blocking most of the solar heat.

Installing reflective window film is a difficult do-it-yourself project. These films require careful installation to an absolutely dirt-free glass surface. The procedure resembles hanging wallpaper, but you have more time to place the film. See "Window Films and Solar Screens" on page 38 for a photo of window-film installation.

Figure 2-5: *Large Adjustable Awnings*

Awnings — Awnings are popular in hot, sunny climates, because they intercept solar heat before it strikes the window.

The three most important considerations to select awnings are these.

1. **Amount of shade desired.** The shade an awning produces is related to how far the awning drops down over the window. This distance is known as the "drop" of the awning.
2. **Cost of the awning.** Awnings are more expensive than solar screens or window films. Costs for awnings vary widely.
3. Importance of maintaining a view out the window. Depending on their drop, awnings can cut off a significant portion of a window's view. Some awnings adjust to allow you to compromise between solar control and view.

Awnings on a home's south side (north side in the southern hemisphere) typically need a drop measuring 45% to 60% of the window height to block solar radiation from high in the sky. Awnings on the east and west need to drop 60% to 75% to block solar radiation coming from lower in the sky in the morning and afternoon.

Shade sails — *Shade sails*, also called *shade canopies*, are a popular way to provide shade to a large outdoor area. Rope or ratchet straps secure these fabric shades to surrounding objects. (Use your imagination.) Shade sails not only shade the outdoor area but also nearby windows. The cooler outdoor temperatures that shade sails provide reduce a buildings cooling load.

Figure 2-6: *Large Canopies*

Exterior shutters — Shutters are solid and fixed to the sides of windows permanently with metal hinges or rails that allow the shutter to slide. Exterior shutters can provide security and storm protection in addition to shade.

Bahama shutters hinge at the top of the window. Sliding shutters slide horizontally to cover the window during the heat of the day. The traditional vertically hinged shutters can also block solar heat effectively. Exterior reflective venetian blinds are very effective where high winds aren't a problem. Inexpensive exterior bamboo rolling shades are also popular in some warm climates.

These movable exterior shading methods require a daily commitment by the residents to operate them. Exterior rolling shutters and shades, controlled indoors by motors or manually, are expensive but convenient.

Exterior blinds — Blinds are more flexible than shutters and are only attached to the top of a window. They're relatively expensive and can be susceptible to moisture and high winds. Exterior Venetian blinds have adjustable slats that can provide reflected light while shading the window from direct solar radiation.

Exterior shades — Exterior shades are the most flexible outdoor movable-shading option. We don't recommend exterior shades for regions with high wind. Some exterior shades feature transparent reflective window film to provide a view and to block most of the solar heat.

Figure 2-7: *Two Types of Shutters*

Figure 2-8: *Motorized Venetian Blinds and Roller Shades*

Indoor window treatments **—** Indoor window treatments with reflective surfaces — metalized or bright white — can block solar heat effectively. An opaque roller shade, with a white surface facing the exterior, rejects more than half of the solar heat entering the window. Roller shades also block most of the light and all the view. White venetian blinds and white slim shades (a smaller-scale venetian blind) also reject approximately half of the solar gain. Venetian blinds and slim shades also block most of the light and view.

If you want to retain some light or view, install roller shades made with metalized plastic window film. Like reflective films applied directly to glass, metalized plastic roller shades preserve the view and transmit some light, while blocking some of the solar heat.

Replacement windows **—** Replacement windows aren't usually a cost-effective retrofit for reducing cooling costs, compared to the other improvements in this chapter. However if you replace your windows for a combination of reasons, the added expense of multiple panes and low-e insulated glass is very cost effective.

The 3 most important window glass specifications for cooling are these.

1. *Solar heat gain coefficient (SHGC)* is the ratio of solar heat passing through the glass to solar heat falling on the glass at a 90° angle. SHGC includes radiant heat transmitted, and also the solar heat absorbed and re-radiated indoors. Single-pane clear glass has a SHGC of 0.87.
2. *Visible transmittance* measures how much visible light is admitted by the window glass. Visible transmittance is important because the window's main job is to give view and admit light. Special coatings — some of which cut visible light to 30% — may be unacceptable in some applications, such as when windows are small and few in number.
3. *U value or U factor* is a numerical rating that reports the heat transmittance of a window because of the indoor-outdoor temperature difference. U-value is less important in mild cooling climates and more important in very hot climates.

For a complete discussion of windows, see"Window Solutions for Cooling" on page 57

Figure 2-9: *Exterior Shades, Shutters, and Awnings*

Figure 2-10: Windows, Shutters, and Louvers for Tropical Climates

Tropical climates, glass, and AC — Glass windows may not be the best cooling strategy for tropical climates because of the greenhouse effect glass creates. Glass traps direct and reflected solar heat. Outdoor temperatures may be comfortable most of the time for acclimatized residents of the tropics. Many traditional windows in the tropics use louvered window shutters without glass. The shutters block solar gain and maximize natural airflow between indoors and outdoors.

When residents retrofit a window opening with a glass window, they may experience little or no improvement in comfort, even after they install air conditioning. Shuttered windows aided by powerful ceiling fans may provide as much comfort as glazed windows with air conditioning.The human body has a wider range of comfort than most first-world citizens realize. In tropical humid climates, air conditioners struggle to create first-world comfort.

Cool Roofs

Buildings with shaded, reflective, insulated, and ventilated roofs can block most solar radiation falling on the roof. The most effective measures for blocking solar heat through roofs are to shade the roof with trees or solar-PV arrays or to install a reflective roof coating. Either trees, PV arrays, or a reflective roof can block most of the solar heat falling on the roof.

If the roofing tiles are a bright white, the roof assembly might block 95% or more of the solar radiation falling on the roof. The red-tile roofs work well for this purpose also but don't block as much solar heat as bright-white roof tiles.

Insulation resists whatever heat that passes the shade and reflectivity. A combination or these improvements typically reduces cooling costs 10% to 35%, depending on climate and R-value (RSI) of attic insulation.

Figure 2-11: *Cool Roofs*

Compared to poorly ventilated attics, good attic ventilation helps to limit attic air temperatures. However, attic solar gain is primarily heat radiation and not conduction.

Ventilation has no effect on heat radiation unless the ventilated space is between the roofing material and the roof deck. In this case the ventilating air removes solar heat from both surfaces — the roofing materials and the roof deck — before their heat can radiate downward into the building.

Figure 2-12: *Composite and Rubberized-Membrane Cool Roofs*

Radiant barriers — A radiant barrier combines aluminum foil with kraft paper or roof sheathing. Radiant barriers have easier consumer acceptance than reflective roofs, because of many consumers prefer dark roof colors. Expect 2% to 10% savings from radiant barriers, depending on climate and insulation level. Radiant barriers fasten to the bottom of roof sheathing or rafters.

Radiant barriers reflect heat radiation coming toward them. They also impede radiant heat flow because they are poor radiant-heat emitters. A radiant barrier's shiny surface must face an air space to be effective. If the shiny surface collects dust, the air barrier's effectiveness declines.

Internal Heat

Internal heat comes from people, refrigerators, lights, cooking appliances, entertainment, and bathing.

People contribute heat to the air by shedding body heat (sensible heat) and by sweating (latent heat). Each occupant of the home contributes about 300 BTUs per hour (88 watts) of sensible heat and about 150 BTUs per hour (43 watts) of latent heat.

***Reduce internal heat* —** Energy-efficient appliances and lights use less electricity and release less heat into the indoors during the cooling season, compared to less efficient models.

Behaviors to reduce internal heat require the participation of residents.

- Minimize indoor cooking.
- Limit hot-water use.
- Use exhaust fans to remove moisture and cooking heat.
- Ventilate with outdoor air, during cool-or-moderate outdoor temperatures.

Air Leakage and Duct Leakage

Air leakage and duct leakage are costly problems for air-conditioned buildings, especially in humid climates. You find the worst air leaks in attics, crawl spaces, and hidden areas.

***Reducing air leakage* —** Duct leakage wastes cooling energy in these three ways.

1. Allows cooled air to escape the supply ducts.
2. Allows hot, humid air to enter the return ducts and enter the building.
3. Creates pressure differences between indoors and outdoors that drive air leakage through the building enclosure.

See "Finding Air Leaks" on page 99.

***Reducing duct leakage* —** Sealing ducts to stop duct leakage greatly reduces air-conditioning costs. Duct sealing and repair should minimize the pressure differences caused by duct leaks and poorly designed duct systems.

Building owners can save 10% to 25% on air-conditioning costs by air-sealing the building enclosure. Technicians should find and seal large, air leaks before smaller ones around windows and doors.

For comprehensive information about sealing HVAC ducts, see "Troubleshooting Duct Leakage" on page 223.

Heat Transmission

Heat transmission through ceiling, walls and windows usually amounts to 10% of the cooling load. Absorbed solar radiation causes heat transmission through the roof and ceiling. However, the indoor-outdoor temperature difference is usually less-important than direct solar gain through windows and the roof.

Dark-colored roofs and walls collect most of the solar heat that falls on them. Thus, they should have high R-values (RSI) to resist solar-heat transfer into the building. Heat transmission is more important in very hot climates than in milder climates.

See "Insulation for Cool Buildings" on page 135.

Moisture Management

Summer comfort in warm humid climates requires the control of indoor relative humidity. The durability of buildings depends on both water barriers and relative-humidity control. If your climate is humid with frequent cloud cover and typical daytime summer temperatures below 85°F or 29°C, humidify could be as much as 50% of your AC-electricity consumption.

If you are only interested in comfort and you live in a dry climate, you could skip this main topic on moisture management. However, moisture and durability are important topics for buildings in most climates. If you're interested in building durability, then continue reading this section on moisture because every sub-section contains important information on both comfort and durability.

See "Dehumidification" on page 239.

Moisture and Durability

Moisture in buildings causes trillions of dollars in property damage and respiratory diseases around the world each year. Moisture condensation problems are common during both in all seasons.

1. Water deteriorates buildings by enabling biological pests to colonize and damage building materials. These pests also threaten human respiratory health.

2. Water corrodes metal, dissolves glue, warps wood, and weakens mortar.
3. Water reduces the thermal resistance of insulation and may permanently damage the insulation.
4. Water expands when it freezes and can damage building materials that absorb water.

Energy-efficiency work can make moisture problems better or worse, depending on the knowledge and skill of the workers. Workers should take all necessary steps to prevent or mitigate moisture condensation, water leakage, and water seepage during energy retrofits. See "Preventing Moisture Problems" on page 52.

Moisture movement — Moisture moves into a building during wet seasons and out during drier seasons. Moisture is a problem when it reaches a level of material-saturation that encourages pests — termites, dust mites, dry rot, and fungus. See "Moisture Movement through Buildings" on page 31 and "Equilibrium moisture content (EMC) and building durability" on page 27.

Figure 2-13: *Foundation Water Leakage: Prevention and Treatment*

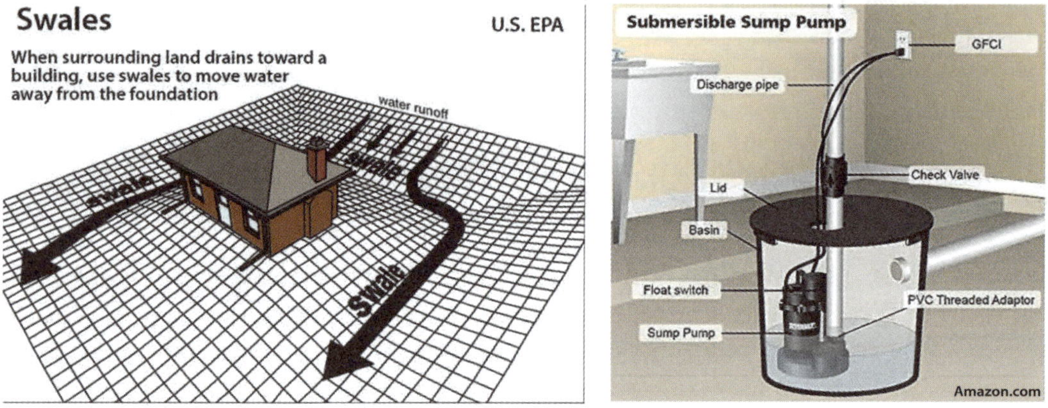

Water Leakage: Sources and Solutions

Problems with water intrusion into the building determine how long the building survives. Rain water leaks in the roof, walls, and foundation are important causes of moisture damage. Poor site drainage lets capillary water seepage into the foundation. Undetected plumbing leaks drip water unnoticed into the building. Water moves easily as a liquid from the ground into porous building materials like concrete and wood through capillary seepage. Snow melting and refreezing on the edge of the roof — known as *ice dams* — can force water underneath the shingles and into the attic and upper walls.

There are four main types of water barriers or *water-control layers*.

1. The *water-resistive barrier* (WRB) stops water penetration into walls and roofs. Manufacturers offer WRBs in a variety of water-vapor permeabilities.
2. *Flashing* protects vulnerable areas in walls and roofs in cooperation with the water-resistive barrier.
3. *Waterproofing* stops water and water vapor from penetrating foundation walls.
4. *Capillary barriers* stop ground water from rising into the building.

Figure 2-14: *Water Barriers: Capillary Barrier, and Flashings*

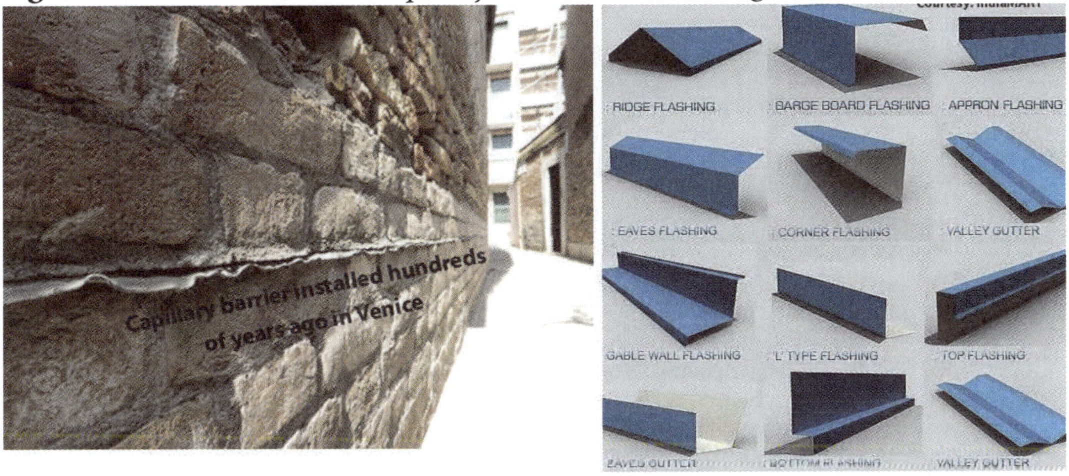

Solutions for foundation water leakage — Crawl-space and basement moisture problems are common in existing buildings. Consider these solutions to below-grade moisture problems, depending on their severity.

- Install a ground-moisture barrier, sealed at the seams and attached to the foundation perimeter in all crawl spaces.
- Install or repair rain gutters and downspouts to drain water 3 feet (1 meter) away from the foundation.
- Correct the slope of soil bordering the foundation walls to drain water away from the foundation.
- In areas where surface water must flow near a building, cap the ground next to the foundation with impermeable clay or concrete.
- Re-position sprinklers to avoid wetting soil near the foundation.
- Install a basement-perimeter drainage system that drains leaking water into a sump with a sump pump or into an exterior gravity drain. Either way, the leaking water should flow to daylight or to a drainage structure.

- Excavate the foundation and waterproof it correctly, if the builders neglected waterproofing during construction.
- Install exterior perimeter drainage to drain water collecting around the foundation's exterior away from the building.

Table 2-1: *Moisture Sources and Their Potential Moisture Emissions*

Moisture Source	Potential Amount (Pints/Liters)
Ground moisture	0–105 pints/0–50 liters per day
Seasonal evaporation from materials	6–20 pints / 3–10 liters per day
Dryers venting indoors	4–6 pints / 2–3 liters per load
Dish washing	1–2 pints / 0.5–1 liters per day
Cooking (meals for four persons)	2–4 pints / 1-2 liters per day
Showering	0.5 pints / 0.3 liters per shower
See "Household Moisture Sources" on page 289.	

Solutions for roof and wall water leakage — Roof-and-wall moisture problems are also common in existing buildings. Consider these solutions to roof-and-wall water leakage, depending on their severity.

- Repair roof leaks promptly. If you fail to stop the leak, try again, or replace the roof.
- Look for areas damaged by water, splashing from the roof. Install or repair gutters, downspouts, diverters, and flashing as necessary to direct water away from walls and foundations.
- Seal air leaks to stop humid attic air from entering the building and causing condensation during the air-conditioning season.
- Improve attic or roof venting if necessary, to keep the whole roof cold in winter and prevent ice dams.
- Add cap flashing to windows if you observe water leakage around windows.
- Apply water repellents to masonry walls or other walls that become water-soaked from rain.

Figure 2-15: Relative Humidity (Rh), Absolute Humidity, and Comfort

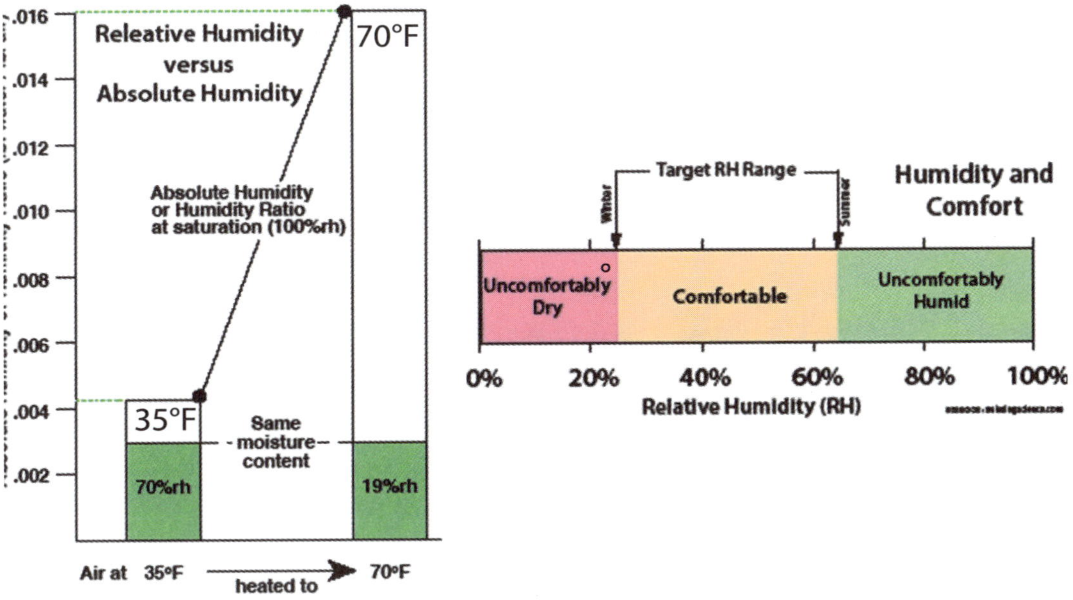

Controlling Water Vapor

High relative humidity (Rh) in indoor air can cause comfort problems in summer and condensation problems in both summer and winter. Experts on cooling say the indoor air should be less than 60% relative humidity for adequate indoor comfort in summer. For more information on the relationship between temperature and condensation, see "Approximate Energy Data of Typical Window-glass Options" on page 58

Sources of water vapor — The average person evaporates 4 pints of water into the air daily through respiration or perspiration. Showers, housecleaning, and cooking can add another 3 pints per person daily. Water tracked into the home on shoes or clothing evaporates and increases Rh. Moist firewood and house plants put additional moisture into the air.

Be sure that clothes dryers and exhaust fans vent outdoors and not into living spaces, crawl spaces, basements, or attics. Avoid using humidifiers unless there is a medical reason to use them. Some air handlers have built-in humidifiers that add moisture to the air in a home without residents being aware of the humidifier's operation.

See "Yearly-Average-Relative-Humidity Map" on page 292 for information about Rh in your climate.

Figure 2-16: *Sources of Moisture*

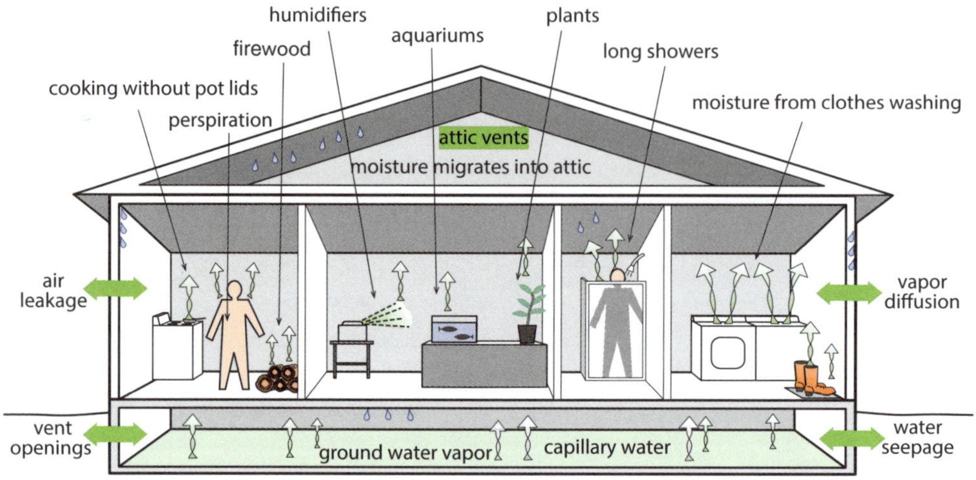

Moisture condensation problems — Moisture condenses on cold surfaces, such as windows and exterior walls, because of a combination of relative humidity and temperature. As air cools, from its contact with cool surfaces, the relative humidity of the air increases until it reaches saturation. When that saturated air is cooled, some of its water vapor turns to liquid water and clings to the cold surfaces that cooled the air. Condensation occurs most frequently on the room's coldest surfaces. Condensation increases as relative humidity increases and as surface temperature decreases.

Effective strategies for reducing moisture condensation include these.

- Reduce relative humidity by reducing moisture sources.
- Equalize pressure between indoors and outdoors.
- Install or improve air barriers and vapor retarders to prevent air leakage and vapor diffusion from transporting moisture into building cavities.
- Ventilate with drier outdoor air to dilute more humid indoor air if possible.
- Remove moisture from indoor air by cooling the air to below its dew point, with air conditioners and dehumidifiers.

Preventing Moisture Problems

Many regions throughout the world experience increasing outdoor relative humidity because of climate change. If you live in such a region, think about preventive measures to stop moisture problems before they begin.

Prevention is the best way to guarantee a building's durability and the respiratory health of residents. Besides the source-reduction strategies listed previously, air barriers, vapor retarders, ventilation, air conditioning, and dehumidification can prevent moisture problems.

Figure 2-17: Outdoor Water Sources and Solutions

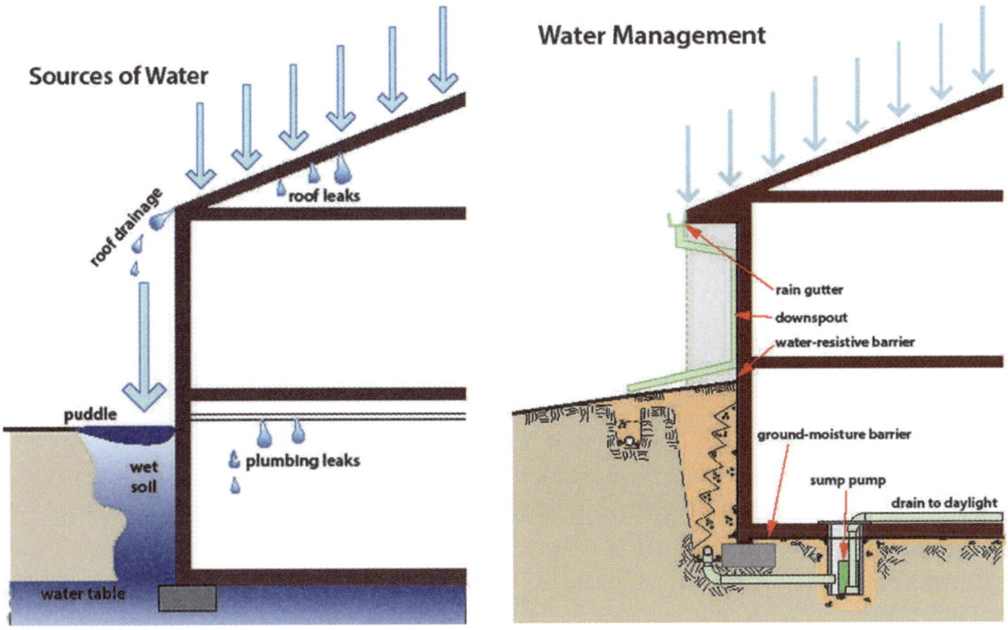

Air-and-water barriers — Air barriers and vapor retarders stop water vapor from migrating into building cavities and then condensing.

Warm, moist air from the building or the outdoors can carry moisture rapidly from a source into the building assemblies. Seal all penetrations in the building's interior and/or exterior membranes to create an *air barrier*.

This air sealing would include the following.

- Identify and plug air leaks like plumbing chases, chimney enclosures, and utility holes.
- Install gaskets on all outlets and switches.
- Seal window and door frames.
- Patch holes and cracks in ceilings and walls.
- Seal the floor-wall joint.

An existing building's air barrier uses caulking, gaskets, and tight joints between building materials. Most buildings employ a water-resistive barrier on the outside of the exterior sheathing that can also serves as an air barrier. This water-resistive barrier is usually vapor-permeable, letting water vapor pass through (but not air).

See "Air-sealing Materials and Application" on page 115.

***Vapor retarders* —** A vapor retarder reduces the rate of water vapor can move through a material. Water vapor's movement through a material is called diffusion. Vapor retarders are important in cold climates, and less important in warmer climates. In hot humid climates, builders may install vapor retarders on the outdoor side of walls and ceilings, but this isn't common.

In buildings with months of air conditioning and months of heating, moisture travels from outdoors to indoors during summer and from indoors to outdoors during winter. In these climates, avoid installing vapor barriers, which stop almost all vapor diffusion. Instead allow the building be vapor permeable by using vapor retarders that are permeable. Vapor permeability allows the building dry to either the indoors or the outdoors. See "Vapor Retarders and Vapor Barriers" on page 159.

***Ground-moisture barriers* —** The ground under a crawl space or basement is a major source of water vapor from airflow through the ground, capillary seepage, and evaporation from the ground. Ground-moisture barriers install over dirt floors in crawl spaces and under basement concrete slabs. Reinforced polyethylene plastic makes a durable ground-moisture barrier where people access that space. *Ground-moisture barriers* stop water vapor rising through the ground and into a building and also prevent evaporation from damp ground.

If the seams of the barrier are sealed with a durable tape and/or sealant, this ground covering may also be an air barrier. Ground-moisture barriers aren't effective in crawl spaces with standing water, unless the ground and the ground-moisture barrier drains water into a sump or drainage assembly.

Some building's employ concrete slabs that stop pests such as rats, rodents, and snakes from colonizing the building through crawlspaces. These slabs should have moisture barriers underneath them.

Figure 2-18: *Crawl-Space Moisture Solutions*

Ventilation for Attics, Roofs, and Crawl Spaces

Attic, roof, and crawl-space ventilation can prevent moisture problems.

Attic ventilation — Attic ventilation removes moisture, deposited in attics by roof leakage and condensation. Most attic moisture problems happen because moist building air leaks through the ceiling into the attic during the winter. The preferred strategy for attic-moisture control is to prevent the moisture intrusion into attics by air sealing the ceiling.

During hot and humid weather, the attic may be cooler than the outdoor air because of ceiling air leaks, attic duct leaks, or poor AC-duct insulation. In this situation, the humid, ventilating outdoor air may condense on rafters or ducts in an attic. Air seal ceilings, seal attic AC ducts, and insulate AC ducts in ventilated attics.

Roof ventilation — To minimize solar gain, roof venting on top of the roof deck may be a more effective solution than venting underneath the roof deck. Battens or furring strips create a 1-to-2 inch (25 to 50 mm) cavity between the roof deck and the roofing. Roof decking and the fibrous insulation underneath can dry effectively with rooftop ventilation.

When we install reflective roof tiles over a vented airspace, this assembly also creates an effective sunblock. The small fraction of solar heat that the tiles collect, they radiate toward the roof deck. The ventilating air carries most of this radiated heat away from the hot tiles and hot roof deck before the heat can radiate from the roof deck to the attic.

Figure 2-19: *Attic and Roof Ventilation*

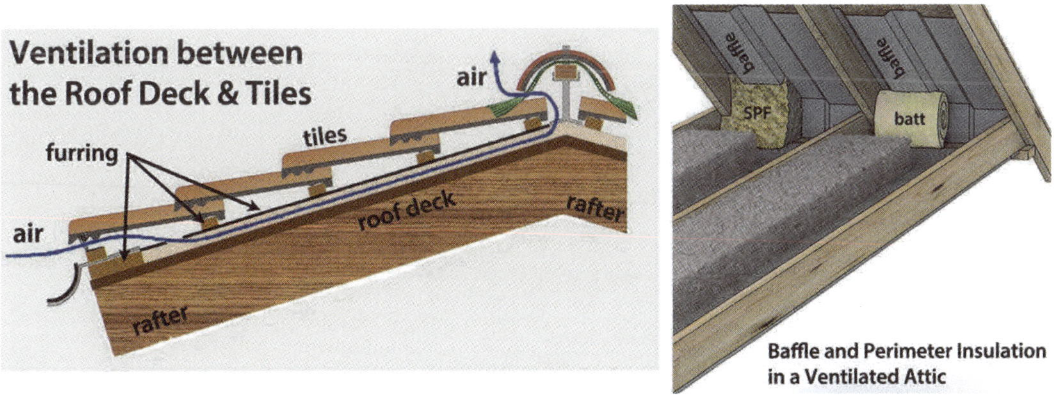

Crawl-space ventilation — Building codes may require you to ventilate crawl spaces with passive vents or small exhaust fans. In cool and dry climates, the passive ventilation option, which complements floor insulation, is a good choice. If the crawl space has an airtight ground-moisture barrier, the code requires vent openings to be a vent-area-to-crawl-space-area ratio of 1-to-1500 square feet or square meters.

Contractors also condition (heat and/or cool) crawl spaces to bring the crawl space inside the building's heated and/or cooled space. This option sometimes serves the same moisture-control purpose as ventilation, but it can waste a lot of electricity and promote condensation especially in summer when air conditioning.

Exhaust ventilation is usually more effective and economical than space conditioning in the crawl space. In summer, an air-conditioned crawlspace, combined with air exchange from the humid outdoors can create a lot of condensation on floor joists and other building materials. Exhaust ventilation can also pull humid outdoor air into the crawl space to replace the air that the fan exhausts, causing condensation on cool floor joists.

The best option may be close-able crawl-space vents or a sealed crawl space, depending on outdoor humidity.

Figure 2-20: *Close-able Crawl-Space Vents*

3 Window Solutions for Cooling

Architects refer to windows and doors as *fenestration*. Windows provide natural light, ventilation, and a view to the outdoors. These functions make windows both a necessity and a formidable energy problem.

Windows are a significant source of transmission heat loss in cold climates and a significant source of unwanted solar heat gain in temperate or hot climates. Windows challenge us to minimize heat loss and heat gain while preserving natural light and view.

Window energy-conservation measures are expensive per square foot or square meter, compared to insulation. To be cost-effective, window energy-efficiency retrofits must significantly reduce thermal transmittance (U), or reduce solar transmittance (SGHC), or reduct both.

Doors have less thermal resistance than walls, but more thermal resistance than windows. Doors' surface area is small and replacement cost is high, so door replacement isn't usually a high priority for buildings cooling improvements. However, doors can be a significant air-leakage and comfort problem because of malfunction or ineffective air seals.

This chapter outlines the most important energy characteristics of windows and doors. We discuss how to retrofit and replace old windows to save energy and improve comfort. We also focus on the challenging terminology associated with windows and doors. Factors that affect both cooling and heating are important to retrofitting and replacing windows, so we discuss both.

***Figure 3-1:** Infrared Scans Show Window Heat Loss*

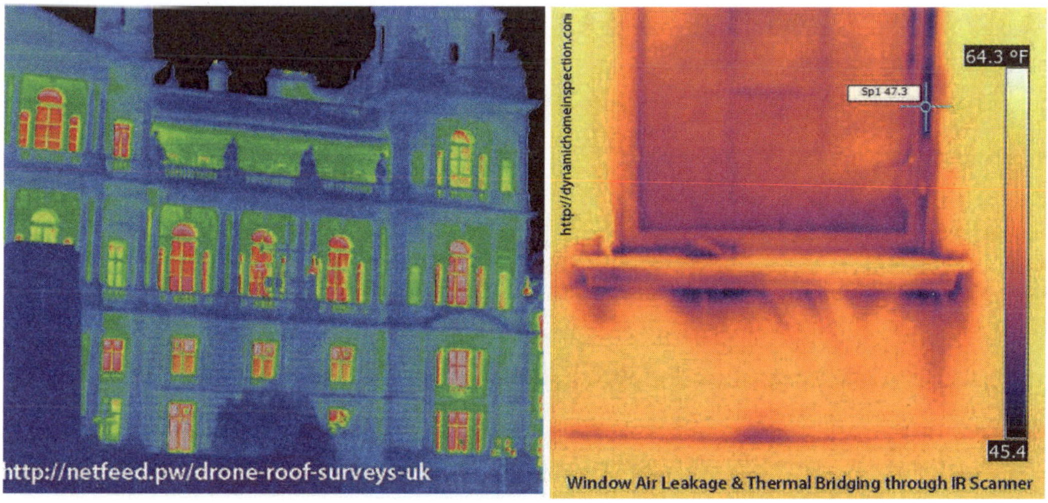

Table 3-1: *Approximate Energy Data of Typical Window-glass Options*

Glazing Assembly	U-factor Imperial	Shading Coefficient	U-value metric	SHGC	VT
Single clear glass	1.1	1.00	6.2	0.87	0.85
Insulated glass, no low-e coating	0.50	0.69	2.8	0.60	0.75
High-SHGC, low-e insulated glass	0.30	0.81	1.7	0.70	0.55
Medium-SHGC, low-e insulated glass	0.26	0.69	1.5	0.58	0.55
Low-SHGC, low-e insulated glass	0.29	0.46	1.6	0.40	0.55
Triple glazed 2 low-e coatings & inert gas	0.16	0.40	0.91	0.35	0.35
4 glazing layers with low-e films & inert gas	0.12	0.35	0.68	0.30	0.25

Window Characteristics

Windows include the following 4 main components.

1. Glass Assembly — One or more glass panes with spacers and gaskets.
2. Sash — The frame around the glass assembly. Sashes are either movable for ventilation, or fixed.
3. Frame — The frame that surrounds the sashes and attaches to the building.
4. Rough Opening — The complete window assembly fits inside the rough opening and attaches to it.

Conduction, convection, and radiation are all important window heat-flow mechanisms. We tolerate the high conductivity of glass because of these useful and unusual qualities.

- Glass is transparent.
- Glass is comparatively easy to clean.
- Glass absorbs most infrared radiation coming from indoors rather than transmitting it (greenhouse effect).
- Ordinary glass transmits most solar radiation — visible light and heat.
- Specialty glass reduces winter heat loss, reduces summer heat gain, reduces glare, and provides other benefits, depending on the type of specialty glass.

Thermal transmittance U-factor (or U-value) and solar heat gain are the most important energy considerations for window selection. The window's air leakage, optical characteristics, frame material, and type of glass assembly also affect window performance and selection.

Radiation, convection, air leakage, and conduction through windows waste energy and reduce indoor comfort. Radiation is more important with windows than with other building components.

Figure 3-2: *Window Rough Opening and Installation Details*

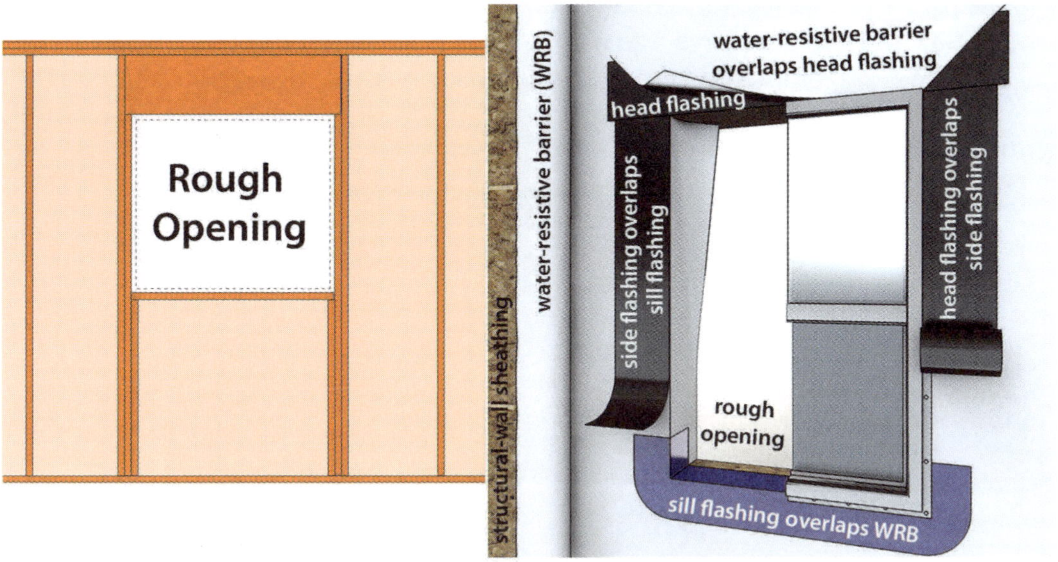

Thermal Transmittance (U-factor or U-value)

The window industry describes and rates its products by U-factor (or U-value), which is the *overall coefficient of thermal transmittance*. The U-factor includes heat transfer by conduction, convection, and radiation through the window assembly.

For both Imperial and metric, the U-factor (or U-value) is the reciprocal or inverse of R-value (RSI) (U=1/R or 1/RSI). As U-factor decreases, heat-transmission rate decreases. Low U-factors transmit heat slower than higher U-factors.

The U-factors of window assemblies are rated using an area-weighted average of different sections of the window that have three (3) distinctly different U-factors.

1. The center of the glass
2. The edge of the glass area
3. The frame and sash

Windows are often a building's most prominent thermal weakness. The best new windows have R-values up to R-20 (RSI-3.5), but most windows remain between R-1 and R-3 (RSI-0.18 to 0.53). And the windows high conductivity creates thermal bridging. If you take an area-weighted average R-value (RSI) combining a wall with the windows in that wall, windows reduce this averaged R/RSI by between 20% and 60% depending on the U-factor (U-value) of the window.

Remember that the U-factor (U-value) is the inverse of the R-value (RSI), so that a U-factor of 0.5 is an R-value of 2, which is the R-value of the average existing window assembly. Compare that to the R-value the typical insulated wall, which is about 15 and then calculate the area-weighted average for that wall assuming 15% of its area is windows. In this example, your average is R-7.6, a 50% reduction. Now calculate the same average for a triple-pane window with an R-value of about 6. In this case, the area-weighted average wall R-value is about 12, a 20% reduction. The same relative averages and percent reductions are the same for metric examples.

Figure 3-3: *Glass Transmittance at Various Wavelengths*

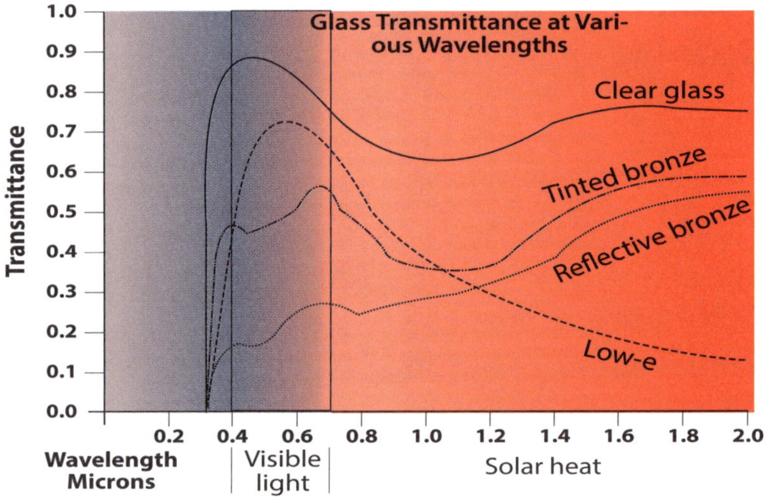

Solar and Optical Characteristics

Solar heat gain is a primary window-energy characteristic. Solar heat gain through windows can be up to 40% of the total unwanted heat removed by an air conditioner. There are three common factors used to measure solar heat gain: *solar heat gain coefficient (SHGC)*, *visible transmittance (VT)*, and *shading coefficient (SC)*. Each of these factors is a ratio expressed as a decimal number between 0.0 and 1.0, such as 0.81 (or 81%).

Solar heat gain coefficient (SHGC) — The ratio of solar heat passing through the glass to solar heat landing on the glass at a 90° angle. SHGC includes the radiant heat transmitted and also the solar heat absorbed and then re-radiated indoors. Single pane glass has a SHGC of 0.87. For buildings in hot, sunny climates, choose window assemblies with SHGC of less than 0.40 for comfort and energy efficiency.

Visible transmittance (VT) **—** Visible transmittance is important because one of the window's main functions is to admit visible light. VT measures how much visible light the window glass transmits. Reflective coatings and dark tints — some of which cut visible transmittance to 30% — aren't acceptable in many residential applications. A window's ultraviolet-blocking ability is important to customers concerned about the ultraviolet radiation's destructive effect on furnishings.

Shading coefficient (SC) **—** This term quantifies the effect of various window-glass assemblies and shading devices. SC is a ratio that compares the solar transmittance of a glass assembly — including its interior and exterior shading devices — to the solar transmittance of single-pane glass. Single-pane glass has a shading coefficient of one (1.0). A glass assembly's shading coefficient is always less than one and approximately 1.15 times greater than the its SHGC.

Air Leakage

Window air leakage is usually a less important energy consideration than thermal transmittance during the heating season or solar heat gain during the cooling season. Exceptions to this generalization include old windows that are falling apart from neglect or poorly installed windows that have significant air leakage around their frames. Window air leakage is often dominated not by frame/sash cracks, but by installation flaws between the window frame and rough opening.

Windows and doors are less an air leakage problem than most people think. Most existing buildings have hidden air leaks in the floor, ceiling, and wall cavities that leak more air than the leaks around windows and doors.

Lab technicians test new windows at a pressure of 1.6 psi or 75 pascals, which simulates a 25 mile-per-hour or 42 kilometers per hour wind. The National Fenestration Rating Council (NFRC) air-leakage rating lists CFM per square foot of window surface area at that test pressure. Test results vary from about 0.3 to 0.04 cfm/ft^2 although the NFRC doesn't give ratings below 0.1.

British and other international test labs measure window airtightness at 100 pascals. The metric measurement of window air leakage is cubic meters per hour per square meter ($m^3/hr{\bullet}m^2$). The range of acceptable air leakage values is 20 to 3 $m^3/hr{\bullet}m^2$ at 100 pascals.

Window air leakage varies according to the window vent's operation. Casement, awning, and tilt-and-turn windows have compression weatherstripping, which is more effective than the pile weatherstrip of sliding windows.

Trickle vents **—** A small amount of window air leakage may help to maintain good indoor air quality in airtight buildings. In fact, some manufacturers make windows with small adjustable vents, called trickle vents, that open and close to provide ventilation air in buildings without balanced ventilation systems.

Figure 3-4: *Trickle Vents Provide Some Ventilation for Buildings*

Resistance to Condensation

Water condensation on glass causes the deterioration of the window and the opening surrounding the window. Condensation is one of the most common consumer complaints about windows. Condensation is a winter problem that gets worse as the outdoor temperature falls. The NFRC now has a optional window rating of condensation resistance with a scale of 1 to 100.

Solving condensation problems requires increasing the thermal resistance of the window or reducing the building's relative humidity or both. Consumers sometimes think that condensed water droplets leak from the outdoors. When the new windows they buy to solve their condensation problem also condense, customers complain.

The energy specialist should know how much thermal resistance must increase and/or how much relative humidity must decrease in order to solve a window-condensation problem. The chart shown here plots the outdoor temperature and relative humidity where condensation occurs for various window R-values (RSI).

Figure 3-5: Glass Condensation and Frost Deposition

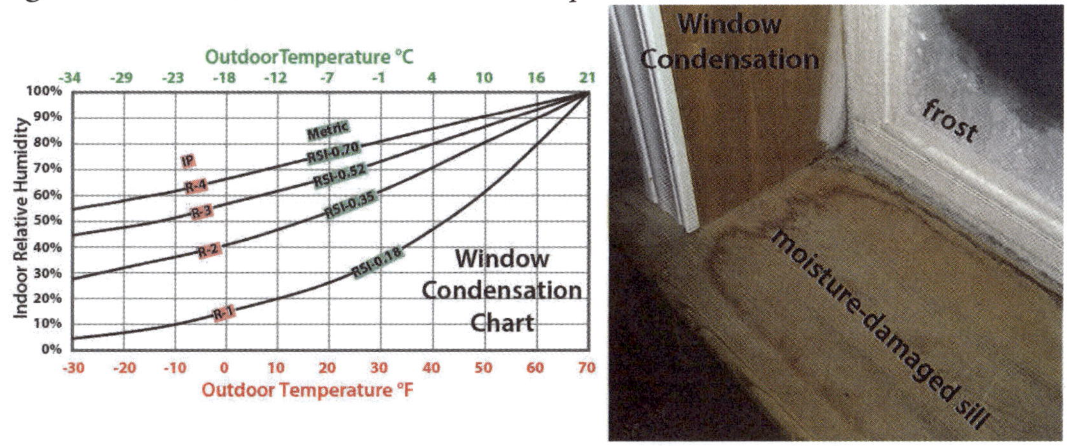

Window Structure

Window frame and sash materials are important for thermal performance, maintenance, and life-span. Movable sashes are very important for cooling and ventilating buildings.

Sash Operation

Windows have two types of sash: fixed sash and movable sash. Movable sashes either slide within their frames or open inwardly or outwardly on hinges. Vertical sliding windows include double-hung windows, where both sashes move, and single-hung, where only the bottom sash moves. Horizontal sliders usually have one sliding sash with one fixed sash. Sliding windows are usually less expensive than hinged windows.

Horizontally hinged windows are called *hopper* or *awning* windows. Vertically hinged windows are called *casement* windows. Hinged windows provide better ventilation and seal more tightly than sliding windows.

Frame-and-Sash Material Type

The four most common types of windows according to their frame material are these.

1. *Metal frames* have excellent life-span and freedom from maintenance, but poor thermal performance. Modern aluminum-frame and steel-frame windows have an insulating gasket inserted in their frame for a thermal break. However, metal remains the most conductive frame. The metal frame is more likely to condense water both indoors in winter and outdoors in summer, compared to less conductive frames.

2. *Wood frames* have good thermal resistance and life-span, but maintenance is an ongoing task. Building owners must repaint the exterior of wood window frames and sashes every 5 to 15 years depending on the climate, paint quality, and paint installation.Wood-frame window manufacturers have improved exterior maintenance by covering the frame's exterior with metal or plastic cladding. For custom new buildings, metal-clad wood windows are popular but expensive. The best wood windows use foam or cork thermal breaks and interior insulation inserts in their frames and sashes.

3. *Vinyl frames* have good thermal performance and freedom from maintenance. Low cost and minimum maintenance explains vinyl windows' popularity. However, with the many manufacturers making vinyl windows, their durability and lifespan differ widely. The most thermally resistive vinyl windows have foam-insulated frames and sashes.

4. *Fiberglass-frames* are energy efficient, strong, and low-maintenance. Fiberglass is stronger than vinyl, and it expands and contracts less with temperature change compared to aluminum or vinyl. However fiberglass is typically more expensive than vinyl or metal. Fiberglass windows have factory-applied baked-on enamel paint. Some window frames combine interior and exterior jamb extensions with the fiberglass frame. The best fiberglass windows have foam-insulated frames and sashes.

Figure 3-6: *Energy-efficient Windows*

***Frame types for installation* —** Be sure to order the right type of frame for the opening and the surrounding building materials in a wall. Here are four (4) common types of window frames for a variety of walls and exterior finishes, such as *siding* or *cladding*.

1. *Nailing fin* — Manufacturers designed the nailing fin for a common type of new installation in a wood-frame wall. The nailing fin attaches flat to the structural sheathing of the building. The nailing fin window protrudes outwardly beyond the nailing fin. This protrusion provides a place where workers can stop the siding or trim next to the window frame.
2. *Brick mold* — Manufacturers designed brick mold window frames to give a finished look to a window installation surrounded by brick or other masonry material. However, many builders use the brick-mold frame for window replacement with wood or composite siding. Vinyl windows with vinyl brick mold are popular for retrofits because of their removable slat that conceals the screws that fasten the brick mold and the window to the rough opening.
3. *Flush flange* — Builders use flush-flange window frames to create a tight seal with stucco exterior cladding. Flush-flange window frames are also appropriate for any kind of flat-surface cladding. (Flush means even or level.)
4. *Block frame* — A block frame window uses no added nail fin, brick mold, or flange. Thus, installers connect the window frame to the inside of the rough opening with specialty screws, plates, or clips. The block frame window installation is the most versatile window-installation option. Also the block frame allows workers to install windows from indoors, which is safe and convenient for heavy windows and tall buildings.

Support the window on the structural sill with wood or hard-rubber shims. Never depend on a nailing fin, brick mold, or flange to bear the window's weight.

Some window manufacturers make it easy to remove the window sashes and glass for installation. This makes triple-pane windows easier to lift and fasten to the rough opening.

Order new and retrofit window frames with interior and exterior *jamb extensions* to align the window jambs with their exterior and interior finish materials. If the window jambs are the correct width, this greatly simplifies window installation and finishing.

***Windows and siding/cladding* —** We use both of these terms because some readers prefer *siding* and some prefer *cladding* to describe wood, vinyl, masonry, steel, and other exterior finishes. **From centuries of experience, we know that some water penetrates the exterior finishes of buildings. Siding or cladding isn't the waterproof layer for a wall. The water-resistive barrier underneath them is the water barrier.**

Figure 3-7: *Window-Frame Types*

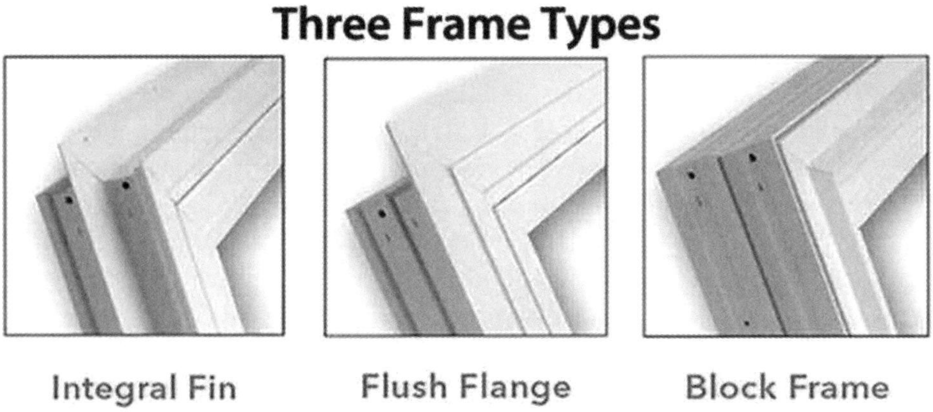

Glass Characteristics and Glass Assemblies

Many types of *insulated-glass units* (IGUs) are available, with one to four glazing layers. Some advanced windows use flexible plastic inner glazings within the outer glass panes.

Energy-efficient windows use three strategies to improve the R-value of glass, besides multiple panes: metalized coatings, gas fillings, and low-transmittance spacers between the panes.

Glass coatings — Low-e coatings coat only one surface of a glass pane in the standard multi-pane or insulated-glass window. The coating faces the air space between the panes on either the inner or the outer pane to protect the coating from damage. The low-e coating works better to resist radiant-heat emittance than to reflect radiant heat. For cold climates, the coating is on the inner pane to resist radiant emittance. In warm climates, the coating is on the outer pane to reflect solar radiation.

During the heating season, that glass absorbs radiant heat from indoors, then the low-e coating resists the heat's re-radiation to outdoors. Low-e coatings also regulate solar heat gain by trapping some of the solar heat and resisting its transmission to indoors.

Low-e glass is available in high, medium, and low solar transmittance for use in various climates and different orientations. Customers can now specify different low-e glass for windows facing different directions.

The low-e coatings come in two varieties: soft coats and hard coats. The soft coats, made of silver, are the most effective at reflecting heat, but they are fragile and must be enclosed in an IGU shortly after the glass is manufactured. The hard coats can function in some single-glass units because they resist mechanical damage better than the soft coats.

Building retrofitters sometimes use hard-coat low-e glass in single-pane storm windows. The low-e surface is more difficult to clean than a standard glass surface. Thus, retrofitters install the storm window's low-e surface toward a the primary window rather than toward the outdoors to keep the indoor surface clean.

IGU gas filling — Gases like carbon dioxide and argon, which have a higher R-value (RSI) than air, fill the space between panes in many high-tech windows. Carbon dioxide and argon are denser than air, which reduces the gas's convection between the glass panes, compared to air.

IGU low-transmittance spacers — Insulated glass manufacturers now use improved edge spacers for assembling IGUs. Older aluminum spacers conduct heat rapidly. New spacers — made from dense-foam plastic or plastic-steel composite materials — reduce the heat flow through the edge of the IGU. These newer spacers reduce thermal-bridging effect, compared to older aluminum spacers.

Figure 3-8: *Insulated Glass Unit (IGU) and Insulated Spacers, Frames, and Sashes*

Selecting New Windows

The energy-selection criteria for new windows, depends on climate and the building owner' budget and maintenance preferences.

Window-selection criteria depend on these factors: the climate's coldness (in heating degree days [HDD]), the solar exposure (both winter and summer), the window's orientation, along with the light and ventilation the window must provide.

U-value or U-factor — U-value or U-factor is the most important factor for cold climates and an important factor very hot climates. Don't replace windows in a very hot climate unless you can make a significant reduction in U-factor. Consider a U-factor of 0.25 (or USI-1.4) as a maximum for very hot climates (or cold climates), but lower is better to minimize hot-weather heat transmission.

Solar gain — In warm, sunny climates, the IGU should have an SHGC of 0.30 or less. The exterior pane's interior surface is the best place for the low-e coating for hot, sunny climates. In cool and cold climates, the low-e coating should be on the exterior-facing surface of the interior pane.

If solar heating from south-facing glass is part of a building's heating strategy, select glass with at least 0.7 SHGC, which requires an insulated-glass unit with a high-transmittance low-e coating. West-facing glass and east-facing glass should have a SGHC of less than 0.35 in warm climates to minimize summer heat gain.

Replacing Windows

Replacing windows often has a small return on investment — less than 10% return annually. However, if a building is well insulated and the heating and air conditioning systems are efficient, window replacement may be the logical next step to optimizing the building's energy efficiency. However, window replacement would ideally precede HVAC-replacement to minimize HVAC-system size.

The two most important considerations for window replacement are to size the windows accurately and to install the windows correctly. Sizing may involve removing casing or drilling holes in the existing window frame to locate the rough opening's boundaries. The installer gives the window supplier either the rough opening size or the window-frame size, depending on the installation method. This book gives brief and general guidance on window installation. Only experienced workers should install windows.

Figure 3-9: Block-Frame Window Installation

Retrofit-installation options — There are three common options for retrofit-window installation.

1. Install a custom-sized window in the existing rough opening. Rough openings must be slightly larger than the replacement window frame (>1 inch or 25 mm). Most window replacements install into the existing rough opening.
2. Install the custom-sized window inside the existing window frame. Building owners choose this method when replacing old wood windows with vinyl or aluminum windows because it is the least expensive option. However, thermal bridging and air leakage, around the old and new window frames, is a common problem.
3. Install a standard-sized window in a re-sized rough opening. This option is the most expensive, because you need to change the rough-opening's size. Also, the siding and interior finish probably require changes with this option. Window installers use his option to down-size existing windows that are unnecessarily large. Resizing the rough opening is a good idea for oversized windows, when you replace siding at the same time.

See "Frame types for installation" on page 65.

Installing a new window in the old window frame is the least expensive and often the least satisfactory option. If the window is worn or rotted, its frame is probably also damaged. Installing a new window in the old window frame isn't the best option unless an existing well-installed frame has a beautiful interior finish that you don't want to demolish or damage.

When window installers use the existing rough opening, this provides a much more original-looking appearance than using the old window frame. A rough-opening installation also promotes better air seals and water seals. In a frame building however, this method requires much more labor than installing the new window in the existing window frame.

Resizing the rough opening, especially enlarging it, increases the labor cost and wastes energy compared to using the existing rough opening. In contrast, reducing the window size saves energy.

Since air leakage around new-window sashes is quite low, leakage around the frame should be low also. Air leakage and water leakage are problems in many recently installed windows because installers fail to connect the building's water-resistive barrier to the new window flashing.

For air sealing, window installers squirt non-expanding one-part polyurethane foam into the gap between window frame and rough opening. Masons use mortar to seal these gaps when they install block-frame windows in masonry buildings. Some contractors install gaskets or tape to seal between the new window and the opening.

Durability issues — The modern approach to window replacement is to provide a low-maintenance window exterior. Vinyl, metal, or fiberglass qualify as low-maintenance materials because you don't need to re-paint them.

Replacing a window often involves repairing water damage to the exterior or interior of the rough opening. Seal the window perimeter with sealants that are compatible with the building materials bordering the joints.

Don't forget these three important window-protection details.

1. Join the existing water-resistive barrier (WRB) to new window flashing you install around the window.
2. Install a waterproof and water-tight sill detail that drains water out and away from both the window and the wall cladding.
3. Install flashing that drains water out and away from the head frame and its joints with the side frames.

See "Selecting caulk" on page 121.

Exemplary window replacement — Energy specialists should encourage customers to replace both windows and cladding (siding) to facilitate the installation of insulated sheathing under the new cladding.

Window replacement presents the best opportunity to add exterior wall insulation to buildings. Combining windows with exterior insulation allows workers to minimize thermal bridging and optimize the exterior finish details around the new windows.

Thermal bridging in wall assemblies is one of the main challenges to making existing frame buildings more comfortable and energy efficient. Windows usually have many linear of thermal bridging around their perimeters.

Another reason to replace windows and cladding at the same time is to integrate of window flashing with the wall's water-resistive barrier (WRB) in wood-frame buildings. Many window replacements and even new window installations cause water leaks and air leaks. Correctly integrating new window flashing with the building's existing WRB, without removing the cladding, is often difficult in frame buildings. Thus, installers should remove cladding if necessary to integrate new window flashing with the existing WRB. Shop for new innovative products to reduce thermal bridging, integrate the water barrier, and reduce labor costs.

Figure 3-10: *Custom Products to Optimize Window Installation*

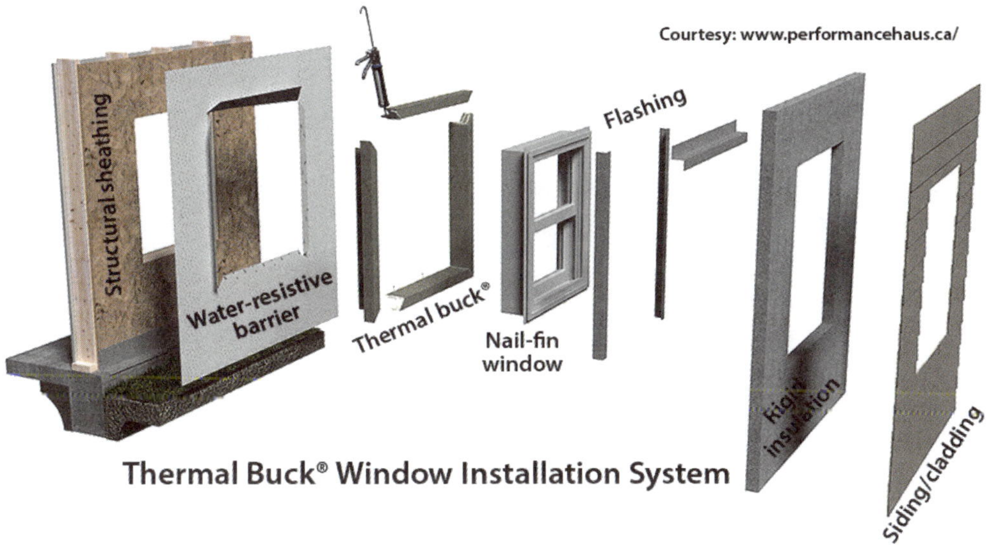

Window Treatments

Consider window treatments including: repairs, retrofits, weatherstripping, storm windows, window coverings, and shading devices before deciding to replace windows. Window upgrades may be a sensitive issue for older structures in the following circumstances.

- Building has a traditional appearance.
- Building is designated a historic landmark.
- Building is located a within an historic district.

Besides property owners, window treatments may involve renovation contractors, lenders, realtors, appraisers, regulating agencies, and insurance companies. These entities consider more than thermal performance and ventilation. Cost benefits, durability, egress, lead-based paint mitigation, privacy, noise control, dust control, ultraviolet light control, personal comfort, water leakage, environmental issues, and historic preservation also affect window-treatment decisions.

Figure 3-11: *Exemplary Window Assemblies to Minimize Thermal Bridging*

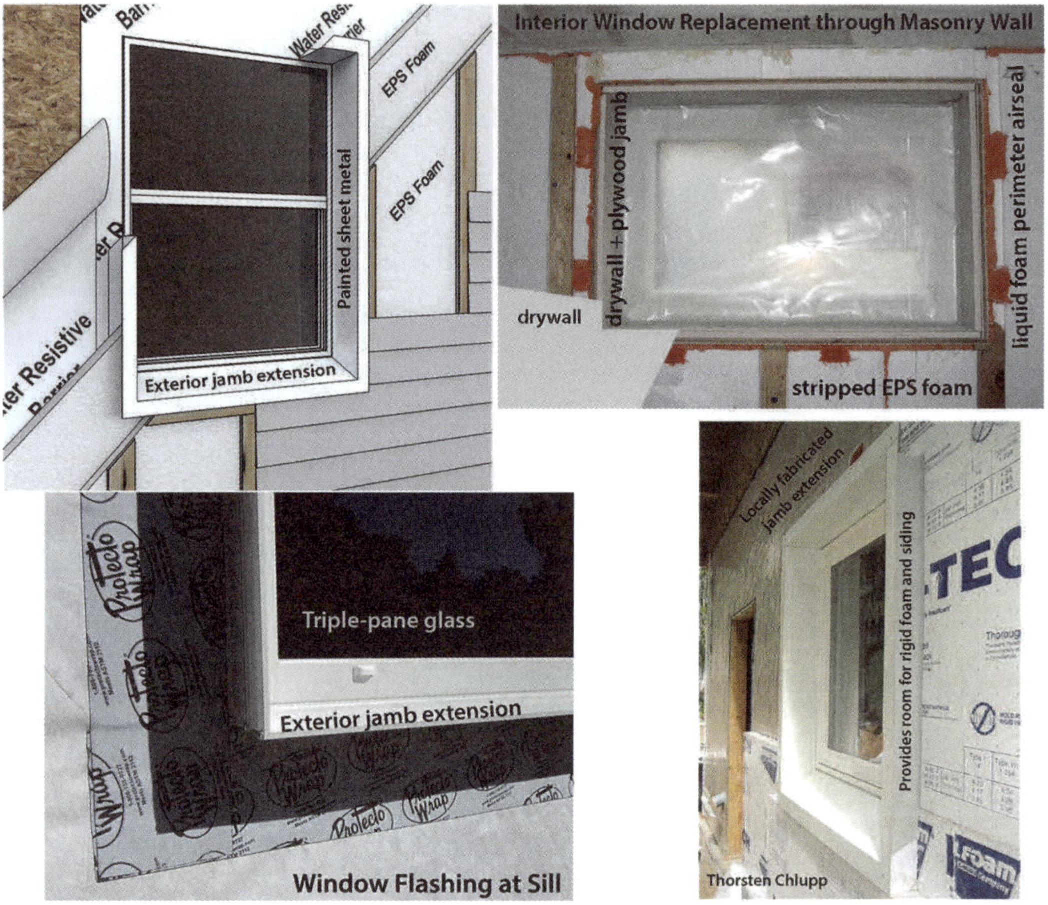

Many property owners have strong feelings and opinions about their windows. The number of window openings, the frequency in which residents operate their windows, and the windows' prominence as architectural features also affect decisions. For all these reasons, energy specialists must balance window energy efficiency with many other factors.

A variety of treatments can enhance the winter and summer thermal performance of windows. Air sealing reduces air leakage and benefits residents all year.

Severe-weather window treatments reduce the window's U-factor or U-value. Choices include storm windows and insulating window shades. Solar-gain reductions include solar screens, awnings, window films, along with shades, blinds, and shutters.

For an extensive discussion of window shading, see "Shading Windows" on page 37.

Figure 3-12: *Storm Windows and Double Windows to Reduce U-Value*

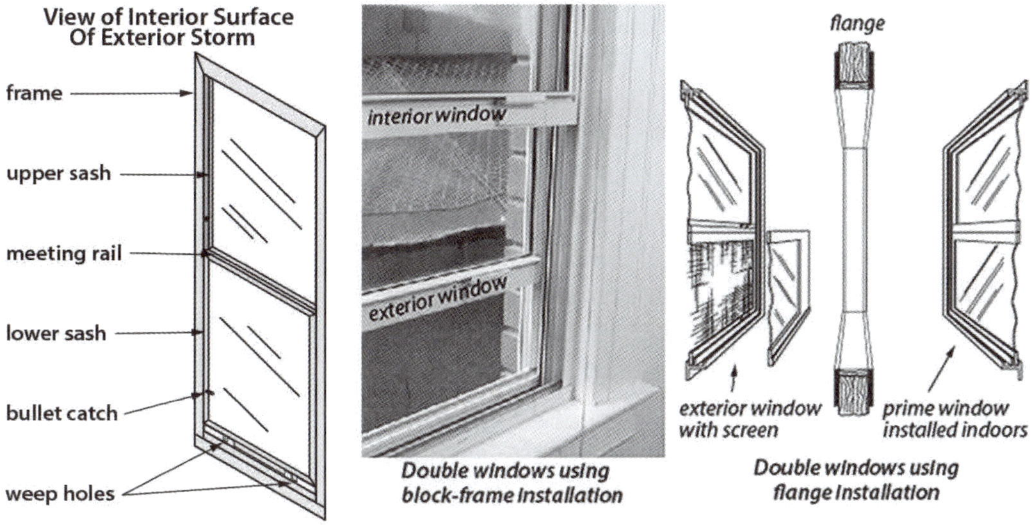

Window Repairs and Weatherstripping

Some renovation companies repair windows (and doors) and even restore them to original condition. Weatherstripping and wood-repair technology requires a wide range of materials and skills. Energy professionals who evaluate traditional homes should know about these window repair and restoration options.

Because this topic of windows is so complex, we can't explain all the options and procedures available to retrofit windows in this book. Energy auditors should know that window repair and renovation are possible in most cases. Decisions to replace or repair depend on these factors.

- The windows' current condition and state-of-repair
- Cost-comparison between new-window installation and window renovation
- Whether the skills to do window repair and renovation are available locally

Common window weatherstrip — Vinyl or polypropylene V-shaped weather-strip, particularly those available in stores, tend to be very thin and may buckle, tear, or peel off.

- Self-adhesive weatherstrip that sticks without fasteners requires very clean surface preparation, but may still may peel off eventually.
- Foam tape is inexpensive, and relatively easy to install. However it can degrade from weather and sun exposure.

- Weatherstrip, packaged in a roll, isn't usually as effective as weatherstrip supported by a rigid metal or wood strip.

Effective window weatherstrips — Pile-type weatherstrips are effective for replacing worn-out pile weatherstrips on sliding windows and sliding-glass doors.

The most durable replacement weatherstrip-mounting method, for both windows and doors, is within a kerf, integrated into the window frame. A *kerf* is a narrow slot that holds the weatherstrip in after installation. Remove the old damaged weatherstrip first. Install the new weatherstrip by pushing it into the kerf. Kerf-type weatherstrip may be available locally or from online stores.

Spring-bronze rigid weatherstrips are very durable, if installed correctly. Spring-metal strips are traditional weatherstrips that are common on older windows and doors.

Exterior Storm Windows

Besides saving energy, exterior storm windows protect the primary window from wind and water. We recommend that you order low-e pyrolytic glass for your storm windows. Low-e glass increases the R/RSI of a single-pane window by about 20% compared to standard glass.

Fixed exterior storm windows are easy to install on fixed primary windows. Attach a glass pane, framed with a metal or plastic sash, to the exterior of the primary window with screws. Use closed-cell foam tape between the storm sash and the primary-window sash or frame for an air seal.

If an open-able primary window is too leaky, moist indoor air may cause fog and ice on the inner surface of the exterior storm window in cold weather. Air-seal the primary window to minimize this fog or ice.

Exterior open-able storm windows appear very similar to one another, making it difficult to judge their quality. The following is a list of features that are important in selecting an exterior metal storm window that opens and closes.

- Order low-e glass for the storm window.
- Frame should be rigid — sashes shouldn't move around in their frame.
- Frames shouldn't deform during installation.
- The gasket that seals the glass should surround the edge of the glass.
- The interlock at the meeting rail should seal tightly.
- The frame should have weep holes or some other feature at the bottom to let water out of the window assembly.
- The window should fit well.
- Sashes should be removable from indoors.

Figure 3-13: *Storm Windows with Low-e Glass*

Interior Storm Windows

Interior storm windows should have an effective perimeter air seal to prevent the warm, moist, indoor air from depositing condensation on the interior side of the cool primary window during winter.

Many interior storm windows have flexible or rigid plastic glazing that gives them a slightly higher thermal resistance than glass interior storms. However, plastic glazing scratches more easily and is more difficult to clean compared to glass.

When you consider interior storms, inspect the frames, sashes, and seals of the interior storm-window assembly to evaluate the storm window's durability. Avoid magnetic tape and other mounting components that depend on adhesives as their only means of attachment to the existing window frame.

Installing an inexpensive metal primary window at the interior side of the exterior primary window is a popular treatment for creating double or triple glazing in hotels, apartment buildings, and mobile homes.

Older buildings often have large single-pane fixed windows. Glazing contractors have retrofitted many of these large windows with interior storm windows that have a sturdy frame to hold the large storm-window sash. Maintenance workers can remove the sash to clean the storm window and the inside of the primary window if necessary.

Figure 3-14: Bahama Shutter and Quilted Window Shade

Insulating Shades and Draperies

Insulating shades and drapes can save energy and improve comfort. They usually cost much less per square foot or square meter than new windows. The residents must operate these window treatments daily to save energy and improve comfort.

Insulating shades, draperies, or shutters are more effective if they're airtight. A good air seal minimizes air leakage and convection and also creates an insulating air space. The air seal also prevents warm, moist indoor air from depositing condensation on the glass. See "Shading Windows" on page 37.

Doors

Like windows, doors have energy problems: air leakage, low R/RSI value, and thermal bridging. Replacing a door is usually a low energy-saving priority because the door's surface area is small, and its replacement cost is high. However when you replace an exterior door for any reason, choose the new doors based on thermal resistance and airtightness.

Door Components

Almost all doors in buildings have similar components. The door frame, which consists of door *jambs, door stops,* and *trim*, surrounds the door and attaches it to the building. Hinges mount the door to its hinge-side jamb and allow it to swing. The door closes against its *door stop*. The *threshold*, at the door frame's bottom, seals out most air and dust. The *knobset* — consisting of door knobs and a *latch* — latches the door in the closed position. The latch inserts into a *strike plate*, attached to the door jamb.

Door Types

Wood doors are still the most common exterior door type for residential buildings. Solid-core wood doors are made of high-quality lumber or are composite-board cores covered by thin wood veneers. Solid-core wood doors have an R-2.5 (RSI-0.44) or U=0.40 (USI-2.3). Wood panel doors, with decorative recessed areas, have an approximate R-1.75 (RSI-0.54) or U=0.57 (USI-3.2).

Energy-efficient residential doors have foam insulation inside a steel or fiberglass exterior shell. Insulated panel doors have R-values of approximately R-6 (RSI-1.1) or U=0.17 (USI=0,95). Many of these new doors have effective weatherstrips integrated into kerfs in their door frames.

Sometimes, carpenters built door frames in place and then hung the door in the frame. Modern exterior doors are pre-hung units with the frame, stop, hinges, threshold, and door packaged together and ready to install into a rough opening.

Door Repairs and Weatherstripping

Main entry doors are important because they are the building's public gateway. Durable materials and expert workmanship produce the best repair-and-weatherstripping results.

Door repair — Many doors need some repair and adjustment before you install weatherstrip. Consider these problems and solutions.

- *Loose hardware:* Tighten loose hinges, knobset, and strike plate. Adjust the strike plate if necessary to provide tight closure.
- *Uneven space between door and its jambs:* Adjust hinges to provide a consistent space around the door perimeter.
- *Settled building:* Level the building, or level the area under the door frame, or plane the door, or reinstall the door and its frame level and plumb.
- *Permanently warped door:* Adjust the door stop to conform to the warped door, and install a flexible weatherstrip that allows the door to move a little while remaining sealed.

- *Damaged door jamb at the strike-plate:* Repair the jamb with a wood insert or wood epoxy, or replace the jamb.

Door-jamb weatherstrips — Select door weatherstrips and thresholds for effectiveness and permanence. If the existing weatherstrip once sealed effectively and simply wore out, replace it with the same type. Ideally, weatherstrip should be continuous and only interrupted by the door-frame corners.Bronze V-strip must bypass the strike plate to allow the door to latch.

Avoid foam tape, felt, and most self-adhering weatherstrips because they wear out quickly and may never seal effectively. An exception is adhesive bulbs and foam tape for steel doors and steel door frames. Steel is smoother and easier to clean than wood, thus a durable adhered weatherstrip should last many years.

Door-stop-mounted weatherstrip is effective and easy to install. Most weatherstrip of this type has a rigid metal or wood retainer to hold the weatherstrips. Weatherstrips include bulbs, flaps, and fabric-covered foam. The disadvantage or door-stop weatherstrip is that it shrinks

Pre-hung exterior doors have a variety of weatherstrips inserted into pre-cut kerfs in the door jambs. These weatherstrips, inserted into the kerf, are easy to replace, and available locally or through the internet.

Certain plastic formulations retain their cold-weather flexibility better than others. For example, silicon weatherstrips have good cold-weather flexibility.

Spring bronze weatherstrips are a versatile and traditional weatherstrip that seal well and are durable if installed carefully. The rigid V-strip is much better than the flexible variety that comes rolled up. The flexible variety is more difficult to install and isn't as durable as the rigid V-strip.

Install corner seals to air-seal the door's bottom corners. Corner seals air-seal where the jamb weatherstrips don't completely seal the bottom corners at the threshold.

Door-bottom weatherstrips — Air, water, and dust may leak through the gap at door's bottom. The combination of a threshold and a door-bottom weatherstrip minimizes this leakage.

Weatherstrips, called *door bottoms* and *sweeps* attach to the bottom of the door and seal the door to the threshold. Some thresholds have weatherstrips mounted to them, but these wear quickly because people walk on them.

Proper adjustment of sweep clearance over the threshold is also important for both air-sealing and durability. Foot-wiping mats require a threshold at least as high as the mat so that the mat doesn't interfere with a door swinging inwardly.

Handicap-accessible doors usually have no threshold. In this case, use a retractable door bottom that seals to the floor.

If the door is exposed to wind-driven rain, the door bottom should divert any water draining from the exterior face of the door. Many door bottoms have built-in flashing to shed rain away from the threshold.

Figure 3-15: *Door Weatherseals*

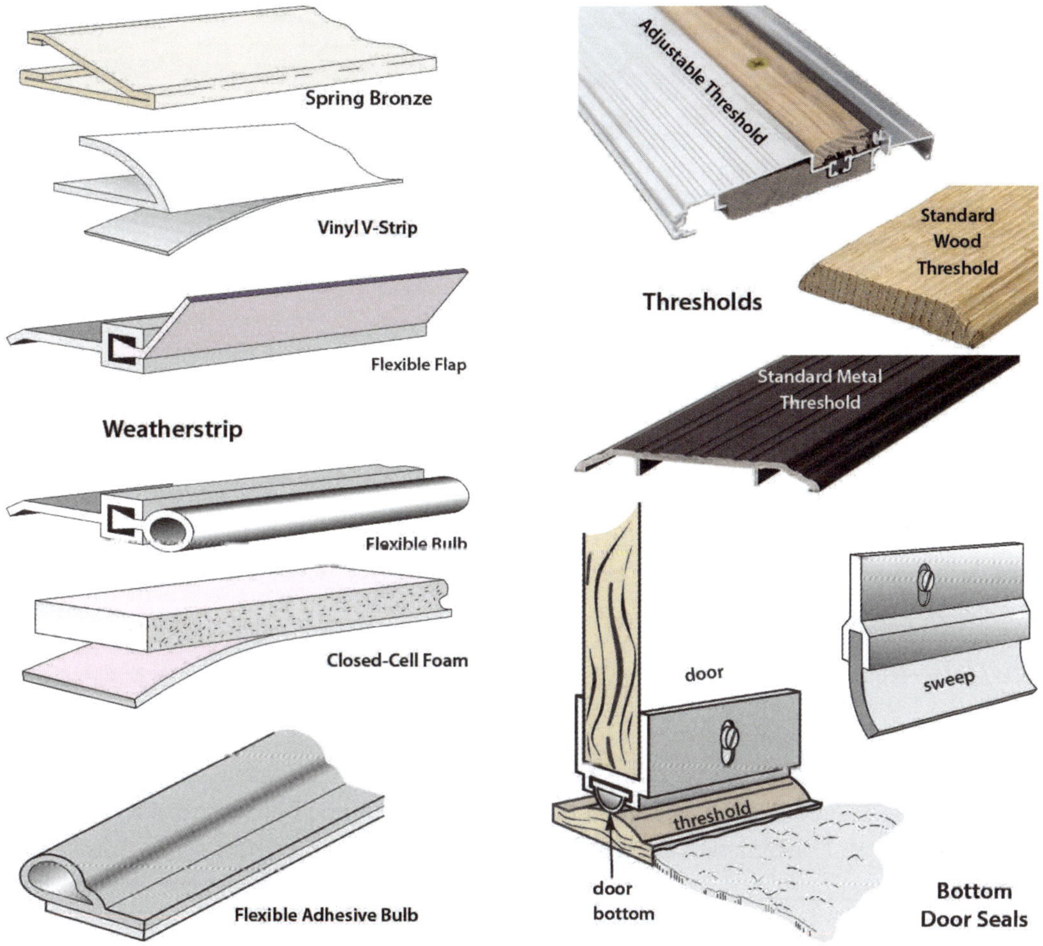

Figure 3-16: *Exterior Shutters: Sliding and Hinged*

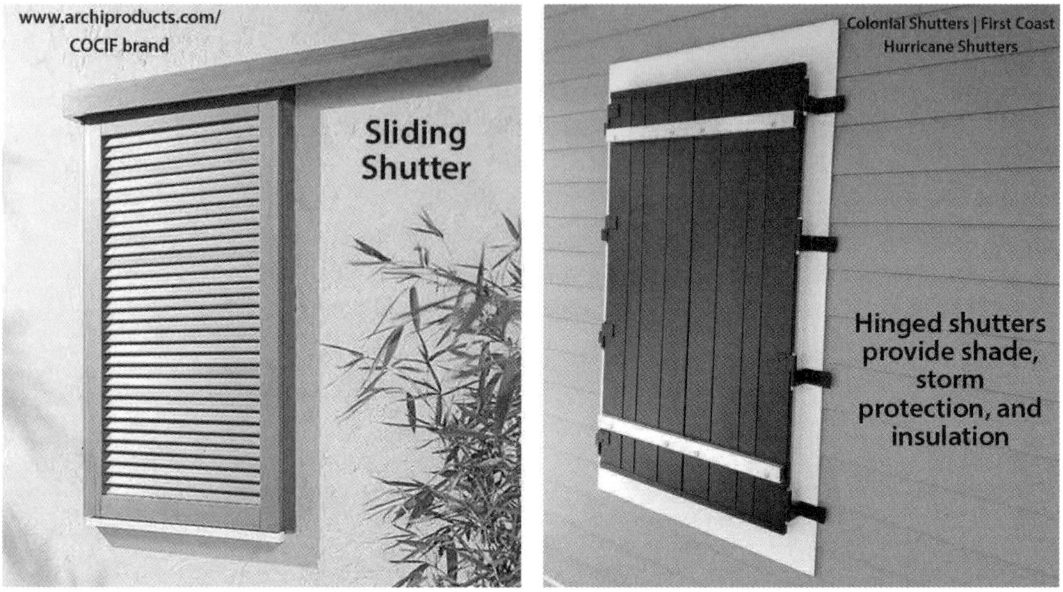

Double-Glass Doors

A double-door's large area of glass transmits a lot of heat into and out of a room. This heat flow takes three forms: indoor heat transmitted through the door and glass in winter, solar heat transmitted through the glass in summer, and air that leaks around the edges of the doors all year. We discuss swinging double doors and then sliding double-glass doors here.

Double-swinging Door Repair and Tightening

To make a double-swinging door (French door) airtight can be difficult. One of the doors is often stationary and held in place by a sliding latch for stability and airtightness. Make repairs and adjustments to the doors first. Understand that a wood door expands and contracts seasonally, so choose weatherstrip that can move a little with the door.

You can buy weather strip designed for French doors. Where the double doors meet, rigid bronze V-strip and a bulb or flap mounted on a rigid strip around the doors' outer perimeter may also work. On the bottom, try a door bottom with a flexible air seal like silicon. Latches on each door should hold the doors tightly into their door-frames.

Sliding-glass Door Repair

Air leakage is a common cause of comfort complaints. Air leaks around the perimeter of the door panels and between the panels. Sliding glass doors often have operating problems, caused by dirt in the tracks and worn rollers at their bottom. Vacuum dirt out of the tracks. Replace broken rollers along with worn and leaking weatherstrip, to improve the operation and the air seal of the doors. You can find replacement parts and repair advice on the internet.

Sliding-glass Door Insulation

Insulating quilted shades, though expensive, can improve winter and summer comfort reduce heat transfer through glass substantially. The best of these products have vertical tracks that seal the edges of the shade. The air seal makes the quilt's insulation effective, and it minimizes air leakage and convection. A roller system, fastened to the wall above the door, allows the shade to open and close vertically.

You can also fasten insulated or quilted fabric to the door frame with velcro, and move the quilt to storage when you don't need the insulation. Insulating fabric is available as either translucent or opaque materials.

Figure 3-17: *Glass-door Solutions*

Sliding-glass door summer solutions — For blocking glare and solar heat, exterior solutions are best. These include exterior reflective shades and exterior solar screens. Fasten the rigid solar screens to the frame of the sliding glass door using hardware that allows their removal in cool weather. Exterior rolling sunshades save energy and improve comfort.

Sliding-glass Doors Replacement

If you choose to replace your old sliding glass door, buy one with advanced insulated glass, improved rollers, and high-quality weatherstripping. A hinged patio-door may be a better option because hinged doors leak less air around their weatherstripping than sliding doors. Consider reducing the size of the sliding door opening and replacing the double sliding door with a single swinging glass door.

Your choice of glass is particularly important when buying a sliding-glass door. Choose low U-factor (U-value) for high and low outdoor temperatures along with low SHGC for superior summer performance.

Storm Doors

Storm doors reduce air leakage, and they increase the thermal resistance of the door assembly. Storm doors also improve the comfort of the area around the doorway. However, the cost for a storm door is high, compared to its modest return-on-investment.

Storm doors can offer outstanding cooling benefits by creating a large screened opening for cooling with ventilation. Order low-e glass in a new storm door if you choose to install one.

Figure 3-18: *Ideas for Storm Doors and Patio Doors*

4 Air Leakage and Ventilation

This chapter first discusses the principles of air leakage, air-leakage diagnosis, construction flaws that cause air leakage, and methods and materials for sealing air leaks. The second topic is ventilation. I put air leakage and ventilation together because so many buildings depend on uncontrolled air leakage for ventilation. I'll explain the necessity for controlled ventilation below.

Air sealing is essential for comfort and controlling electricity costs. However, you must control the humidity originating inside the building enclosure before or at the same time that you seal air leaks. Otherwise, indoor humidity could increase because the tighter enclosure traps the humidity, originating from indoors.

See "Moisture Management" on page 47 and "Household Moisture Sources" on page 289.

The Problem with Air Leaks

Air leakage in buildings represents from 5% to 40% of the air-conditioning costs. Air sealing is one of the most important goals of building renovation and new construction. Air sealing is usually the most challenging energy-conservation task because it requires skill, persistence, and hard physical labor.

An air barrier is a building component that prevents air leakage. The air barrier combined with insulation and solar control defines the *thermal enclosure,* also called the *thermal envelope.*

See "Finding Air Leaks" on page 99.

The Solution: Air Sealing

Air sealing stops hot-humid-air leakage into buildings during high outdoor relative humidity in summer. With high outdoor relative humidity and air leaks, air conditioners and even dehumidifiers often fail to remove enough water vapor from indoor air to provide sufficient comfort. If you plan to air condition a building that currently lacks AC, we suggest that you air seal the building before AC installation. Air sealing is particularly important for air-conditioned buildings in humid climates to proved comfort and prevent condensation.

See "Air-Sealing Methods and Materials" on page 113.

The Necessity: Controlled Ventilation

Most buildings employ accidental air leakage to provide ventilation. This accidental ventilation was good enough when outdoor temperatures were predictable, energy prices were stable, and outdoor air quality was healthy enough. Now, if we value comfort, healthy indoor air, and affordable energy, ventilation is a necessary part of changes we must make to our building.

The problem with using air leakage for ventilation is that the air-leakage rate is usually either too high or too low. Air leakage doesn't ventilate all areas of the building equally, leaving some areas drafty and fresh, and others stagnant and polluted. Also, air leakage is highest during severe weather, especially in winter, because wind and temperature differences maximize air leakage. The last topic in this chapter discusses mechanical ventilation and how to achieve controlled ventilation for air-sealed buildings.

See "Whole-Building Mechanical Ventilation" on page 126.

Air-sealing Goals and Benefits

The main benefits of air sealing are these.

- Save energy and money.
- Optimize comfort.
- Protect insulation's thermal resistance.
- Reduce direct heating of people and building components by hot outdoor air leaking through the enclosure.
- Stop moisture movement into building cavities to prevent condensation.

Air sealing may provide these additional benefits.

- Reduce pest's access to indoors.
- Reduce air pollution and dust from outdoor sources.
- Reduce rainwater leakage.
- Increase fire safety by preventing moving air from feeding flames.
- Promote controlled whole-building ventilation.

Figure 4-1: *Locations of Typical Air Leaks*

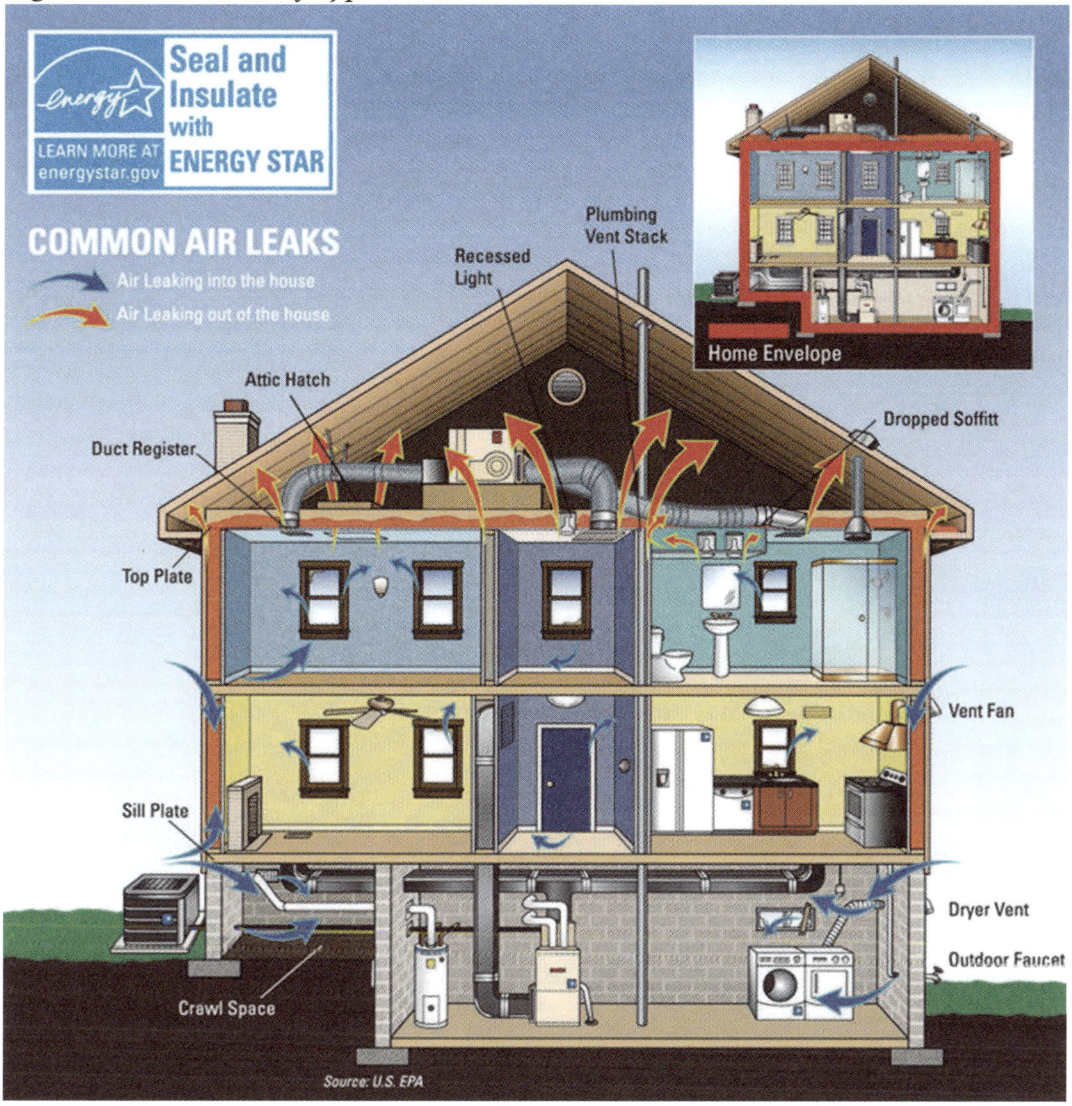

Air-sealing Research

Experts used to think that existing buildings were relatively airtight, except for joints where building materials meet, especially around windows and doors.

From 1975 to 1985, scientists and technicians developed instruments and methods to evaluate air leakage, including blower doors testing and infrared scanning. Using this technology, we now know that the building's hidden air leaks usually leak more air than small visible joints between building materials. As a result, technicians discovered new ways of finding and sealing these hidden air leaks.

In the 1990s, Canadian building scientists measured the air permeability of various building materials and joints. They also evaluated the durability of air barriers in high winds and found that high winds can disable weak air barriers like house wrap and polyethylene.

***Table 4-1:** Approximate Air Leakage per Unit of Surface Area at 50 Pascals Pressure*

Building Material	$CFM_{50}/1000\ ft^2$	$l/s_{50}/100\ m^2$
5/8" (16 mm) oriented-strand board	0.9	0.45
1/2" (13 mm) drywall	2.6	1.3
4-mil (0,1 mm) air barrier paper	2.6	1.3
1/8" (5,7 mm) hardboard	3.7	1.85
1" (25 mm) EPS (high density)	15	7.5
(Materials below not considered air barriers)		
Perforated asphalt felt	50	25
Standard concrete block	80	40
1" (25 mm) EPS (low density)	1700	850
5/8" (16 mm) tongue & groove boards	3000	1500
6" (150 mm) fiberglass batt	4900	2450
3" (75 mm) vermiculite	9300	4650
1.5" (38 mm) spray-on cellulose	11,600	5800

These values represent the approximate air leakage through each square unit of material during a 50-pascal blower-door test.

Based on: "Air Permeance of Building Materials" research report by Canada Mortgage Housing Corporation. Units converted from all-metric by the author.

Natural air leakage ventilates most homes and small multifamily buildings. Ventilation affects decisions about how airtight to make a building. Scientists, engineers, and technicians debate whether or not air leakage is an acceptable strategy to ventilate buildings.

Some building codes now require strong and effective air barriers. Many builders, architects, and engineers understand the need for air barriers because of the moisture problems and discomfort that leaky air barriers cause.

See "Air Leakage @ 50 Pascals for Different Types of Building Enclosures" on page 95, and "Air-leakage Rates @ 75 Pascals for Materials, Assemblies, and Buildings" on page 95.

Air-sealing Principles

Air leakage from one zone to another (for example indoors to outdoors) requires a hole between the zones and pressure to push air through that hole. Air that leaks in is called *infiltration*, and air that leaks out is called *exfiltration*. The airflow rate through a hole or group of holes depends on two factors: the cross-sectional area of the holes and the pressure difference (ΔP) between the zones.

Natural zonal airflows are usually too small and variable to measure. A blower door's purpose is to amplify the air-leakage rate in order to measure air leakage. Measuring pressure and airflow with a blower door allows you to estimate both the air-leakage rate and the surface area of holes in the air barrier.

Direct air leakage occurs at windows, doors, and other concentrations of joints, where air leaks directly through a place in the building enclosure. Indirect air leakage enters the enclosure in one location, flows through building cavities, and exits at a different location.

Energy auditors find many indirect air leaks in intermediate zones like attics and crawl spaces. One seldom-recognized air-leakage source is airflow through building materials themselves. Concrete block, brick, perforated felt, and fibrous insulation have high air permeability. We don't consider these materials to be air barriers.

Pressures driving air leakage come from wind, exhaust fans, *HVAC blowers*, chimneys, and the stack effect. (The HVAC blower is a large fan used to distribute heating, cooling, or ventilation air.) The stack effect is pressure caused by density differences between warm and cool air masses.

Chimneys, exhaust fans, and clothes dryers remove air from the building and thus create a slight vacuum, that we call *depressurization*. The wind, HVAC blower, and stack effect pressurize some areas of a building and depressurize others.

Beyond air leakage, air also circulates inside building cavities, increasing the rate of heat transmission. Air circulates inside building cavities, carrying heat from one surface to another. Air can circulate over the insulation's surface, convecting heat away. Or air can penetrate through an insulation material, reducing its thermal resistance.

Ideally, an effective air barrier, adjacent to the insulation, surrounds a building on all sides. An effective air barrier prevents most air leakage and convection. However, most buildings have ineffective air barriers that need air-leakage diagnosis and air sealing to become sufficiently airtight.

Figure 4-2: Measuring Pressure the Old Way

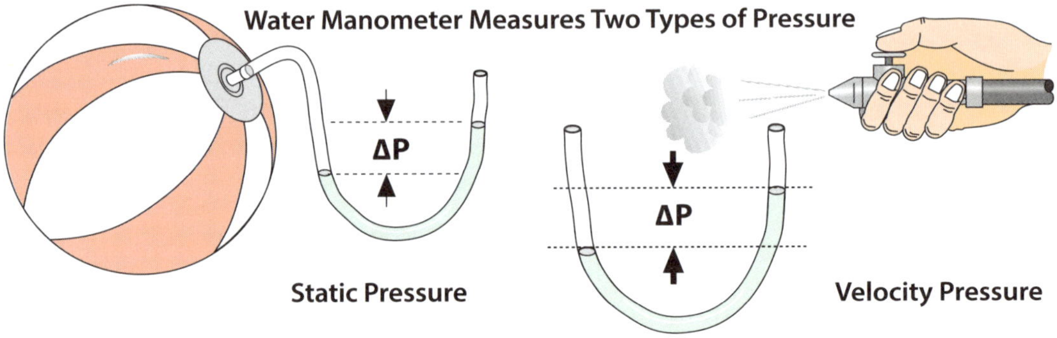

Air Pressure and Flow

Air pressure, airflow, and the size of air leaks are directly related to each other. A pressure difference on opposite sides of a hole causes airflow through the hole. Greater pressure causes greater airflow. Bigger holes allow more airflow than smaller ones at the same pressure.

Instruments called *manometers* measure pressure. Manometers come in three common types.

1. A transparent tube filled with water
2. A gauge with a needle indicating the pressure
3. A digital manometer, providing a digital readout

The air inside an inflated beach ball is denser than the atmosphere outside. Measure this pressure difference by attaching a water-tube manometer to the beach ball's valve. The lighter atmosphere presses on one side of the water manometer, and the denser beach-ball air presses on the other side. The distance that the beach ball's denser air moves the water column off level — measured in inches or millimeters — is a unit of air pressure.

We measure the small air-pressure differences caused by wind, blower doors, HVAC fans, and chimneys in inches of water column (abbreviated IWC) in the Imperial measurement system. The more common metric pressure unit is the pascal — 249 pascals equal 1.0 IWC.

When we discuss pressure differences between two zones, we say that the zone having denser air is pressurized, or is the high-pressure zone. The zone with less dense air is depressurized or is the low-pressure zone.

An analog manometer is a gauge with an arc-shaped scale and a needle for measuring pressure. This manometer has a high-pressure tap and a low-pressure tap. The manometer's needle moves in direct proportion to the pressure difference. The two taps usually connect to hoses that sample the air in the two different zones.

The digital manometer measures pressure with sensors, called pressure transducers. Digital manometers give their readings on a digital screen, and some can measure pressure and calculate airflow simultaneously because they have two channels.

A manometer measures airflow by sensing the pressure between the manometer's low-pressure tap and an opening parallel to the airflow stream. Airflow perpendicular to an opening creates vacuum inside the tube. This vacuum's strength is directly proportional to the airflow rate. Manometers convert this pressure to airflow by an airflow scale shown on the gauge face or by an electronic calculator inside the digital manometer.

Figure 4-3: *Manometers Measure Air Pressure*

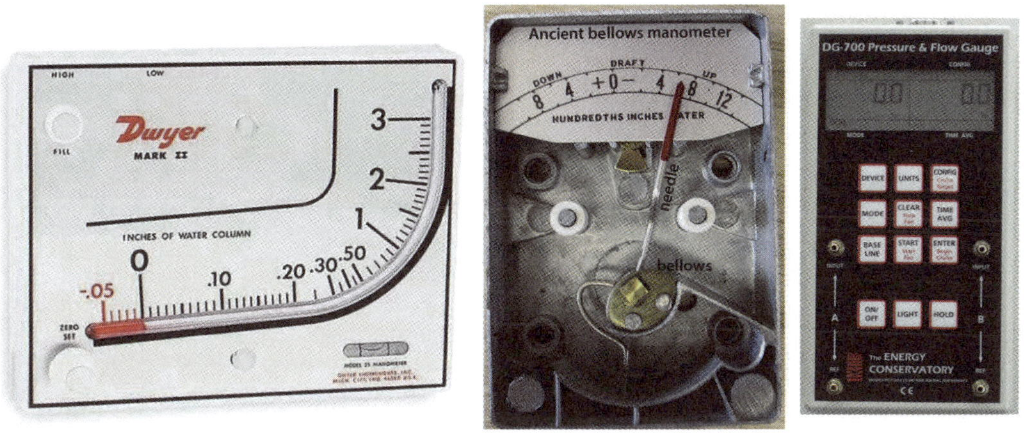

Fluid Manometer

Bellows Manometer

Digital Manometer

Pressure Drives Air Leakage

Air flowing inside or outside a building creates pressures that affect the building. Different buildings have different internal pressures for different reasons. Because of differences in pressure, one building may have two-to-four times the air leakage of another.

Homes with forced-air distribution systems, fireplaces, and large kitchen exhaust fans have larger pressures than homes without these features. Homes on hilltops in high-wind areas may have twice the air leakage of similar homes in less windy regions.

This section describes how a building's height, chimneys, wind exposure, exhaust fans, and HVAC blowers affect its indoor air pressures.

Stack-effect pressure — Cooler air is denser than warmer air, and this density difference creates pressure that moves air. A building's indoor air stratifies in layers because of density differences, caused by air-temperature differences. During both the heating and cooling seasons, hot air rises to the top and cooler air falls to the bottom. If the building has air leaks, warm air leaves through higher openings in the enclosure. The building's lower openings suck in cooler air. This air-leakage pattern is called the *stack effect* because it resembles airflow in a stack or chimney.

A building's natural pressure difference between indoors and outdoors varies depending on location and indoor-outdoor temperature differences. Somewhere near the building's vertical midpoint is a boundary region, called the *neutral pressure plane* that separates the building's negative-pressure and positive-pressure zones. A hole near the neutral pressure plane allows little or no leakage because there is no indoor-outdoor pressure difference there.

The pressures, created by stack effect, are greatest at the highest and lowest points in the building. Thus, a hole in a basement or attic leaks more air than an equal-sized hole near the neutral pressure plane.

Figure 4-4: *Causes of House Pressures*

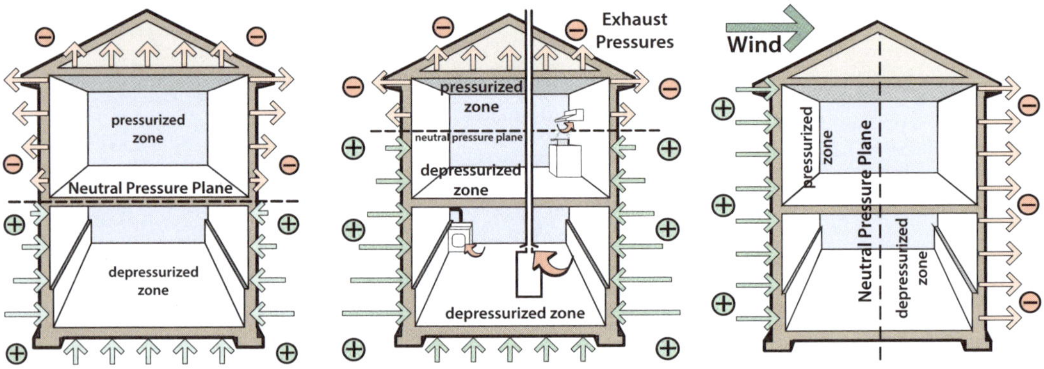

Wind pressure — Wind blowing against a wall creates high pressure that drives outdoor air into that windward wall. Wind blowing parallel to a wall or roof sucks air out of that building assembly. The building characteristics and surrounding features — overhangs, corners, fences, hills, depressions, other buildings, or vegetation increase or decrease the wind's pressure.

The wind's speed is greater the higher from the ground that you measure it. As building height increases, wind's pressure against the building increases. Wind speed varies widely between geographic regions and even within local areas. Strong wind gusts can even damage air barriers and permanently increase air-leakage rates.

Chimney and exhaust pressures — Chimneys, exhaust fans, and clothes dryers create a slight vacuum indoors because they exhaust air out of the building. They also move the neutral pressure plane up because more of the building's interior space is under negative pressure. A fan that forces air into the building moves the neutral pressure plane down, putting more of the building under positive pressure.

Replacement air, for the air that exits exhaust devices, is called *make-up air* for exhaust fans or *combustion air* for combustion appliances. Make-up air and combustion air enter through air leaks, intentional openings, fans, or ducts. Make-up air or combustion air may accidentally come down the chimney or a combustion appliance like a heater. If indoor make-up air does come down a chimney, a dangerous situation called *spillage* (temporary) or *backdrafting* (continuous) may occur.

Duct pressure — The HVAC blower circulates air through the *air handler* with its supply and return ducts. An air handler is a cabinet with a blower inside that moves air through a forced-air HVAC system. Supply and return ducts move air through supply registers and return grilles. Supply registers blow air into a room, pressurizing nearby areas of the room. Return grilles suck air out of rooms, depressurizing areas near those grilles.

If the ducts are leaking, or interior doors are closed, rooms may have positive or negative pressures. These pressures are often large enough to double or triple the building enclosure's air leakage, compared to its air leakage when the air handler is off.

Figure 4-5: *Blower-door Components and Air-leakage Measurements*

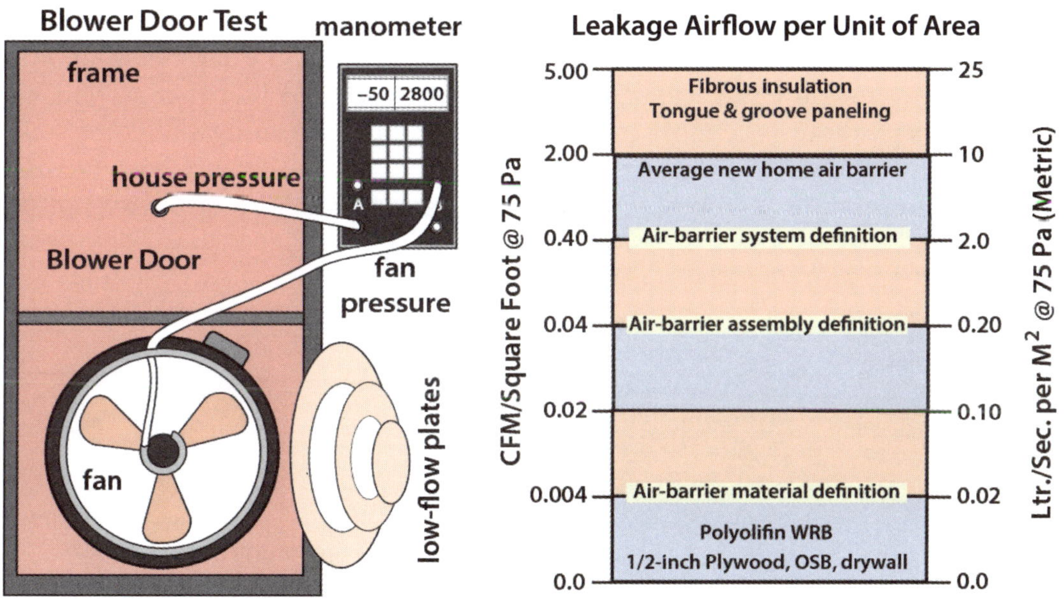

Blower-Door Testing

The blower-door includes these four components.

1. A frame and flexible panel to block an open doorway
2. A variable speed fan to create pressure and airflow
3. A manometer to measure the pressure difference between the building and outdoors or the pressure created from the fan
4. Hoses for attaching the manometer to outdoors and to the fan

The blower door depressurizes the building by sucking indoor air out of the building or pressurizes a building by blowing outdoor air into the building. Depressurization, which creates a vacuum indoors, is the most common procedure. Depressurization sucks outdoor air through leaks into the building. You can feel the leaking air coming in better than indoor air flowing out driven by pressurization.

Blower-door testing is the most practical way to predict energy savings or to verify compliance with airtightness standards. Technicians can evaluate existing air-leakage levels. They can also measure their air-sealing performance. Building codes now require specific levels of airtightness for code-compliance for new buildings.

The combined area of the building's holes and the pressure difference between indoors and outdoors determine the airflow rate through the blower door. An energy auditor measures air leakage in cubic feet per minute, liters per second, or cubic meters per hour. The most common indicator for evaluating a building's air leakage is the number of complete building-volume air changes per hour (ACH) through a blower door at 50 pascals of building pressure — abbreviated ACH_{50}.

Figure 4-6: *Digital Manometers for Blower-door Tests*

Preparation and Set-up for Blower-Door Tests

Preparing the building for a blower-door test includes these 4 steps.

1. Close primary windows and storm windows.
2. Open all interior doors.
3. Disable heaters and water heaters by turning down their thermostats.
4. Shut fireplace dampers, fireplace glass doors, wood-stove dampers, and wood-stove air intakes.

Blower-door tests require these 6 steps to set up the equipment and do the test.

1. Prepare the building for the test.
2. Cover ashes in wood stoves and fireplaces with damp newspaper, to prevent the blower door from sucking ashes into living areas.
3. Set up the blower door in a doorway.
4. Connect the manometer.
5. Turn on the blower door.
6. Read the pressure and airflow rate on manometer.

After preparation and setup, the blower-door slowly increases the building pressure to 50 pascals with the fan's variable-speed control. The operator inspects the building's interior during the test's first moments to ensure that dust isn't entering through a fireplace, wood stove, suspended ceiling, or other dusty area. With the building pressure at 50 pascals, the operator notes the air changes per hour at 50 pascals (ACH_{50}) from the manometer's digital display.

***Figure 4-7:** Air Changes Per Hour Estimates for New and Old Buildings*

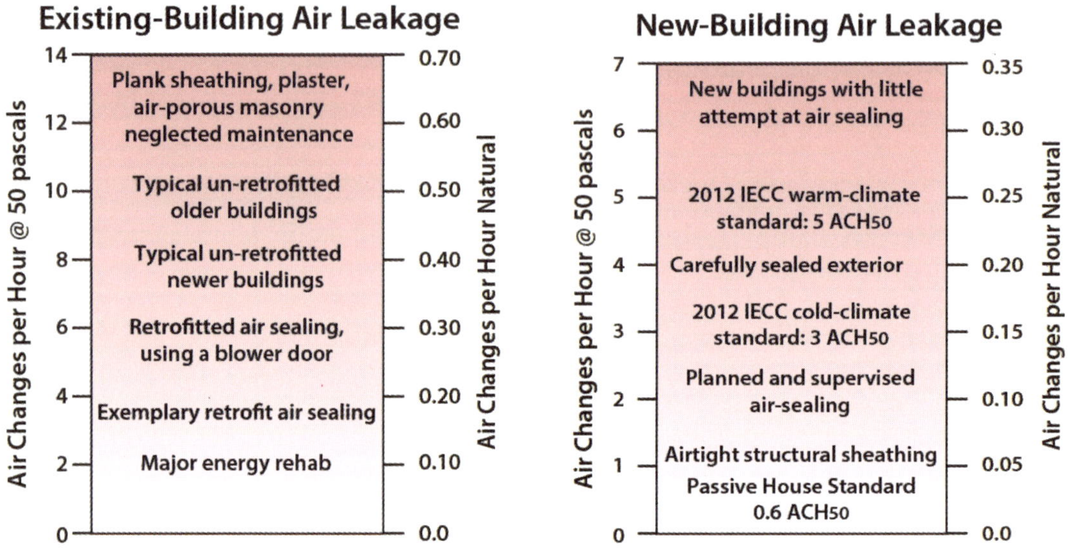

Blower-door Measurements

There are four common factors used to quantify air leakage with the blower door.

1. *The 50-Pascal Airflow Rate* — A blower-door reading, expressed in cubic feet per minute (CFM_{50}), liters per second (l/s_{50}), or cubic meters per hour (CMH_{50}) is the airflow rate, measured at 50 Pascals of building pressure. This airflow rate is the simplest and the most direct measurement of a building's airtightness.

2. *The 50-Pascal Air Change Rate (ACH_{50})* — The number of air changes per hour is a universal and common way of rating a building's air leakage. To calculate air changes per hour at 50 Pascals, multiply the airflow rate at 50 pascals by 60 min/hour and then divide by the building volume in cubic units (Imperial or metric). To calculate ACH_{50} from l/s_{50}, multiply l/s_{50} times 60 min/hour times 60 seconds/minute and then divide by the building volume in liters. The ACH_{50} is the same measurement in both Imperial and metric.

3. *Natural Air Change Rate (ACH_n)* — We express the natural air change in air changes per hour *(ACH_n)*. The *ACH_n* is the 50-pascal air-change rate (ACH_{50}), modified by numerical factors that account for wind speed, shielding of the building by external elements, and the building's height and size.

4. *Area-Specific Airflow Rate* — Energy auditors analyze large-building air-leakage with an air-leakage rate per unit of building surface area — cubic feet per minute per square foot (CFM/ft^2), liters per second per square

meter ($l/s \cdot m^2$), or cubic meters per hour per square meter ($m^3/h \cdot m^2$). The typical reference-building pressure for these blower-door tests is 50 pascals or less frequently 75 pascals. The airflow at 50 pascals is about half of the airflow at 75 pascals for the same test specimen. **You can estimate the natural air-change rate by dividing the 50-Pascal air-change rate by 20.**

Table 4-2: *Air-leakage Rates @ 75 Pascals for Materials, Assemblies, and Buildings*

Tested Specimen	CFM/ft^2	m^3/h•m^2	l/s•m^2
Air-barrier material	0.004	0.07	0.02
Air-barrier assembly	0.04	0.7	0.2
Air- barrier system	0.4	7	2

CFM/ft^2 means cubic feet per minute per square foot.
m^3/h•m^2 means cubic meters per hour per square meter.
l/s•m^2 means liters per second per square meter.
These air leakage rates would be about half of the listed values at 50 Pa.

Table 4-3: *Air Leakage @ 50 Pascals for Different Types of Building Enclosures*

Building Enclosure	Normal / Existing		Best Practice	
	CFM/ft^2	m^3/h•m^2	CFM/ft^2	m^3/h•m^2
Naturally ventilated dwelling	0.38	7.0	0.27	5.0
Mechanically ventilated dwelling	0.27	5.0	0.055	1.0
Office building	0.33	6.0	0.14	2.5
School	0.50	9.0	0.17	3.0
PassivHaus Standard			<0.055	<1.0

CFM/ft^2 means cubic feet per minute per square foot of floor area.
m^3/h•m^2 means cubic meters per hour per square meter of floor area.

Multifamily blower-door-test Strategies

Here we discuss three different strategies for blower-door testing a multifamily building: the whole-building test, the compartmental test, and the guarded test.

Energy auditors also measure zonal pressures to evaluate air barriers in multifamily buildings, like they do in single-family homes.

Air leakage from zone to zone in a multifamily building causes more problems than in a single-family building. The larger the multifamily building, the more important it is to seal leaks between the building's many zones. Tall buildings often have a very strong leak-driven stack effect that could total 2-4 pascals for each floor. When competing with this stack effect, forced-air heating, cooling, and ventilation systems sometimes can't provide the airflow necessary for comfort and ventilation.

Zone-to-zone leakage also spreads smells, microbes, dust, and chemicals. This zonal leakage causes variable temperature, comfort, and air quality among the apartments. You can't usually solve these problems without blower-door testing and effective air sealing. **With our warming climate, air-leakage tests and air sealing will become more common, if we hope to maintain current comfort standards.**

Whole-building test — Although difficult, the whole-building blower-door test is a preferred strategy in large buildings. Technicians usually need multiple blower doors for this test along with blower-door computer software.

The whole-building test provides an air-leakage measurement for the entire building. Among these three tests, this whole-building test gives the best information about the location of large leaks to outdoors.

Compartmental test — The compartmental test measures one apartment and requires only one blower door. This test measures air leakage to both the outdoors and indoors. The compartmental approach gives the energy auditor a sample of the air leaks in a single apartment. The leaks in this single apartment may inform the auditor about typical leaks in all apartments or in apartments with the same characteristics.

Guarded test — Energy auditors use the guarded test when they measure a single apartment's air leakage to outdoors. Similar to the compartmental test, the guarded test is another way to measure a single apartment's air leakage, but only leakage to outdoors. The guarded test requires multiple blower doors to pressurize surrounding apartments with the same pressure applied to the test apartment.

Zonal pressures — Creating and measuring zonal pressures is one of the most effective ways to evaluate air leakage in multifamily buildings. Testers observe the building's assemblies and architectural features. Then they formulate assumptions and questions to guide their air-leakage testing and air sealing.

The testers may decide to isolate various zones and assemblies to determine their individual leakage to outdoors or to adjacent zones. The goal is a testing and air-sealing process that is cost-effective. The tests may be quantitative or qualitative. Here are some zones and assemblies that you may want to pressure-test.

- Stairwells and elevator shafts
- Laundry chutes and garbage chutes
- Ventilation-and-HVAC ducts, duct joints, and duct chases

- Overhangs and balconies
- Basements, crawl spaces, and penthouses
- Floor, ceiling, and wall cavities

Figure 4-8: *Multifamily Blower-door Tests*

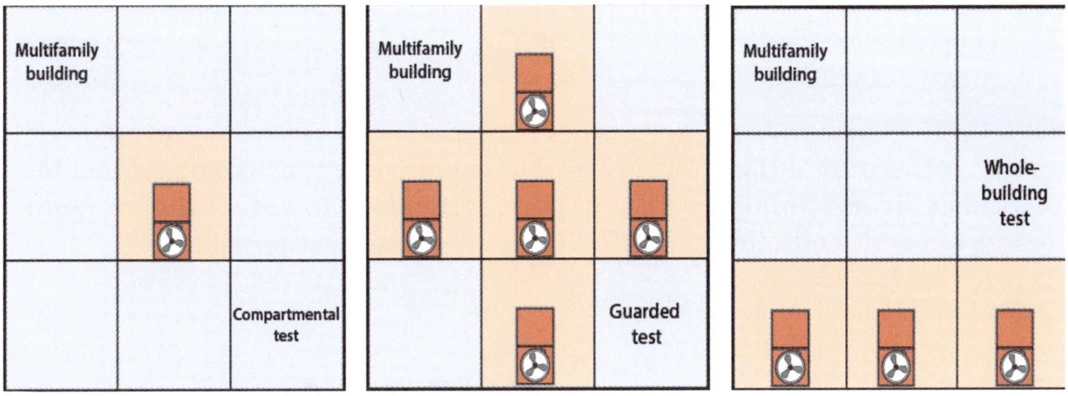

Ventilation and Air Leakage

Either whole-building ventilation system or air leakage must provide acceptable indoor air quality according to various codes and standards. ASHRAE and other organizations set ventilation-airflow requirements to ensure acceptable indoor air quality for residential buildings in the USA and in other countries where local jurisdictions adopt ventilation standards.

Ventilation adjustments for air leakage — The current ASHRAE Standard 62.2-2019 doesn't consider air leakage to be a legitimate ventilation strategy, with a few exceptions. ASHRAE 62.2-2019 specifies ventilation fan airflow according to floor area and number of bedrooms. However, the standard allows an infiltration credit to reduce the required ventilation airflow for existing homes if a blower door measures the air leakage. Climate is a factor in the infiltration credit, which makes existing buildings in mild climates more likely to qualify compared to buildings in severe climates.

Ventilation and air-leakage strategies — There are three strategies for reducing air leakage while providing adequate ventilation in existing homes.

1. Seal the building to the point where air leakage ventilates the home adequately under the infiltration credit of ASHRAE 62.2-2019, which requires no ventilation fan when the calculated airflow requirement is less than 10 CFM or 5 l/s (single family).
2. Seal the building as tightly as possible, and install a whole-building ventilation system using exhaust-only or supply-only ventilation (ASHRAE 62.2-2019).

3. During new construction or a major building retrofit, seal the building very airtight, and then install a heat-recovery ventilator (HRV) or an energy-recovery ventilator (ERV).

The first option allows air leakage to ventilate if the ASHRAE 62.2-2019 infiltration-credit calculation reports enough air leakage. The last two options comply with the ASHRAE Standard requirement for mechanical whole-building ventilation in most homes. The third option recognizes that HRVs and ERVs work best in very airtight homes.

Many residential buildings have indoor air pollution caused by combustion heating, tobacco smoking, moisture problems, and other environmental pollutants. Don't air seal buildings without considering how to reduce indoor pollution. Install a whole-building ventilation system whenever possible.

Figure 4-9: *Balloon-frame Problems*

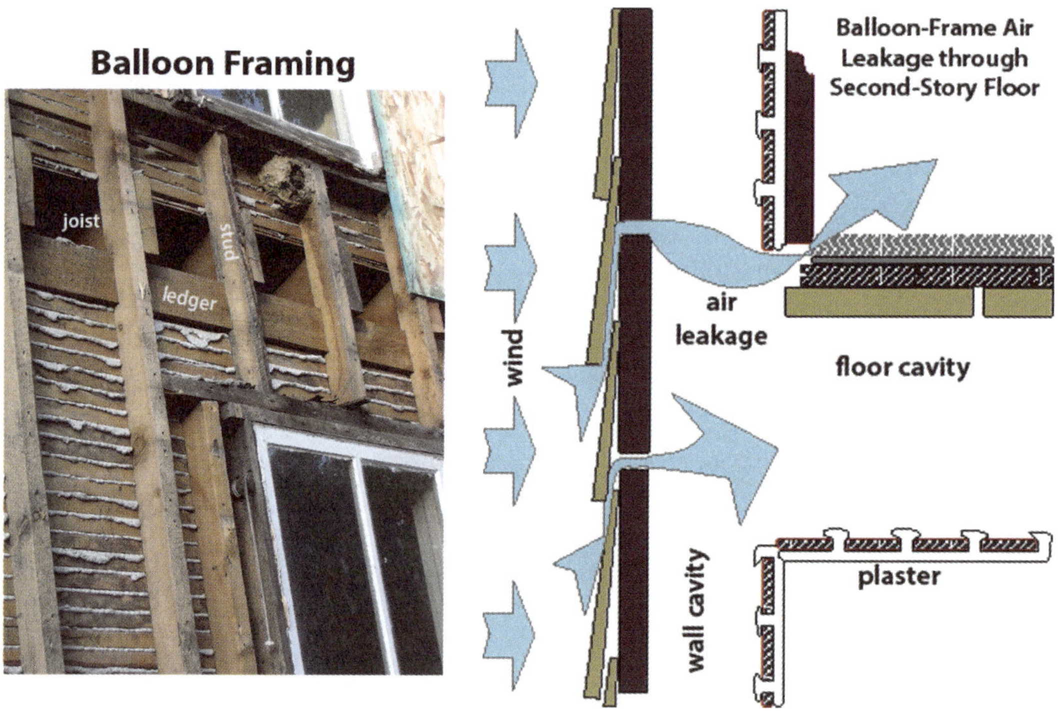

Finding Air Leaks

Energy specialists inspect many buildings where the architect and builder failed to plan and install an effective air barrier. Any gaps or holes in the building's exterior floor, walls, or ceiling contribute to air leakage. Damage, weather, remodeling, and other building changes create additional air leaks.

When a blower door depressurizes a building, small leaks around doors and windows become obvious. However, most buildings have less obvious but larger air leaks in attics, roof structures, crawl spaces, and basements.

Air-sealing Strategies

Look for air leaks in unexpected places. Technicians use diagnostic procedures to avoid guessing, unnecessary searching, and unsuccessful air sealing.

To control air leakage, try to create a continuous air barrier around all sides of the building that separates the indoor air from the outdoor air. Use a blower door to find the large air leaks where building components join like floor-wall joints or wall-roof joints.

Another secret to effective air sealing is knowing the relative leakiness of building materials and building assemblies. Plywood or drywall, installed carefully and sealed at joints, is extremely airtight. Polyethylene vapor barriers and house wrap can also be very airtight if sealed at all joints and protected from wind and sun during construction. Segmented building assemblies like lap siding or wood shingles over plank sheathing are very leaky. Masonry block, brick, stone, and tongue-and-groove paneling are also very permeable to airflow.

HVAC contractors should install ducts inside the thermal enclosure, making duct leakage a minor problem. However, more often ducts are located in intermediate zones, where duct air leakage wastes energy and causes building-pressure problems. Technicians use a variety of special procedures to locate air leaks in ducts using a blower door, duct blower, or pressure pan.

Simple Air-leakage Diagnostic Methods

The diagnostic procedures described here help yow find the best existing air barrier and to seal its high-priority air leaks. Give large air leaks and leaks at the building's high and low points (which experience high pressure) the highest air-sealing priority.

Use your hand to sense air leaks inside the building during a blower-door depressurization. Smoke from a *smoke generator* can locate leaks when you're working from an attic or crawl space because the smoke moves into air leaks during depressurization. A smoke generator is a chemical-smoke dispenser.

You can also pressurize the building, instead of depressurizing it, and then observe the attic floor from the attic. Look for signs of air blowing through the ceiling and into the attic (such as dust clouds or wavering cobwebs). Smoke works well during blower-door pressurization because you see the smoke leaving the conditioned zone through leaks.

Technicians sometimes use interior doors to test parts of a building's air barrier. Three common methods, used during a blower-door test, give you clues about which areas of the building are leaky and which are tight.

1. Compare rooms to one another by closing the door to a 1-inch-wide (25 mm) opening, and then feel how much air is moving, through the opening, from each room.
2. Close interior doors one by one, and measure the pressure difference between the home's main body and that room. The greater the manometer's measured pressure during depressurization, the more air is leaking through that room.
3. Open and close doors to rooms and intermediate zones, noting the effect on blower-door airflow (CFM_{50} or l/s_{50}).

Measuring pressure differences across interior doors allows you to evaluate that air leakage. If the room has no pressure difference across its door during depressurization, then there are few air leaks between that room to the outdoors. If the pressure difference is –30 pascals, then that room connects to the outdoors through some large air leaks.

Using the third technique, close the doors to two rooms, one after the other, and note the reduction of airflow through the blower door. If closing one room door reduced the airflow significantly, and then closing the second reduced the flow much less, you know that the air barrier in the first room leaks more air than the air barrier in the second room.

Figure 4-10: Blower-door Zone Pressures and What They Mean

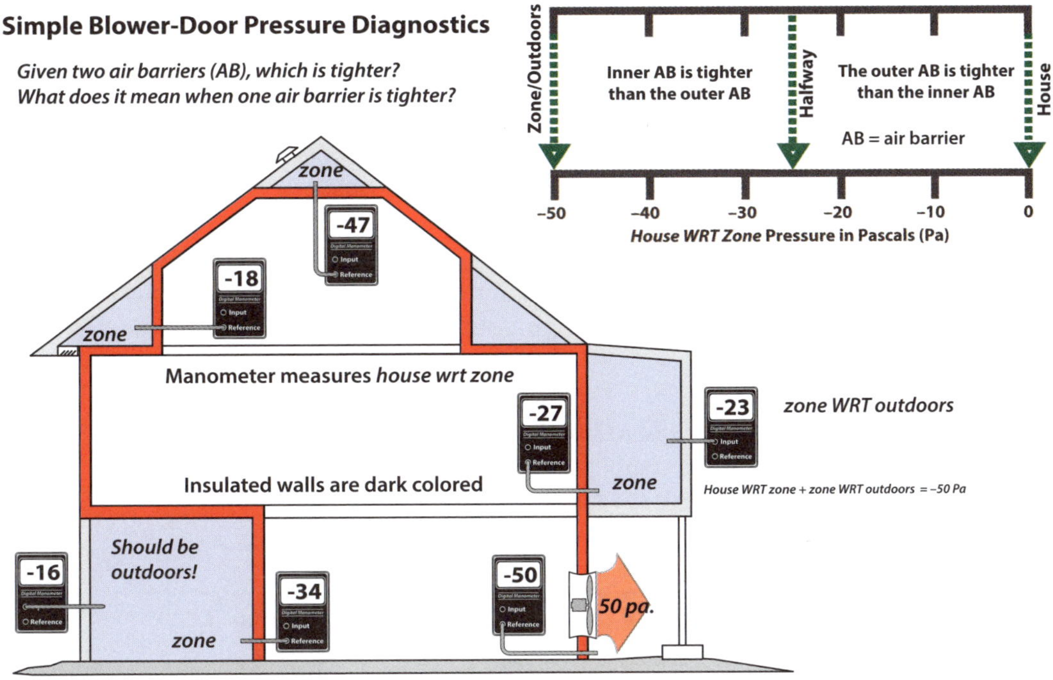

Attic air-leak testing — Opening basement doors, crawl-space hatches, or attic hatches can give you an indication of whether the floors or ceilings (between living spaces and intermediate zones) are effective air barriers. If the blower door shows little airflow difference between an open or closed basement door, then the foundation wall, not the floor, is the primary air barrier. This comparison helps decide how much time and material to spend air sealing a particular air barrier.

For example: To test the airtightness of a home's ceiling, depressurize the building to −50 pascals, and then insert a manometer hose through a small hole in the ceiling so that the hose's tip reaches above the insulation. Below are three possible pressure readings, with a probable cause for each.

1. *48 pascals:* The attic has almost the same pressure as the outdoors, indicating the ceiling is the tightest air barrier. The attic is well-connected to the outdoors, and the roof isn't an air barrier.
2. *25 pascals:* The ceiling is a partial air barrier, and the roof is a partial air barrier. They are equally airtight or leaky.
3. *5 pascals:* The building is almost completely connected to the attic. The roof is the tightest air barrier, and the ceiling is a leaky secondary air barrier.

You could assume that the attic connects to outdoors, but a 5-pascal pressure difference across the ceiling disproves that assumption. Air leaks connect the attic to the conditioned space through that ceiling. Of course, we could consider the attic a conditioned zone if it contains an HVAC system, or if the roof is airtight and insulated.

Technicians may employ a similar procedure to test the airtightness of the floor, foundation, or walls. If you need this information to prioritize air sealing, drill holes, insert the manometer tube, then read the manometer during a blower-door test.

You measure the largest pressure difference across the tightest air barrier. A 50-pascal pressure difference indicates that the air barrier is very airtight compared to any alternative air barrier. A 25-pascal pressure difference indicates that two air barriers are equally airtight.

Figure 4-11: *Measuring Zone Pressures during a Blower-Door Test*

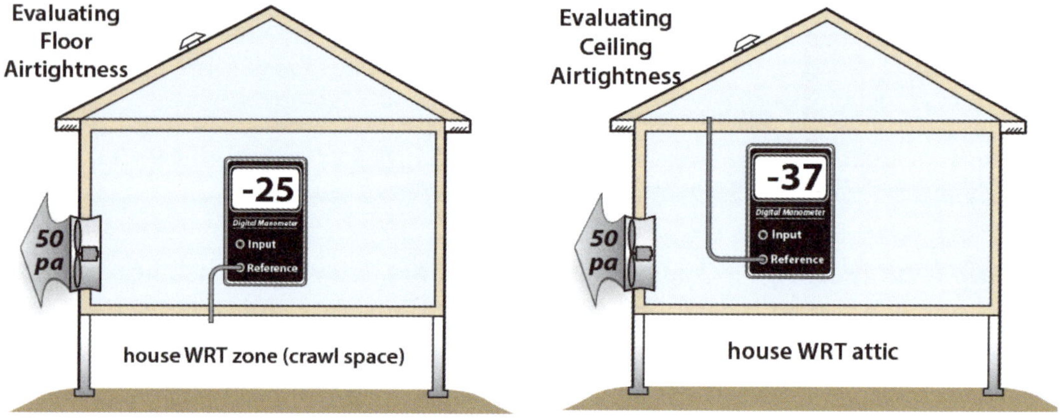

Air-barrier testing — Technicians prefer to seal one air barrier — ideally the one next to the existing or proposed insulation. Knowing which of two possible air barriers is tightest may not provide enough information for a satisfactory decision. If you can measure the amount of air leakage through an air barrier, this helps you determine the amount of effort worth making to seal that air barrier. Technicians sometimes use advanced diagnostic techniques that require them to open a hole of a known size in one of the two air barriers. That opening may be an attic hatch, crawl space hatch, basement door, or the door to an attached garage.

Using the surface area of these openings, their effect on inter-zone pressure difference, and the opening's effect on blower-door airflow, technicians can calculate how much airflow in CFM_{50} or l/s_{50} moves through a section of the air barrier. That measured airflow helps determine how much time and money to spend sealing the air barrier. The calculations, related to these diagnostic methods, are beyond this book's scope.

***Figure 4-12:** Visualize Air Leakage with an Infrared Scanner (Winter Test)*

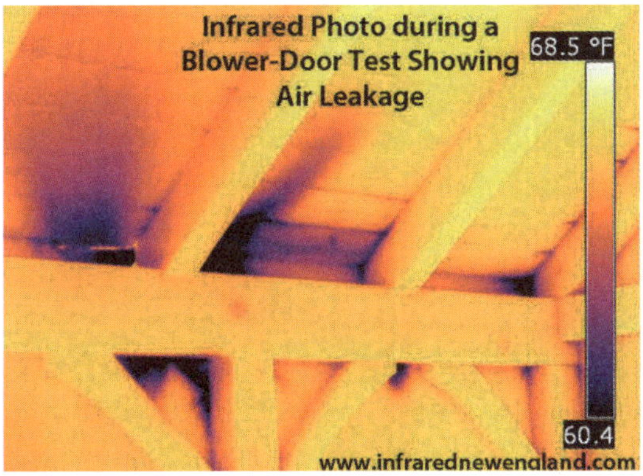

Infrared Scanners

Infrared scanners let you observe temperature differences visually. Energy specialists save time, using the scanners to find thermal flaws in the building enclosure.

Energy specialists usually inspect buildings with infrared scanners from indoors because solar heat interferes with outdoor infrared images. When there's a considerable temperature difference (at least 30°F or 17°C) between indoors and outdoors, the scanner shows the temperature difference of thermal weak points as colder (in winter) or warmer (in summer) compared to their surroundings.

Infrared scanners, used with blower doors, are a quick and effective way to find air leak. Energy auditors view interior surfaces, without the blower door to locate thermal bridges, and with the blower door to observe air leaks. Hot air, pulled in by the blower door during summer testing, heats surfaces near the air leaks.

***Figure 4-13:** Removable Airseals*

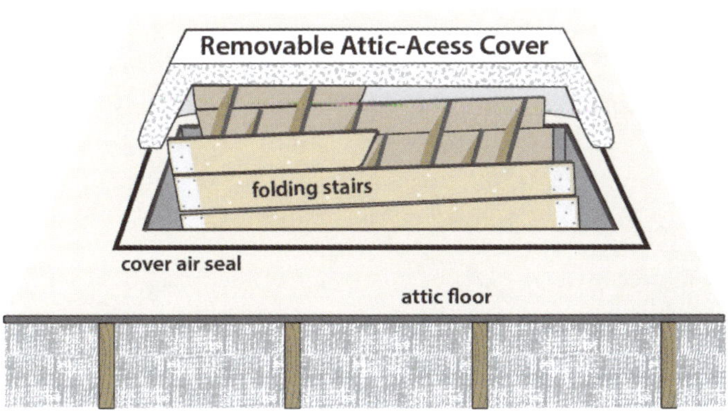

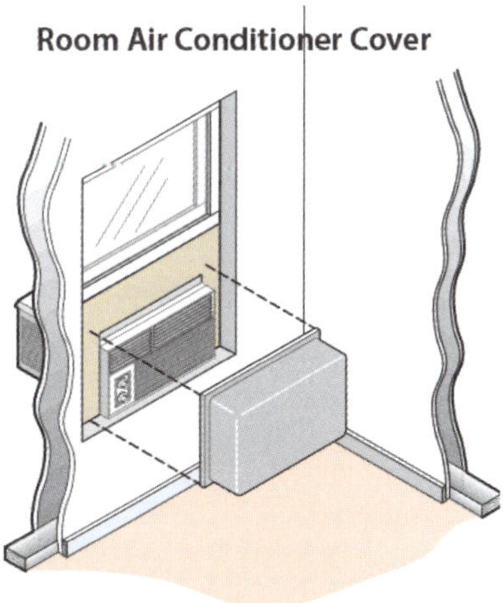

Construction Flaws and Air Leakage

The ideal building, from a thermal-performance viewpoint, is a simple 6-sided box. This ideal building has no protrusions or indentations. The insulation and air barrier, located next to each other, create a thermal enclosure that surrounds the conditioned space. Vertical shafts and horizontal building cavities are airtight. These cavities don't connect to one another and don't connect the indoors to the outdoors.

Air leaks in existing buildings — Unfortunately, existing buildings rarely have these ideal characteristics. Most builders, architects, and the general public don't yet understand the benefits of a continuous air barrier. The building professionals who do understand air barriers use one or more of these construction details as the air barrier.

- Exterior sheathing, like plywood, supported and sealed at all joints and at their perimeter
- A vapor retarder or exterior water-resistive barrier that also functions as a durable air barrier
- Drywall applied with gaskets around the perimeters of assemblies and openings indoors
- Concrete or masonry blocks, sealed airtight at their exterior surface

The construction flaws outlined in this section aren't necessarily energy problems, although they create air-sealing challenges. Builders can air-seal them during construction. However, since builders frequently fail to seal these potential air leaks, renovators can seal them later, when they weatherize or renovate the building.

Figure 4-14: *Large Attic Air Leaks*

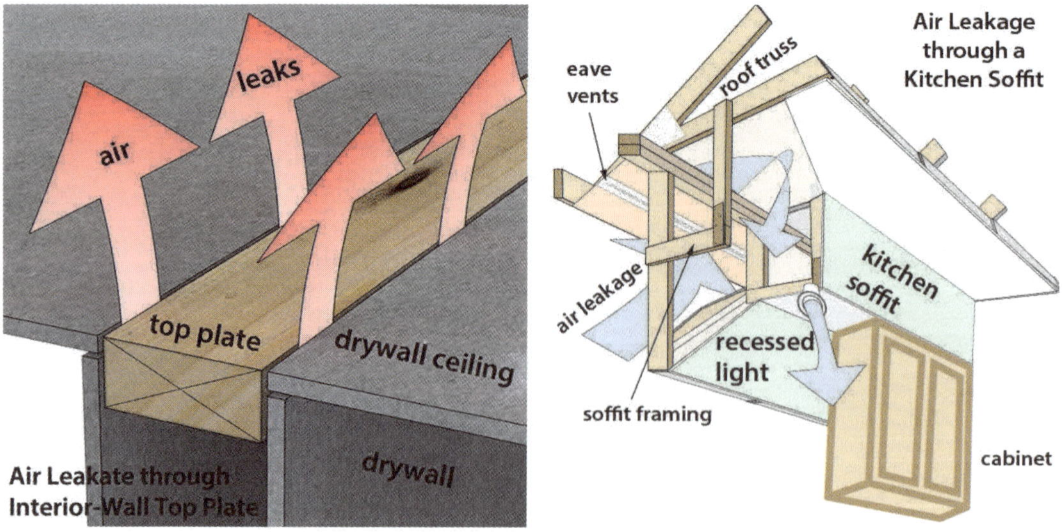

Single-family Structural-Leakage Sites

Homes and small multifamily buildings have typical air-barrier flaws caused by common construction methods. Effective air-sealing prioritizes the building's big leaks first. Retrofitters seal air leaks at the surfaces, joints, and edges of the building assemblies surrounding the air-conditioned space. Retrofitters seal air leaks, hidden inside cavities, with large airtight patches or by installing dense insulation, as a last resort.

Dropped ceilings — Dropped ceilings create invisible horizontal cavities that give convection and air leakage a path to move air-conditioned air outdoors and to move hot outdoor air indoors. Dropped ceilings in older homes frequently hide falling plaster above them. The space above dropped ceilings connects directly to ventilated attics. The suspended ceiling's panels don't create an air barrier because they have unsealed joints, and the ceiling panels themselves aren't airtight.

Soffitted ceilings (narrow sections of dropped ceiling) above bathtubs, cabinets, and stairways often connect directly to unconditioned attics or unconditioned floor cavities above. Even if the cavity created by the soffit is airtight, convecting air from the unconditioned attic moves in and out, increasing heat transmission.

If the interior surface of a suspended ceiling cavity is leaky, indoor air leaks into that cavity, and others such as: connected walls, attics, or floor cavities. If outdoor air flows into any of those cavities, it mixes with indoor air through leaks in the cavities. Besides energy waste, this indoor-outdoor air mixing can cause condensation and indoor-air pollution.

Cathedral-and-vaulted ceilings — Builders often create thermal flaws during original construction of cathedral ceilings, and these ceilings are difficult to repair later. Cathedral ceilings are thermally weak at their eaves because of wind washing and airflow through fibrous insulation's porous edges at the top of the wall. Infrared scanning usually shows both air leaks and thermal bridging through the interior wall-ceiling junction — an obvious thermal weakness.

Indoor air leaks into the cathedral ceiling through cracks in tongue-and-groove wood ceilings, around recessed light fixtures, and through joints at beams. Condensation, caused by humid indoor air or outdoor air, can dampen insulation and other building materials. In many cathedral-ceiling cavities, ventilation to remove this moisture is absent or ineffective, especially if the roof has no continuous ridge vent.

Scissor trusses create the sloping ceiling desired for aesthetics during original construction. The vaulted ceilings, created by scissor trusses, are an effective way to avoid the problems of cathedral ceilings. Vaulted ceilings are much easier to ventilate and insulate to modern thermal-resistance standards, compared to cathedral ceilings.

Figure 4-15: Cathedral-ceiling Air Leakage and Moisture Problems

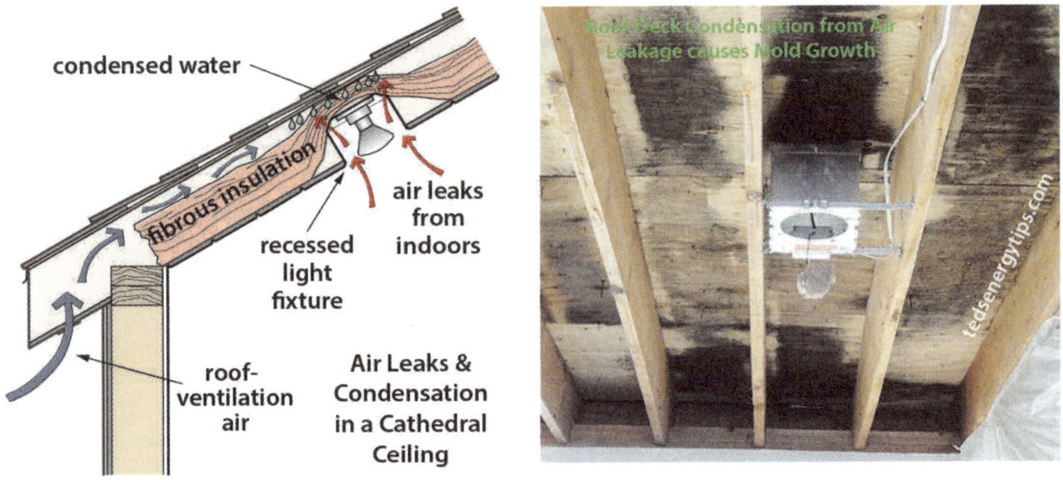

Frame-wall cavities — Balloon framing creates a continuous airway for air convection and air leakage around the conditioned space. Even if the building enclosure has an effective air barrier, this convection allows rapid heat transmission. Most balloon-framed walls don't have bottom and top plates (although some do). Exterior or interior walls both may be open to the unconditioned basement or attic.

If the balloon-frame construction connects to outdoor porches and bay windows, air leakage may be a building's dominant energy problem. The porches and bay

Figure 4-16: Vertical Leakage Paths: Basement and Attic

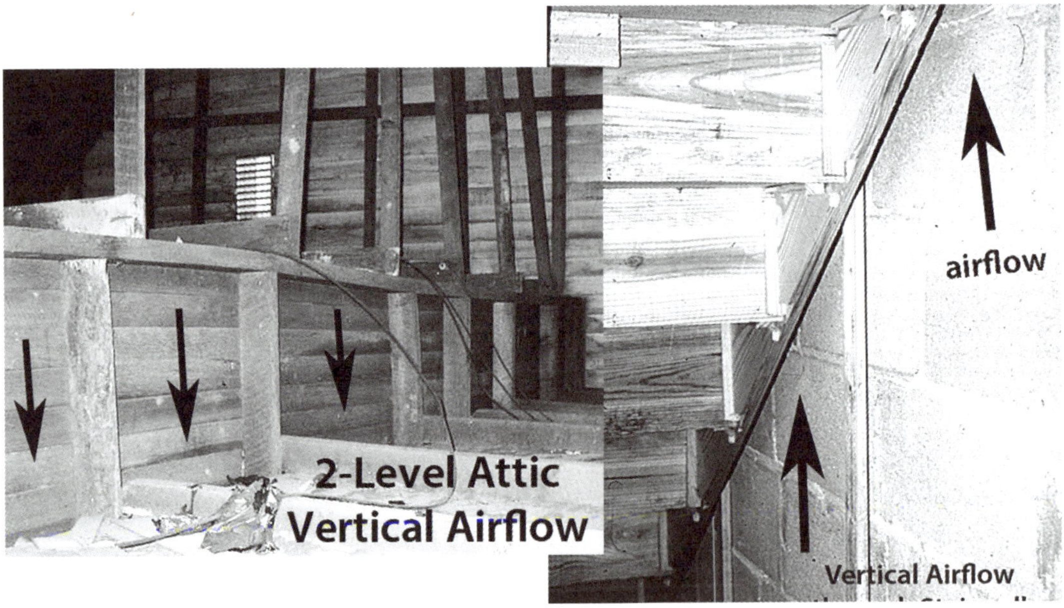

windows may connect the continuous balloon-framed floor, wall, and ceiling cavities directly to the outdoors through porch lights or any concentration of exterior joints.

Either balloon-framed or platform-framed walls may be open behind bathtubs, porch roofs, and interior soffits, without interior or exterior sheeting. The missing air barrier allows outdoor or indoor air to enter these wall cavities.

The wall's junctions to the ceiling, roof, and floor, especially where pipes and wires penetrate, are also large air-leakage sites in many buildings, even new ones.

***Figure 4-17:** Air Sealing Kneewall Floor Cavities*

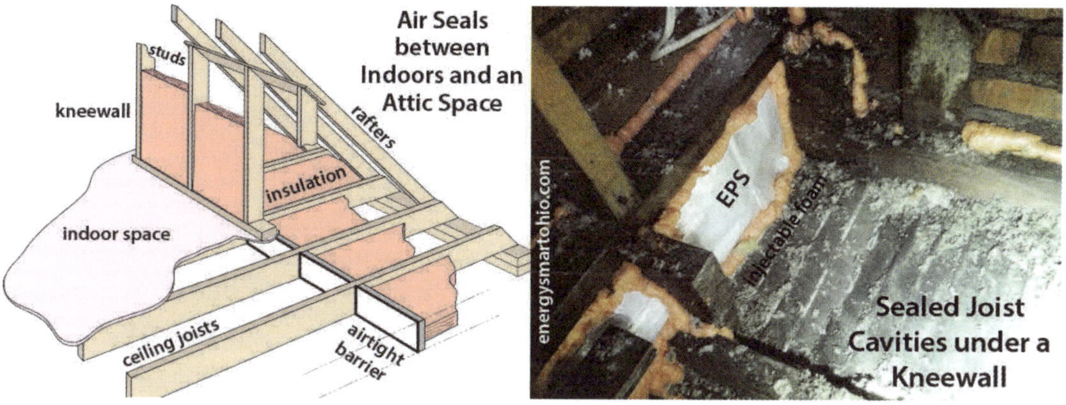

Floor cavities connected to outdoors — *Cantilevered* floors over-hang the supporting wall below to add floor space to an upper floor or to provide framing for a balcony or porch roof. This overhang often allows outdoor air to enter the floor cavity. If the soffit underneath the cantilevered floor is leaky, air leaks into that floor cavity from outdoors, and indoor air from the floor cavity leaks out.

The finished attic of a one-and-one-half-story (Cape) home is a common example of a floor cavity that connects to the outdoors. The triangle-shaped section of attic behind the second-story knee-walls usually ventilates to outdoors, or is at least a leaky intermediate zone. The floor-joist spaces underneath the kneewalls are usually unsealed. These rectangular holes combine for a large surface-area of air leaks into a floor cavity. Also, the outdoor air convects through the cavity, causing rapid heat transmission. Outdoor air also leaks indoors (and indoor air leaks out) if the floor cavity has air leaks into conditioned spaces above or below.

Figure 4-18: Workers Removed a Porch Ceiling to Install an Airseal and Insulation

Roofs joining walls — Joints between roofs and walls can be large air leaks. Porch roofs, in particular, can create a connection between indoors and outdoors through joints in the porch's ceiling or cracks around its roof perimeter. The porch's roof cavity may connect directly to the second story's floor cavity or exterior wall cavities.

The protrusion of the porch roof, formed by overhanging floor joists, can allow a leaky porch ceiling or soffit to bring outdoor air into the second-floor cavity. Another possibility is that the porch roof rafters were nailed to the second story wall studs before the builders applied wall sheathing. This practice connects the second-floor cavity and the exterior wall to that porch-roof cavity. A third possibility is that carpenters built the porch after they installed sheathing, but before the siding. In this case, there is sheathing behind the porch roof. Cracks between and around the sheathing may still allow air leakage into wall-and-floor cavities.

Plumbing penetrations — You can find air-leakage pathways in many buildings by following the plumbing. Bathrooms, in particular, create air leakage through bathtubs sitting against unsheathed wall studs, holes in the floor underneath bathtubs, and plumbing walls with large holes in their tops and bottoms.

Plumbers often enclose the main vertical drain pipe (called the soil stack) in a wall or in its own vertical cavity. This cavity can move air between the basement, attic, indoor spaces, and outdoors.

Figure 4-19: *Sealing Large Air Leaks in Ceilings and Floors*

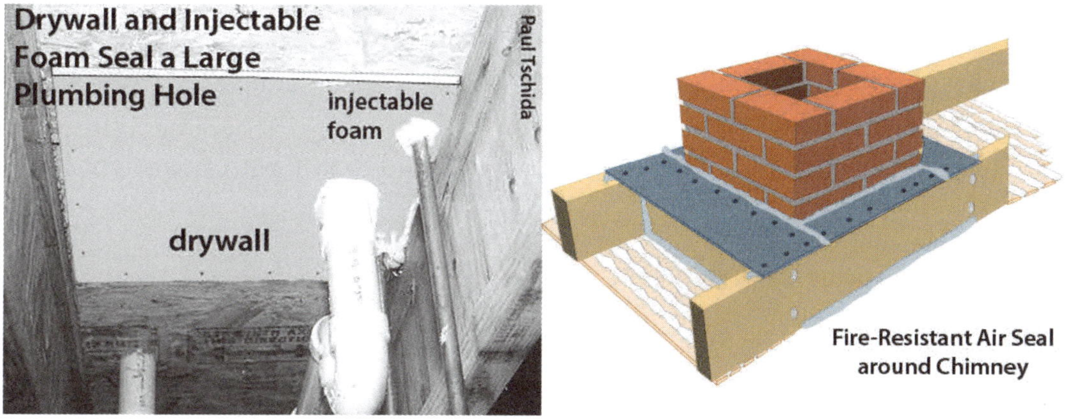

Attic and basement stairways — A stairway is a rectangular hole in one floor with stairs leading to another floor. On one or both sides of the stairway are walls. Stairways leading to unconditioned spaces offer two choices for air sealing and insulation.

1. Workers may install an insulated and weather-stripped hatch cover or door at the top of the stairs.
2. Or, retrofitters may air seal and insulate the stairway's walls and under the stairs. An airtight door at the bottom completes the isolation of the unconditioned stairway from the conditioned space surrounding it.

Figure 4-20: *Air Sealing Recessed Light Fixtures*

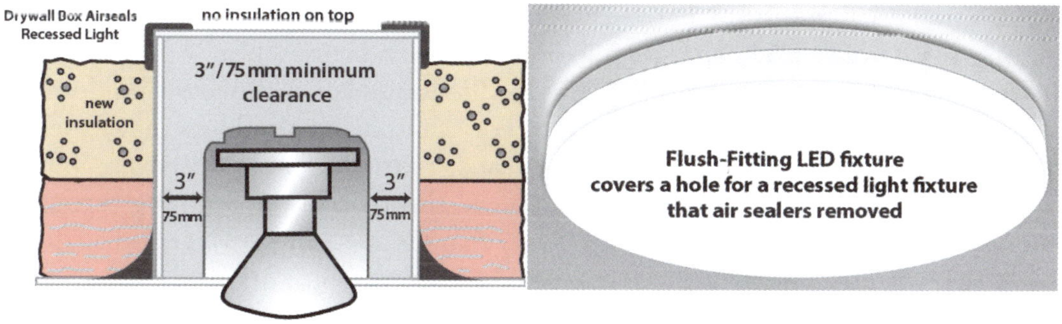

Recessed light fixtures — Recessed light fixtures often leak air through them. These fixtures, when they contain incandescent bulbs, must move air through holes in their enclosure to carry heat out of a fixture. If you replace the incandescent bulb with an LED bulb, you can insulate and air seal the fixture as shown above.

Installed in soffits, cathedral ceilings, and suspended ceilings, recessed light fixtures connect the conditioned space to attics, floor cavities, or roof cavities.

Recessed light fixtures allow warm, moist indoor air to reach cold roof decking, causing condensation. Or hot humid air may leak through the fixture and condense inside the air-conditioned indoors.

One remedy is to build an airtight drywall or sheet-metal enclosure around the existing recessed fixture. This enclosure must be both an air barrier and a fire barrier. Another remedy is to replace the existing recessed fixture with an airtight LED fixture that needs no ventilation.

***Chimneys* —** Older residential buildings, especially large ones, may have multiple chimneys. Any fireplace chimney can be a very large air leak if it has a leaky damper or no damper. Heater or water-heater chimneys also exhaust conditioned air from the building, depressurizing the building. Chimneys with weak draft can backdraft continuously, pulling outdoor air and combustion gases into the depressurized building, during a coal or wood fire in the fireplace.

Chimney dampers almost never fit tightly. Inflatable pillow-like bags or flexible foam plugs may be effective for sealing unused fireplace chimneys. Seal the tops of unused chimneys with airtight metal caps to prevent the entry of both water and air.

Multifamily Buildings — Air Leakage

Multifamily buildings often have relatively tighter enclosures than single-family homes. They also have less exterior surface area per unit of floor area. Also, the multifamily-building exterior is often quite weather-tight to resist rain and wind. This is important because wind is stronger at higher elevations above the ground.

The multifamily building experiences greater pressures than the single-family building. High pressure encourages both air leakage and convection inside cavities. This airflow and convection wastes energy, redistributes conditioned air randomly, reduces comfort, and distributes pollutants. For example, higher floors may experience uncomfortably high temperatures because warm air migrates from lower floors. Pollutants from the basement can spread throughout the building through duct leaks and through zone-to-zone air leaks. Unintended air pressure and airflow can overpower ducted HVAC systems, creating energy, comfort, and indoor-air-quality problems.

Vertical shafts and horizontal cavities circulate air within them and frequently have leaks to the outdoors. Air seals for multifamily-building cavities must be structurally strong to resist strong air pressures. Stack effect combined with wind can send blasts of air rushing through vertical shafts, blowing lightweight air seals out of their holes.

Air-sealing materials installed between floors and through shafts of large multifamily buildings should also be fire-resistant. Fire-resistant air seals prevent vertical shafts and horizontal cavities from becoming an air source or pathway for a spreading fire. Among the fire-resistant materials builders use are drywall, masonry block, mortar, non-combustible caulk, fire-resistant foam, and sheet metal.

Ceiling and floor cavities — Horizontal cavities in multifamily buildings may be deep, giving air convection and intruding outdoor air a large airflow pathway. The same cavities may function as return-air ducts for large air handlers that heat, cool, and ventilate the building. Indoor air enters cavities through grilles, ceiling joints, recessed light fixtures, plumbing penetrations, and other openings in ceilings and floors. These cavities are sometimes connected to the outdoors through joints in balconies, bay windows, or overhangs. If floor and ceiling cavities contain outdoor air, that air can mix with the cavity's indoor air and also enter conditioned spaces.

Vertical shafts — Most larger buildings have vertical shafts for elevators, stairs, pipes, wires, and structural columns. Vertical shafts also include laundry chutes, garbage chutes, ventilation shafts, and incinerator shafts. Openings between indoor spaces and these shafts cause major air leaks. Openings at the shafts' bottoms and tops create pressures that pull air into the shafts and push air out of them.

Contractors often renovate many formerly open stairways with fire walls and fire-rated doors to prevent a fire from spreading rapidly through the open stairways.

Ventilation systems — During the early part of the 20th century, many building designers and health experts believed that a building needed overly large amounts of ventilation air. Older buildings may have natural ventilation systems or power ventilators that include shafts from the basement through the building, terminating on the roof.

The most common type of shaft links the kitchens and/or bathrooms with the outdoors. These shafts usually provide excessive ventilation and contain air leaks. Diagnostic methods may be necessary to measure and adjust the airflow rate of existing ventilation systems and to measure leaks in the ventilation shafts.

Figure 4-21: *Multifamily Air Leakage and Air Sealing*

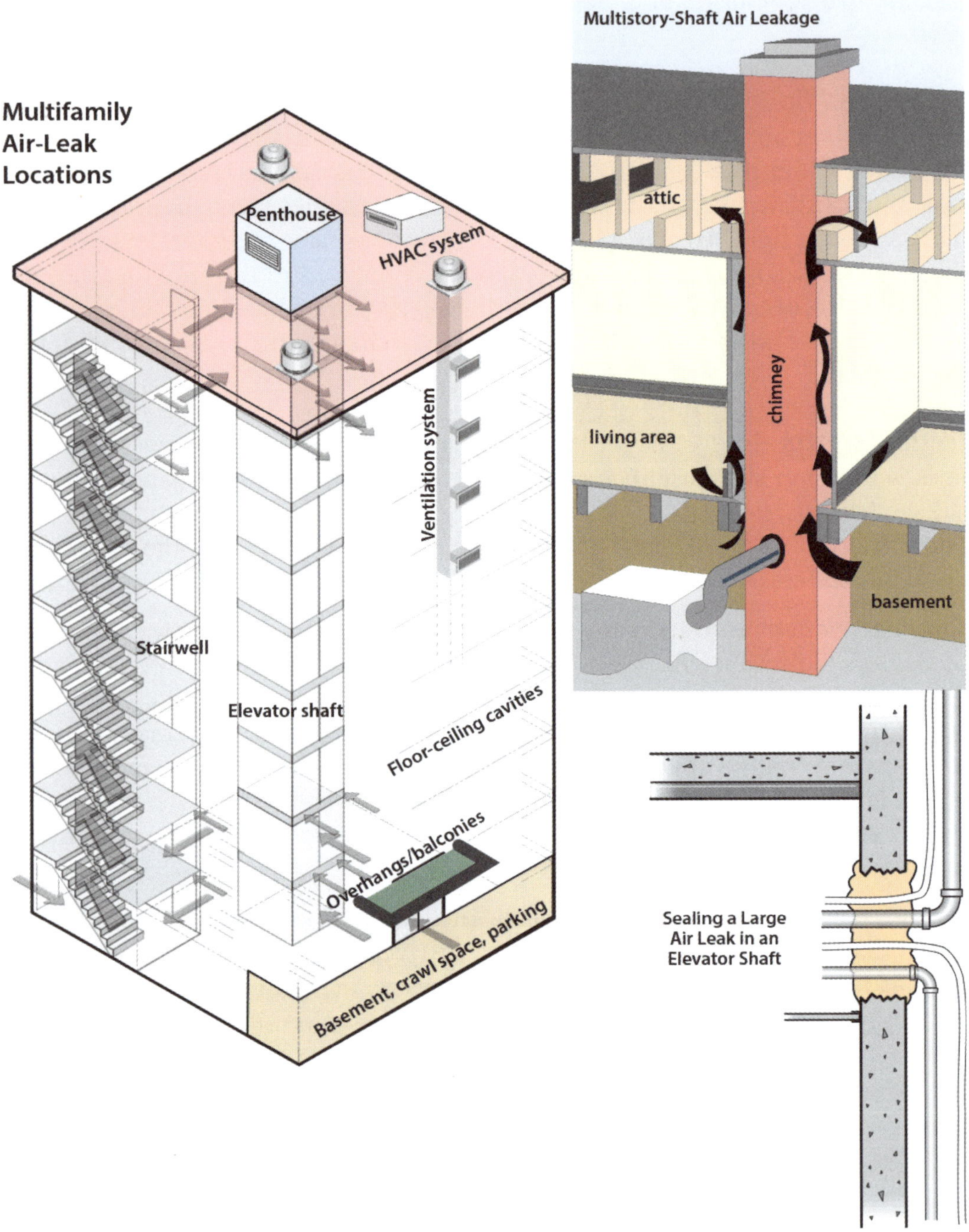

Figure 4-22: Suspended Ceilings: Problems and Solutions

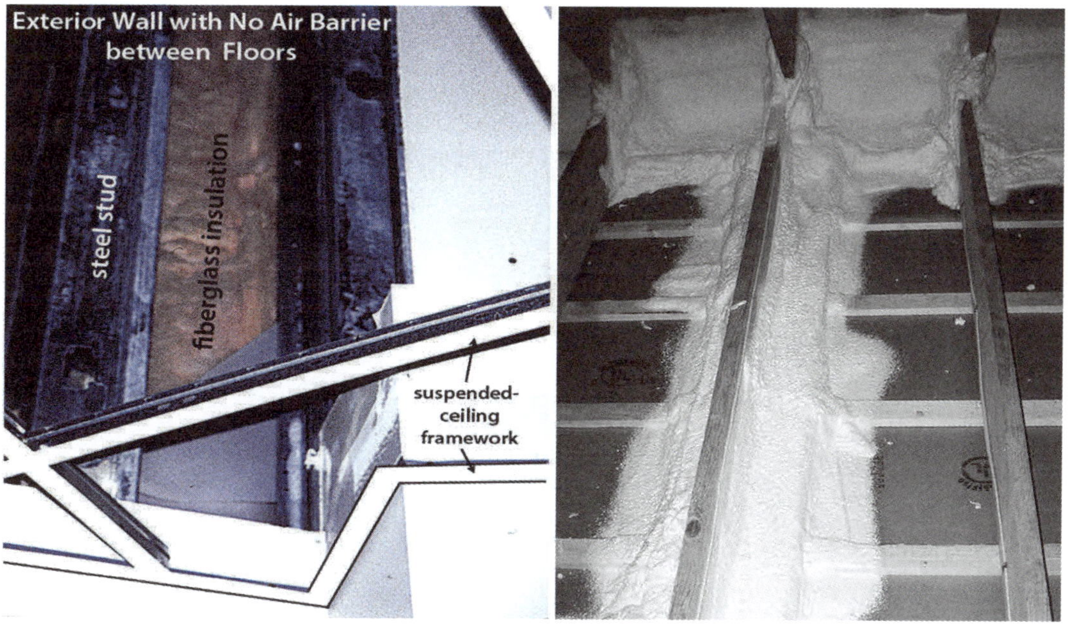

Air-Sealing Methods and Materials

A building's design, its materials, and their installation determine the size, shape, and location of air leaks. The choice of air-sealing methods and materials depends on the following.

- The air leak's size, shape, location, and the location's visibility to residents.
- The air-sealing material's compatibility with nearby materials.
- The pressure difference that the patch or sealant must resist.

Seal the biggest leaks first. The most effective air-sealing methods target large air leaks at the building's surfaces and inside its cavities. Seal leaks around plumbing, wiring, chimneys, and joints in the building sections. Joints in supply ducts are very cost-effective to seal. Experienced air sealers find and seal holes in attics, under floors, above suspended ceilings, and in many similar locations.

Air-sealing Strategy

Some technicians believe that sealing all visible cracks in the building's interior and exterior creates an effective air barrier. Of course, maybe every tube of caulk and every piece of weatherstrip increase a building's airtightness. However, there are usually so many parallel air-leakage paths that caulking and weatherstripping may not have measurable effect. The building's hidden air leaks usually leak more air than obvious cracks and small holes. You find the bigger leaks around penetrations and collections of joints where the building enclosure changes direction.

In climates with cold winters, you can seal air leaks from indoors, establishing the air barrier at the interior surfaces of exterior walls and ceilings. Interior wallboard and plaster are good air barriers.

Figure 4-23: *Rigid Foam and Liquid Foam Combine for Air Sealing and Insulating*

During a major renovation, you can install a better air barrier on the building's exterior, which avoids a mess indoors and can be tighter, because of less obstacles to work around compared to indoor air sealing.

In warm, humid climates where residents use central air-conditioning, the air barrier is more effective on the building's exterior surface. This prevents humid outdoor air from entering cavities of an air-conditioned building, where the moisture could condense.

Wood-frame walls are often leaky, and they're almost impossible to seal effectively with caulking, tape, or gaskets. When workers retrofit a building with densely packed fibrous insulation, this often reduces wall air leakage, even after the workers were unsuccessful at air sealing by other methods. The blown insulation fills the cavities and resists airflow through the cavities. Densely packed insulation's fibers plug many of the air leaks. Besides walls, densely packed insulation can plug floor cavities, porch roofs, and other difficult-to-seal building cavities.

Sealing duct air leakage can be tricky, especially in multistory buildings. Often the ducts are hidden behind walls and ceilings — inaccessible except by demolishing interior finish. In these cases, air-sealing specialists may choose to seal the building materials around the ducts, including them inside the building's thermal boundary.

See "Sealing Duct Leaks" on page 226.

Air sealing and renovation — Energy specialists should encourage building owners, contractors, and building managers to focus on air sealing whenever they renovate. Air sealing can be highly effective when it is a planned part of renovation because technicians can demolish finishes to access air leaks if necessary.

For example, renovators can remove wallboard or ceilings. They can drill through the interior wall surfaces for blowing dense-pack wall insulation and through flooring to dense-pack a rim-joist cavity between floors. Also, renovators can patch, repaint, and install new materials in an airtight manner during the planned renovation.

Repainting gives you an opportunity to caulk all cracks in wall and ceiling surfaces for airtightness and visual appeal. New floor tile or carpet gives workers the opportunity to seal the floor-wall junction, which is often a large air leak.

Siding (cladding) replacement offers a good opportunity to install or repair the water-resistive barrier (WRB) and air barrier under the new siding. Workers can also install wall-cavity insulation, insulated sheathing, and new windows when they replace siding.

See "Retrofitting Insulation" on page 161.

Air-sealing Materials and Application

Here we discuss the most common types of air-sealing materials and their applications. Air sealing saves energy, improves comfort, enhances health and safety, and aids in fire containment.

Masonry Air-sealing Strategy

Masonry materials must be airtight and have airtight joints between building assemblies to function as air barriers. Masonry walls, floors, and roofs, may or may not be airtight. Concrete block and brick aren't an air barriers. Mortar joints between masonry blocks usually leak air. Cavities in masonry blocks may connect air-permeable soil, floor cavities, attics, and roof cavities to conditioned areas of a building. Effective air sealing requires sealing leaks in the block and filling empty block cavities or at least air-sealing them.

For buildings built with only masonry materials, these materials should somehow form an air barrier. This requires construction knowledge and attention to detail for new construction, repair, and retrofit air sealing. Mortar, plaster, and stucco are potential air barriers. Workers who repair concrete, masonry block, plaster, and stucco may achieve significant reductions in air leakage and improvements in structural strength.

***Figure 4-24:** Masonry-wall Air Leakage*

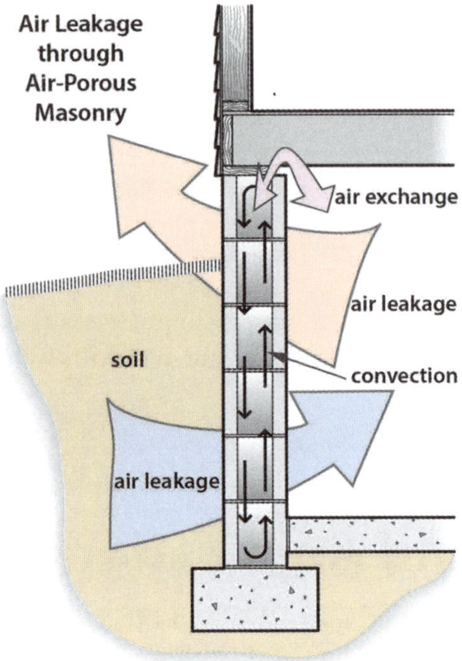

Masonry Air-sealing and Repair

Air-sealing and structural strengthening are complementary renovations to masonry buildings. Seek guidance from masonry experts when planning any kind of wall retrofit to a masonry building.

Strengthening existing masonry buildings is a priority for many regions, especially those threatened by earthquakes. Masonry buildings lose strength over time from damage by moisture, ground movement, and wind. Repair crews install masonry retrofits to the exterior of existing masonry walls and beams. If these crews understand the importance of air-sealing, they're careful to make sure that their repairs result in an airtight wall assembly.

Masons should know the composition of the masonry materials they repair. Repair materials may be incompatible to original materials, causing repair failure. The repair material's composition, water-vapor permeability, and water porosity should be compatible to the existing masonry. For example, use lime mortar to patch or re-point lime mortar between bricks and blocks. Use cement-based mortar or stucco to patch existing cement-based materials. Prime the damaged areas with a masonry primer or adhesive before repair.

Figure 4-25: *Air-sealing Masonry Walls and Supporting their Structure*

There is no substitute for knowledge and experience when repairing and air-sealing masonry building assemblies. If the repair must strengthen the building against earthquakes, a structural engineer with expertise in masonry should advise the building owner and the masonry contractor. With the right specifications and workmanship, an ancient masonry building could last another century or more after renovation.

Masonry air-sealing and repair materials — Consider these materials and methods for repairing and air-sealing masonry walls.

- Masonry bonding agents of various formulations (depending on required strength) that workers apply onto prepared surfaces and push into cracks. Masons also add these liquid bonding agents to mortar — for example, liquid epoxy.
- Glass, steel, carbon, basalt, fiberglass, and plastic fabrics applied to mortar to strengthen walls, columns, beams, and vaults in a process called *jacketing*.
- Polymer fibers to reinforce mortar and stucco to jacket deteriorated masonry wall.
- Traditional mixtures of lime, water, and sand with or without Portland cement, for patching mortar, stucco, and interior plaster. Lime surface mixtures work well to match the texture of existing surfaces.
- Dry mixtures in bags that combine ingredients for specific repair purposes.
- Wet mixtures in buckets that combine ingredients for specific repair purposes.
- Backer rod or 1-part foam to fill large cracks and polyurethane caulk to seal the surface cracks.

- Masonry adhesives that may repair broken masonry block and brick better than common mortar.
- Steel rods, welded wire, and other metal profiles can reinforce mortar joints and support masonry walls and lintels above windows and doors.

Figure 4-26: *Water-resistive Barriers as Airseals, Water Seals, and Structure*

Masonry WRBs and air barriers — Masonry specialists can install a variety of vapor-permeable coatings to masonry materials that function as air barriers, vapor retarders, and water-resistive barriers (WRB). The best of these coatings come with compatible caulking and sealing materials. Comply precisely with manufacturer's instructions for surface-preparation and the coating's applied thickness. WRBs resist water penetration but aren't always "waterproof" by building-code definition.

Pay attention to the vapor permeability of the coating. With most masonry assemblies, we want a class II or class III vapor retarder, and not usually a class I vapor retarder (vapor barrier), except when specified by an expert. We want the water vapor to travel in both directions, but also want an air sealant to improve airtightness of the leaky masonry walls. Besides liquid-applied WRBs, some special masonry paints may also function as air barriers and vapor-permeable water repellents.

Masonry coatings and repellents — Water repellents are usually class II or class III vapor retarders. They may reduce airflow through masonry materials slightly, depending on the thickness of the coat, but they don't usually reduce air leakage significantly.

Water repellents are similar to WRB's but not as airtight or water resistant. Some manufacturers use water repellents mixed into their masonry block. Masons sometimes mix water repellents into mortar and stucco. Workers spray transparent water repellents onto the surface of masonry walls with variable results. Many of the less expensive water repellents aren't very durable.

For below-grade application, a thick waterproof material is best. Waterproof coatings usually contain asphalt, silicon, or clay.They coat below-ground walls in a thick layer (~60 mils or 1.5 mm). Most waterproofing coatings are air barriers and vapor barriers.

Whenever a vapor barrier coats an above-ground masonry wall or a masonry floor, water vapor flowing through the masonry assembly may condense on the coating's backside and dislodge it from the masonry surface. Be sure to avoid this common mistake, whenever vapor pressure could push water vapor against the coating.

Figure 4-27: *Repairing and Sealing Mortar Joints*

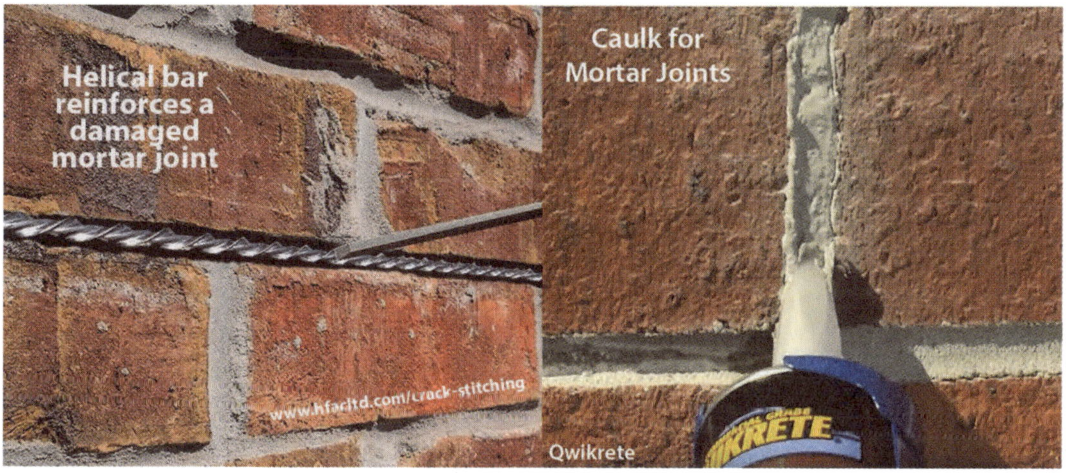

General Air-sealing Materials

Air sealers use a variety of materials to seal holes and gaps in the building enclosure. The characteristics of these materials are very important. Consider their airtightness, strength, flexibility, adhesion, and water resistance.

Rigid panels — Rigid patching materials include composite wood and drywall. These materials are stronger and more damage-resistant than flexible materials. Rigid panels can also cover openings leading to cavities, attics, crawl spaces, and other unconditioned zones. The building codes prefer rigid panels for air sealing cavities, and the codes require rigid panels for *fire stops* in new buildings. See "Fire Barriers" on page 160.

Rigid materials are the best for sealing large air leaks permanently. Rigid patches resist pressure and mechanical damage better than flexible materials. If workers need future access, screw the panel to its surroundings and seal it with a gasket or non-adhesive caulk.

Figure 4-28: *Self-sealing Fasteners Create a Tight Air Barrier*

Film or flexible sheeting — Polyethylene film and cross-linked polyethylene paper, are two common types of flexible air barriers. Polyethylene film is also a vapor barrier. Cross-linked woven polyethylene is a vapor-permeable water and air barrier that allows water vapor pass through it.

Paper, films, and tape are the easiest air-sealing materials to cut and fasten, but wind can tear them or pull them away from their fasteners. Wind can pull flexible sheeting through staples, and the staples themselves leak air. Use self-sealing fasteners for a airtight flexible air barrier.

Stuffing — Insulators and air sealers stuff foam rubber and fibrous insulation into spaces to resist air movement. Foam rubber and fibrous insulation can fill cracks, crevices, and cavities. By filling the cavity's whole space, the material resists airflow.

Fibrous materials and lightweight foam rubber aren't air barriers and must be packed densely enough to provide sufficient resistance to air movement. Enclosing fibrous insulation in a plastic bag can provide a more effective air barrier than the fibrous insulation alone. Workers can seal over the stuffing material with rigid materials, spray foam, caulking, or mastic to complete the air seal.

Blown insulation — Cellulose, fiberglass, and mineral wool loose-fill insulation resist airflow inside inaccessible building cavities that are air-leakage pathways. Technicians can reduce air leakage in areas where they can't crawl or reach, by blowing densely packed insulation into those cavities.

Fibrous insulation isn't an air barrier even when blown at high densities. However it does resist airflow through the cavities. High-density fibrous insulation also acts as a fire-block when installed in closed cavities because it blocks airflow and the spread of flames.

Selecting caulk — Caulk fills and seals narrow and consistent joints between building materials. Selecting the best caulk requires recognition of the *substrate* materials (the materials bordering the joint), the gap size, weather exposure, and joint movement from temperature changes.

Common caulks, like acrylic latex and siliconized acrylic latex work well for indoor gaps less than $^3/_8$ inch (10 mm) between wood and other common building materials. Caulk has a limited shelf life, so buy your caulk from a supplier who sells a lot of caulk.

Silicone, polyurethane, and butyl offer superior durability, adhesion, and flexibility compared to less expensive acrylic latex caulks. When in doubt about selection, especially when replacing failed caulk, use a high-performance caulk that the manufacturer specifies for adhesion and flexibility to your particular substrates.

Suppliers sell caulks and other sealants in a wide variety of formulations. Read the specifications carefully, especially when choosing a sealants for joints between different materials. Many available caulks defy classification because they mix the ingredients of the well-known caulks and adopt a unique name. For these formulations, read the specifications before you buy them.

Ask these questions about your caulking needs. Then consult the comparison table for some preliminary advice.

- Is the substrate porous like concrete and wood or non-porous like glass or metal?
- How wide is the joint?
- How experienced are the installers at working with caulks that are difficult to apply and clean up?
- How much movement do you anticipate in the joint? This depends on material and temperature variation.
- Will you paint the joint?
- How exposed is the joint to water?
- How exposed is the joint to heat and UV radiation?
- How exposed is the joint to abrasion?

Installing caulk — Preparation and installation are just as important as selecting the right caulk. Exterior joints, designed to seal out air and water, must tolerate moisture, ultraviolet light, joint movement, and air pressure. These joints may require careful preparation, to include cleaning, priming, packing, and tooling the caulk after installation.

Porous materials like wood and masonry usually need scraping, wire-brushing, and dusting. Priming both sides of the gap is sometimes necessary, depending on the materials and caulking formulation. Non-porous materials, like aluminum and steel, usually require cleaning with a solvent. Use rubbing alcohol and water for common dirt and methyl-ethyl ketone for oil and grease.

Polyethylene-foam backer rod or bond-breaker tape can prevent three-sided adhesion. Three-sided adhesion may tear the caulking bead when the joint moves. Tool the caulk with a caulking tool if necessary. Various tools create various shapes to provide flexibility, strength, and appearance. An hour-glass shape and slender profile that optimizes the caulking bead's flexibility. Un-tooled beads provide greater strength.

Joint failure is a major problem in new construction. For challenging joints in important locations, designers should specify the materials and preparation of joints. Then supervisors should enforce these specifications during installation.

Liquid foam — Liquid closed-cell polyurethane foam is a versatile air-sealing material in addition to an insulation material. Closed-cell foam is packaged in a one-part injectable variety and a two-part spray-able variety. Liquid foam has a high R-value per inch and works well for insulating and air sealing small, poorly insulated, and leaky areas in a single application.

1. One-part high-density foam comes in expanding or non-expanding varieties. It is superior to caulk for filling large cracks with varying width and depth. The cracks should be at least 1/4 inch (6.4 mm) in width. This one-part foam sticks to skin and fabrics, so wear gloves and long sleeves. Installers improve their neatness and waste minimization with practice.

Two-part foam is also known as low-pressure spray polyurethane foam (SPF). Two-part foam has good adhesion and structural strength. SPF bridges over irregularities and seals the both surfaces and joints. Two-part foam helps rigid patching materials to seal large openings by covering joints and gaps. Cut structural sheeting or foam board to as closely a s possible around obstacles, then fill the edges and gaps with the SPF.

Spraying or injecting liquid-foam is easier compared to installing other materials to do the same air-sealing work. However, cleanup is difficult enough that you should avoid multiple applications and cleanup on the same job. Instead, identify all the spots needing foam application, make a list, and foam them one after another.

Liquid foam's durability as an air sealant isn't perfect. Cracks in the foam may occur at cracks in the air-sealed materials.

Gaskets — A gasket is a flexible material, designed to seal a gap between two less flexible materials. Gaskets seal joints between metal and glass in window systems. They also serve as thermal breaks between conductive building materials.

Many windows, doors, and manufactured wall panels use gaskets, instead of caulking because the gasket allows more movement while maintaining its seal. Some window installers use gap gaskets instead of foam or caulking to seal between window frames and the rough opening. The gap gaskets expand after installation to seal this important gap.

Table 4-4: *Advantages and Disadvantages of Common Caulks*

System	Advantages	Disadvantages
Acrylic latex	Inexpensive Paintable, fast curing Easy to apply and clean up, Best caulks for interior use	Shrinks and cracks Limited durability Inconsistent viscosity and product quality
Urethane	Bonds to almost any material More flexible than acrylic latex Resists abrasion in floor applications Paintable: paint improves durability Water-based products clean up with water	Solvent-based products are difficult to apply and clean up Some products are expensive Degrades from UV radiation Limited durability Slow curing Toxic odors from solvent-based products
Pure Silicon	Waterproof and mold-resistant Superior durability and UV resistance Extremely flexible Good for highly variable temperature applications	Expensive Not paintable Bonds poorly to porous materials Difficult application and cleanup Many products contain acids that deteriorate masonry
Butyl	Bonds to almost any material Never really hardens Waterproof and durable Survives random movement Resists sound transmission Good for low-temperature applications	Difficult to apply Expensive Very messy; difficult cleanup

***Figure 4-29:** Caulking Joints*

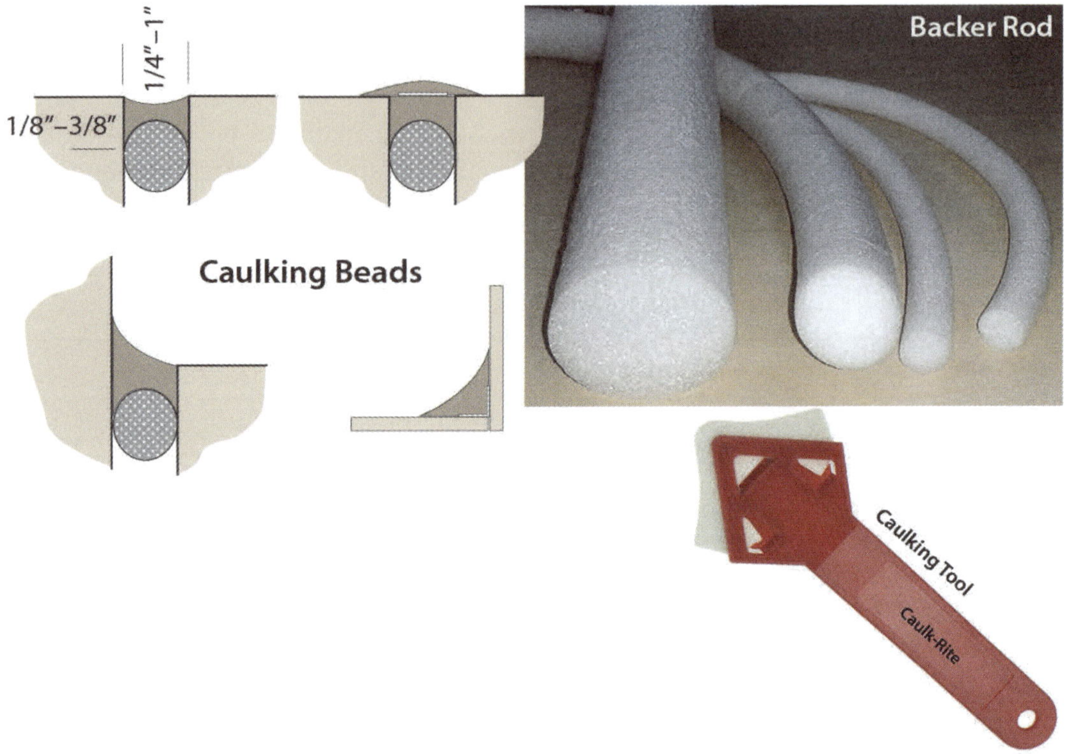

***Figure 4-30:** Liquid-Applied Foam*

Figure 4-31: *Gaskets and Tape for Air Sealing*

Tapes **—** Tapes can be effective air-sealing materials when used exactly as the manufacturer specifies. Various builder's tapes are versatile for air sealing a variety of materials. Air-barrier tape is made by the manufacturers of air-barrier material to seal that material at its joints.

Tape is a popular air sealant for joints in smooth material like plywood, OSB, and foam board. New adhesives have made tapes more reliable compared to the past. Contractors use tape to seal joints in new construction and renovation work for materials with clean, dry surfaces.

Tapes had a reputation of failing more than other types of air-sealing materials because their thin adhesives are susceptible to drying, material movement, or other causes. However new air-sealing tapes have proved reliable when used as the manufacturer recommends.

Adhesives **—** Adhesives adhere a variety of construction materials together. Manufacturers formulate specialized adhesives for specific purposes like gluing foam insulation to masonry materials.

Some caulks and sealants have sufficient adhesion to use as structural adhesives. These include polyurethane foam, polyurethane caulk, duct mastic, acoustical sealant, and siliconized acrylic-latex caulk.

Safety concerns **—** Consult the Safety Data Sheet (SDS) to learn the health hazards of caulk, foam, and adhesives. Use ventilation, and wear an organic-vapor respirator when you install caulk, SPF, or adhesives.

Figure 4-32: Ventilation and Dehumidification Controls Moistures

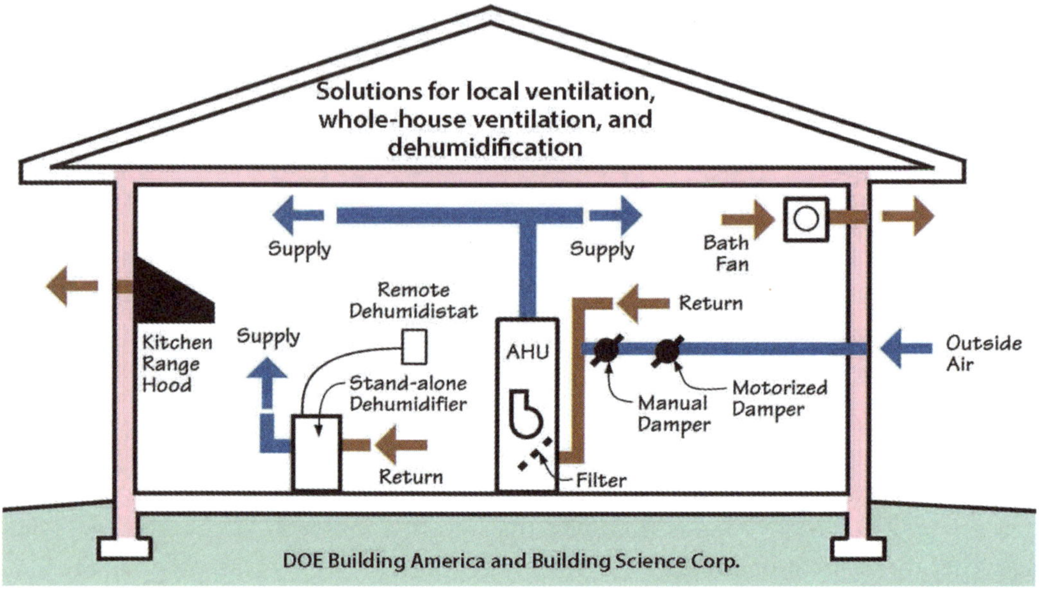

Whole-Building Mechanical Ventilation

Whole-building ventilation is one of the most important issues in residential energy efficiency. Three major strategies for whole-building ventilation are these.

1. Ventilation with air leakage
2. Ventilation with a whole-building mechanical-ventilation system
3. Ventilation with a ventilating dehumidifier

The problem with using air leakage for ventilation is that the air-leakage rate is usually either too much or too little. Air leakage doesn't ventilate all areas of the building equally, leaving some areas drafty and fresh, and others stagnant and polluted. Also, air leakage is highest during severe weather, especially in winter, because wind and temperature differences magnify air leakage.

ASHRAE (American Society of Heating, Refrigerating, and Air Conditioning Engineers) publishes residential-ventilation standards. The current standard 62.2-2019 requires fan-powered ventilation in all single-family homes and multifamily buildings with the exception of very leaky buildings or buildings that require little heating or cooling. If you air-seal homes during energy-retrofit work, ASHRAE 62.2 usually requires you to install whole-building mechanical ventilation and to improve local ventilation in kitchens and bathrooms as needed.

Notice: The numerical information in this chapter about ventilation and dehumidification equipment is approximate. Always use the information provided

by the manufacturer if it's available on paper or online. See "Online Energy-Information Sources" on page 285.

The four types of whole-building ventilation systems are these.

1. *Exhaust ventilation* pulls stale air out of the building and brings make-up air through air leaks.
2. *Supply ventilation* pushes fresh outdoor air into the building, and stale air moves out through air leaks.
3. *Balanced ventilation* uses one fan and duct to bring fresh air in and another fan and duct to exhaust stale air out.
4. *Ventilating dehumidifiers* move outdoor air through a central dehumidifier with a filter. The filtered and dehumidified air then enters either the return or supply side of the air handler, or it goes directly to the conditioned space. See "Dehumidification" on page 239.

ASHRAE Ventilation Standard

The current ASHRAE Standard 62.2-2019 specifies ventilation-fan size according to floor area and number of bedrooms. The standard describes how to size the ventilation system's fan using either the following formula or table to find the total ventilation rate (TVR).

$$TVR\ (CFM) = (0.03 \times \text{floor area}) + 7.5 \times (\#\ \text{bedrooms} + 1)$$

From ASHRAE Standard 62.2-2016

ASHRAE 62.2-2019 also requires 50 CFM local ventilation in bathrooms and 100 CFM in kitchens. An openable window contributes 20 CFM. These fans and windows remove moisture and pollutants at their source. If there is a deficit of local ventilation, add 25% of that CFM deficit to the whole-building ventilation fan's capacity to comply with the standard.

See "Ventilation and Air Leakage" on page 97.

Adjusting the required airflow — When you test a building with a blower door, a calculated infiltration credit may reduce the required ventilation rate under the ASHRAE Standard. The procedure to qualify for the infiltration credit is complex and beyond the scope of this book.

If you ventilate a building to the ASHRAE 62.2 Standard in a hot, humid climate, the building may become too humid. In a cold dry climate, the Standard's required airflow might make the building too dry. You may want to decrease the ventilation rate to avoid these ventilation side effects.

Figure 4-33: *Fans for Local Ventilation and Whole-Building Ventilation*

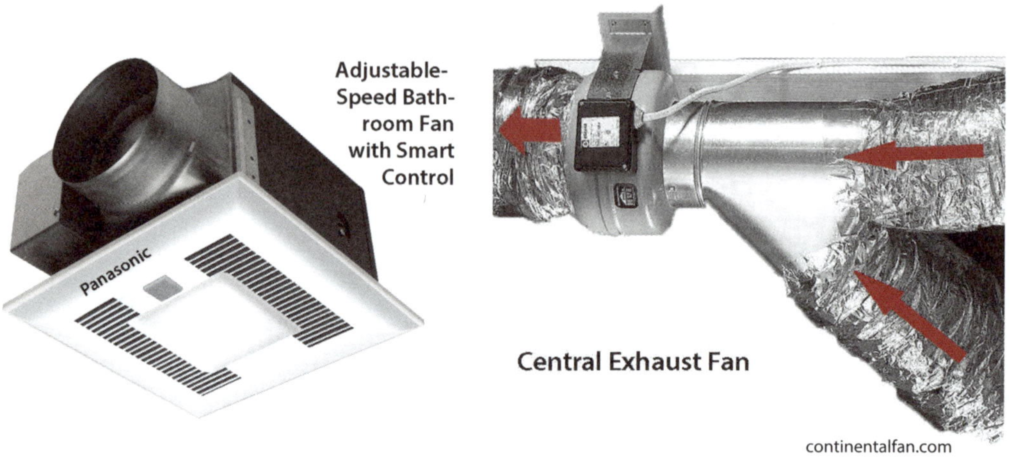

Exhaust Ventilation

Exhaust ventilation systems use a fan to exhaust indoor air, which leaked outdoor air then replaces. You can achieve better air distribution by installing a remote fan that exhausts air from several rooms through ducts. In existing homes, install a high-quality ceiling exhaust fan, to operate continuously. If you decide to replace a bathroom exhaust fan with an exhaust-ventilation fan, the new fan should run continuously on a low speed with a high-speed control to remove moisture and odors quickly. ASHRAE 62.2-2019 requires that new fans be no louder than 1.0 sone during continuous operation.

Exhaust ventilation systems are inexpensive and easy to install, but they don't control the source of incoming air or recover heating and cooling energy. Since most exhaust-ventilation systems don't have filters, dust collects in the fan and ducts. Residents must clean the fan blades every year to maintain the design airflow. Exhaust ventilation systems create negative pressure, so they may risk backdrafting in buildings with open-combustion appliances.

Exhaust systems create negative pressure within the home, drawing air in through leaks in the building enclosure. Exhausting air keeps the moist indoor air from moving into building cavities, which reduces the chance of moisture condensation in cold climates. In air-conditioned buildings however, this depressurization can draw moist outdoor air into building cavities where it can condense.

Supply Ventilation

Supply ventilation uses a supply fan or the blower inside an air handler to deliver outdoor air to the building. The building becomes pressurized, which forces indoor air out through air leaks. Using a supply fan or the building's air handler makes it possible to filter incoming air.

Figure 4-34: Exhaust Ventilation Options

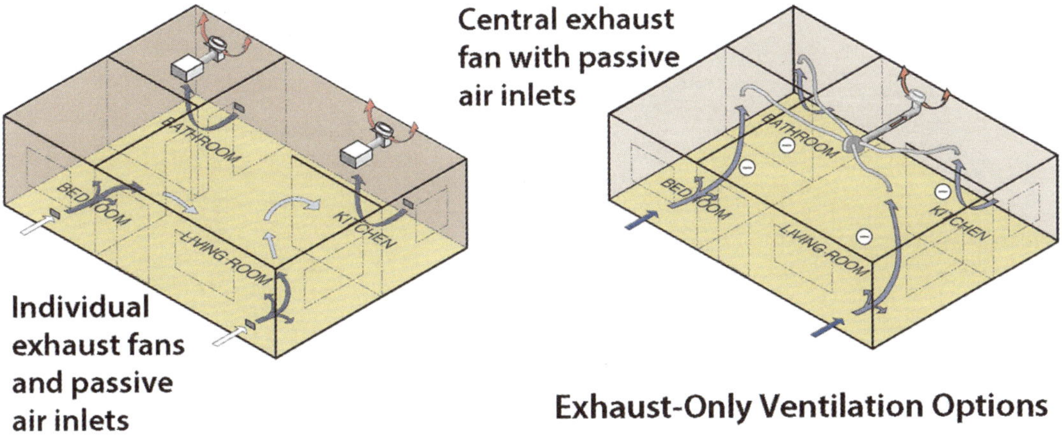

Supply Ventilation

Supply ventilation uses a supply fan or the blower inside an air handler to deliver outdoor air to the building. The building becomes pressurized, which forces indoor air out through air leaks. Using a supply fan or the building's air handler makes it possible to filter incoming air.

Air-handler supply ventilation — The fan may distribute incoming air to the bedrooms and living areas through ducts. A supply duct provides outdoor air to the air handler's return plenum. This supply duct may have a motorized damper that opens and closes with air-handler operation. When the air-handler's blower operates, the duct brings outdoor air in, mixing it with return air, and then heating or cooling the air. Ventilation specialists may add an existing exhaust fan, electrically connected to the blower or supply fan, to balance a supply-only system

Ventilation only happens when the air handler operates, so the air handler must operate even when heating and cooling aren't required. An electronic control can activate the blower, timing its cycles to deliver the design ventilation rate. Using the air handler for ventilation is inefficient because forced-air blowers usually move far more air than the building needs for ventilation.

Dedicated supply fan — Supply ventilation doesn't require an air handler. You can use a correctly sized fan with its own small duct system, which operates continuously. Outdoor air is filtered, keeping most dust out of the duct system, which is an advantage of central supply ventilation over central exhaust ventilation.

Supply ventilation doesn't work for very cold climates because it pushes warm, moist indoor air through exterior walls, floors, and roofs where moisture can condense on cold surfaces.

Figure 4-35: Supply-only Ventilation

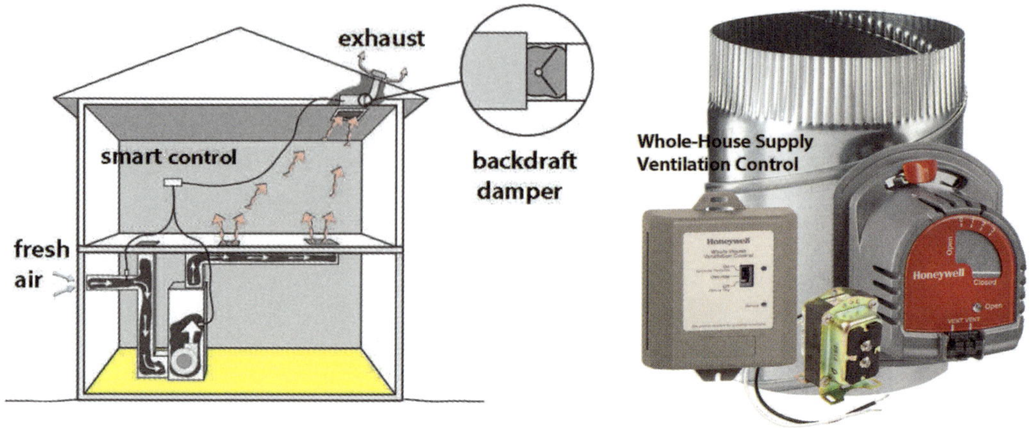

Balanced Ventilation Systems

Balanced ventilation systems exhaust stale air and provide fresh air through a ducted distribution system. Of all the ventilation schemes, balanced systems do the best job of pollutant control and drying.

Balanced systems move equal amounts of air into and out of the building. Most balanced systems use heat recovery ventilators (HRVs) or energy recovery ventilators (ERVs). HRVs exchange only heat. ERVs exchange both heat and moisture from the supply-and-exhaust air streams.

Balanced systems, when operating properly, reduce many of the safety problems and moisture-induced building damage that may occur with unbalanced ventilation. These ventilators aren't trouble-free, however. Many buildings with "balanced" ventilation systems experience pressure imbalances and poor air quality due to design, installation, or maintenance problems.

These complicated systems can improve the safety, comfort, and energy-efficiency of a building. However residents or building maintenance staff must maintain these ventilators to assure that they operate correctly. Inspection and testing are vital during both the initial installation and periodic service calls.

Heat and Energy Recovery Ventilators

The difference between an HRV and an ERV is that HRVs transfer heat only, while ERVs transfer both sensible and latent heat (moisture) between the two air streams.

Balanced ventilation systems can provide indoor air quality (IAQ) and comfort, but they require exemplary installation and maintenance. Installers must test and commission the ventilation system during both the initial installation and periodic service calls. At least once a year, residents must clean the two filters inside the balanced ventilator along with grills and termination fittings in the ducts.

Figure 4-36: *Heat-recovery Ventilators and Energy-recovery Ventilators*

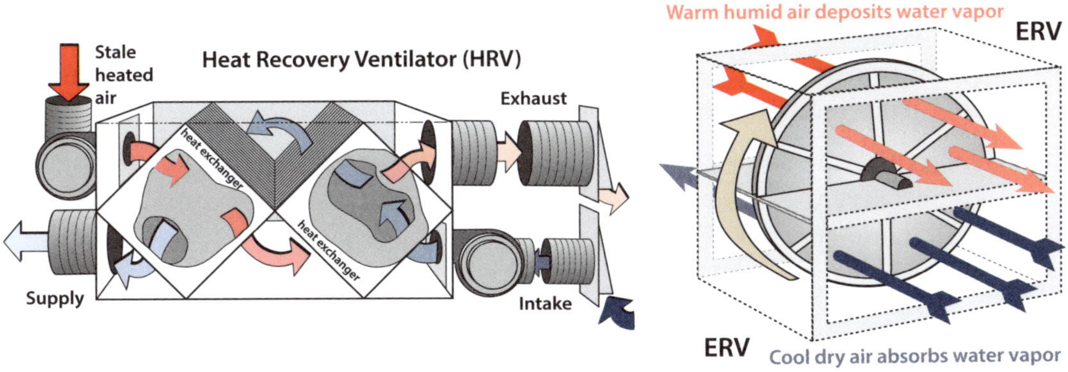

Currently, a lack of trained installers and educated residents causes problems with existing balanced ventilation systems. The most common problem is insufficient airflow because of faulty ducts, dirty ducts, or dirty filters. Other problems include these.

- Residents turn the ventilator off.
- Operation time is too long or too short.

Heat-recovery ventilators — The HRV core is usually a *flat-plate* aluminum or polyethylene air-to-air heat exchanger in which the supply and exhaust air streams pass one another with little or no mixing.

Heat travels through the core, exchanging heat by conduction and convection, from the warmer air stream to the cooler one. During the heating season, the heat contained in the exhaust air warms the incoming supply air. During the cooling season, the cooler exhaust air cools the warm incoming fresh air.

In cold weather, outgoing moisture condenses when it passes the cold incoming air in the HRV core. This condensate drains out of the unit. HRVs may require defrost mechanisms in cold climates to thaw frozen condensate in the heat exchanger. The most economical defrost method is periodic recirculation of warm indoor air, though some HRVs with aluminum cores use electric heat strips to melt the ice.

Energy-recovery ventilators — ERVs reclaim moisture and heating or cooling energy, depending on the season. Energy recovery ventilation exchanges both sensible and latent heat between the outgoing air stream and incoming air stream.

Energy recovery ventilation removes between 20% and 50% of the water vapor from the more humid air stream and transfers it to the less humid air stream. In comparison to an HRV, ERVs add less outdoor humidity to the supply in warm humid weather and subtract less indoor humidity from exhaust in cold dry weather.

The ERV's exchange of water vapor has no significant dehumidifying effect on the building. Instead this water-vapor exchange merely reduces the amount of water vapor that the ventilation air adds to the building. However in very humid weather, the moisture-transfer medium may become saturated and fail to remove much water vapor from incoming air.

ERVs also reduce the amount of water vapor leaving the building. During winter, the ERV can prevent excessive dryness in dry climates, but may increase indoor humidity in damp climates.

Experts disagree about whether the ERV has significant advantages compared to the HRV. In persistently humid regions, neither HRVs nor ERVs reduce indoor humidity enough. Dehumidification would require a whole-house dehumidifier or else a ventilating dehumidifier that both ventilates and dehumidifies using the refrigeration process.

See "Whole-house Dehumidifiers" on page 241.

Figure 4-37: *Simplified and Fully Ducted Ventilator Distribution Systems*

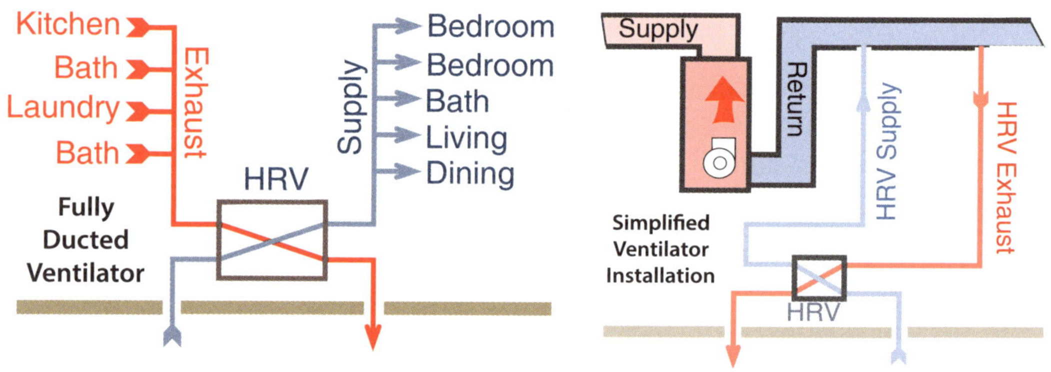

Ventilation-air distribution — Whole-building ventilators are most effective when installers equip them with their own system of correctly sized supply and return air ducts. The duct system may have the same problems of leakage, inadequate airflow, and balancing as other forced-air distribution systems.

Fully ducted systems are popular for buildings with hydronic heating, or for very energy-efficient homes conditioned by space heaters, room air conditioners, and ductless mini-split heat pumps.

In homes with central forced-air systems, installers often connect the ventilator to the central duct system. The outdoor ventilation air moves through the ventilator and into a main HVAC return duct.

When balanced ventilators use forced-air ducts for distribution, called a simplified distribution system, the ducts should be very airtight. To distribute ventilation air, the air handler's blower operates at its lowest speed whenever the ventilator operates without heating or cooling. Operating the blower increases energy use compared to just operating the ventilator. Electronically commutated (ECM) blowers, make the simplified ventilator installations more practical and energy efficient, by keeping the ventilation airflow consistent.

Similar to forced-air heating and cooling systems, many installers often don't measure system airflow or room airflows for whole-building ventilation systems. Lack of testing allows many ventilation systems to perform inadequately. At minimum, installers should measure supply and exhaust airflow to verify those design specifications.

See "Solve Obvious Air-handler and Duct Problems" on page 222.

Figure 4-38: *Balanced-ventilation Controls*

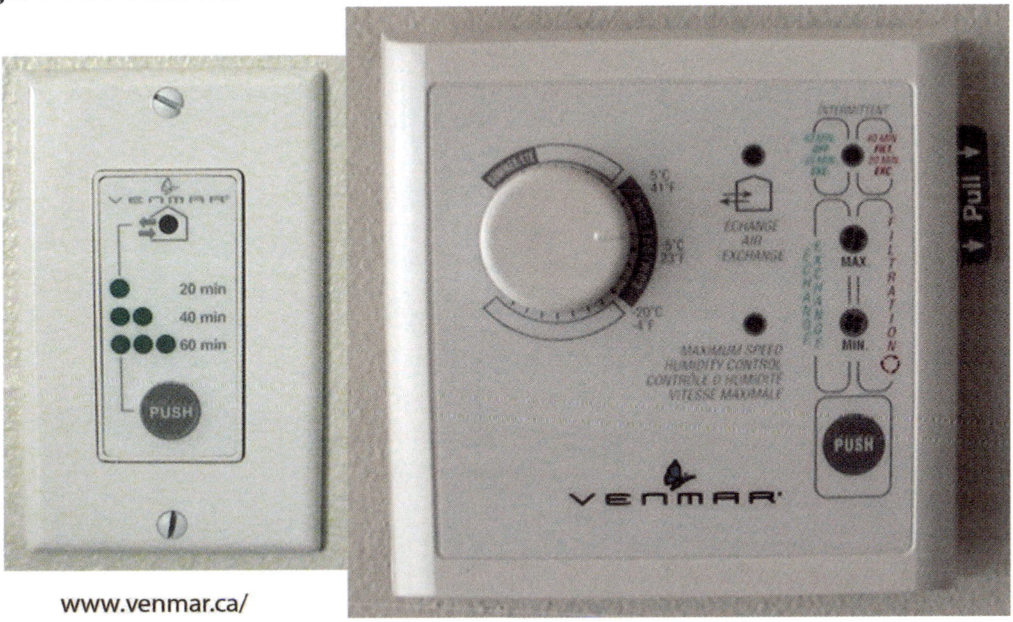

Ventilator controls — Humidistats, timers, and manual switches control the operation of HRVs and ERVs to provide ventilation as needed. Ventilation engineers, led by Lawrence Berkeley National Laboratory, are developing controls that limit ventilation run time to the minimum necessary for acceptable IAQ.

The innovative current and future control strategies include these features.

- Minimize ventilation during unoccupied times.
- Shift ventilation time to ventilate less when energy costs are highest.
- Shift ventilation time to ventilate less during high pollution (rush-hour traffic) or high humidity (above 70% Rh). This strategy prevents pollution and humidity coming from outdoors.

- Use timers or temperature sensors to shift ventilation from times of high temperature difference to times of low temperature difference. This strategy reduces both energy consumption and peak-demand.
- Ventilate more at favorable times to compensate for unfavorable times when controls reduce ventilation.
- Adjust ventilation run-time to account for the operation of other air-moving equipment, such as kitchen fans, bathroom fans, and clothes dryers.

Engineers and manufacturers are working together on smart controls that use some or most of these ventilation-control strategies.

Figure 4-39: *Evaluating CO2 Levels*

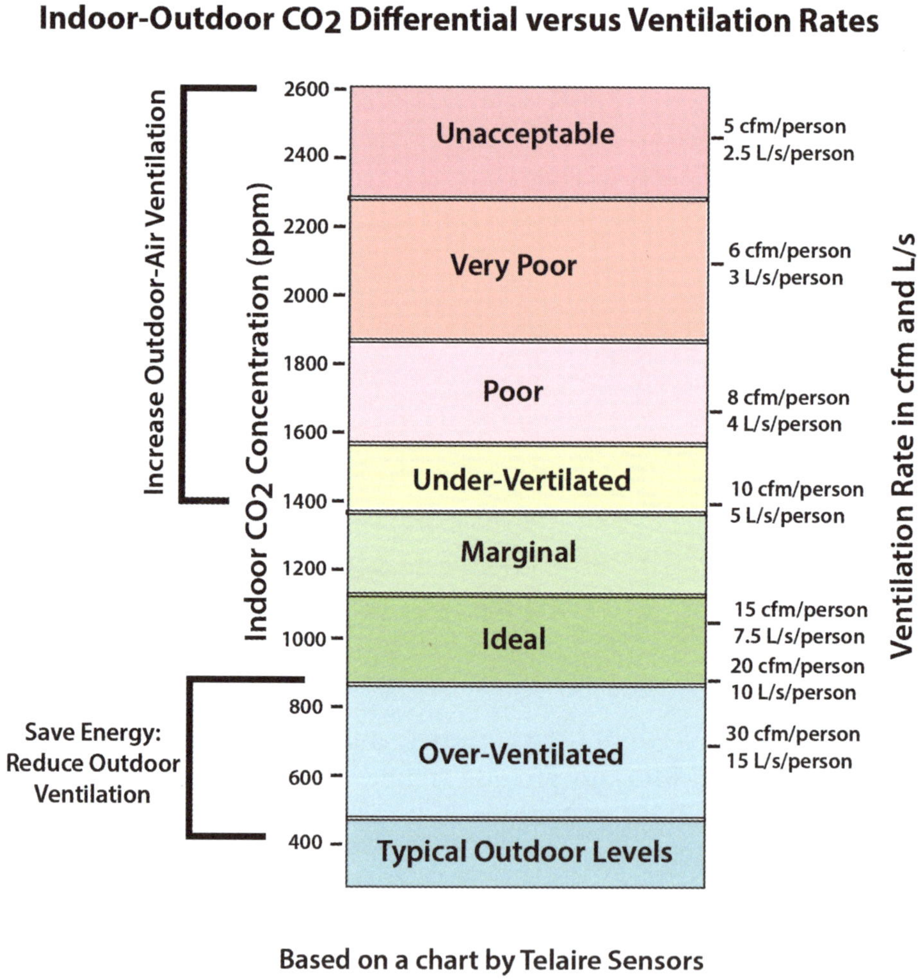

See "Air Filtration" on page 233.

5 Insulation for Cool Buildings

This chapter explains how insulation works and describes insulation types, characteristics, and installation methods. Insulation and air sealing are the two most effective and also the two most neglected solutions discussed in this book. Older buildings, especially in warm climates tend to be under-insulated or not insulated at all.

If installed correctly, insulation reliably saves energy and provides comfort without any maintenance, for decades or even centuries. In contrast, cooling, ventilation, and dehumidification equipment ultimately fails. Some equipment may become obsolete because it requires too much maintenance, which cooling systems may or may not receive.

Roof insulation and wall insulation are very cost-effective investments in hot climates. Solar-heat gain through the roof and windows dominates summer heat gain. Floor insulation, and foundation insulation are less important in hot climates compared to cold climates.

Insulation performs these thermal functions.

- Conserves energy by resisting heat transmission.
- Enhances comfort by minimizing temperature variations inside the conditioned space.
- Keeps the wall and ceiling surface temperations cooler compared to uninsulated assemblies.
- Minimizes the required size of cooling equipment, depending on thermal resistance or R-value (or RSI) of the insulation.

Insulation may also offer these secondary benefits.

- Reduces noise and vibration.
- Resists air leakage and water vapor transmission.
- Provides fire resistance to the building.
- Fills building cavities to resist airflow that feeds building fires.
- Adds structural strength.

Figure 5-1: *Effect of Density on Thermal Resistance*

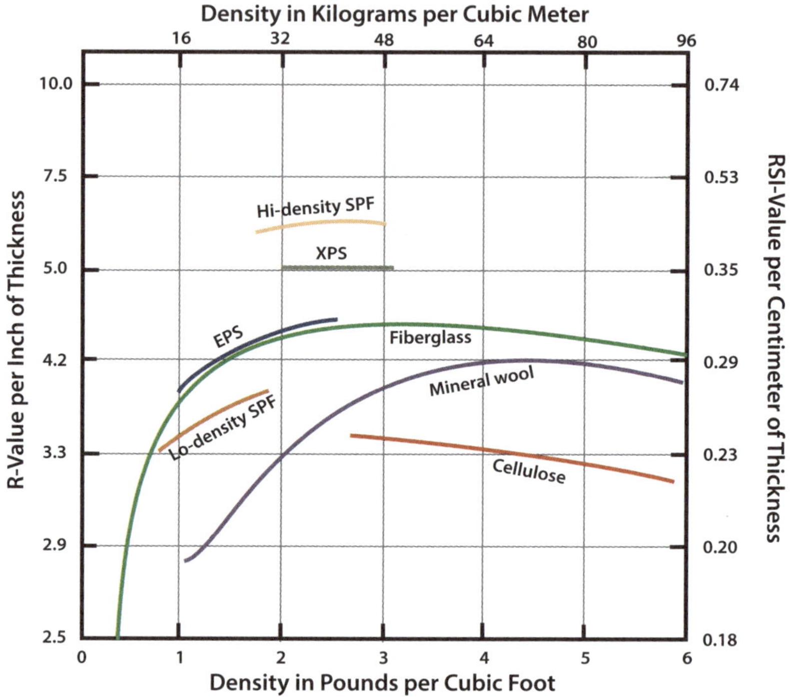

Insulation Characteristics

Insulators install insulation in building cavities, or they attach insulation to the building interior or exterior surfaces of building assemblies.

Insulation slows heat transmission in two ways.

1. Forces the heat to conduct through a gas or through very small connections among fibers or foam bubbles. Gases are poor heat conductors compared to solids.
2. Minimizes heat radiation and air convection inside cavities that contain insulation.

Insulation harbors millions of tiny air pockets within fibrous insulation or bubbles in plastic-foam insulation. Heat transmission happens slowly through insulation compared to other building materials because of the slow-conducting air or gas pockets in the insulation.

Insulation Thermal-performance Factors

The information in this section can help you decide which type of insulation is best for your climate and your building.

R-value expresses insulation's ability to resist heat flow. "R" stands for "resistance." Like many numerical values, used in this book, there are two R-values. The metric R-value is the RSI. SI means: "Systeme International" — French for the metric system. The RSI is the R-value divided by 5.678.

Table 5-1: *Clear-Wall R-Values from Full-scale Tests*

Wall Type	C-W R	C-W RSI
Standard 2-by-4 with R11 fiberglass-batt	9.7	1.7
2-by-6 perfect R19 fiberglass-batt installation	13	2.3
2-by-6 poorly installed R19 fiberglass-batt installation	11	1.9
Steel frame wall C-stud with R11 fiberglass-batt	5.6	1.0
Steel stud wall w/ R11 fiberglass-batt & EPS sheathing	10.5	1.8
Structural 6-inch EPS-insulated panel	22	3.8
Stucco-covered straw bale ~24 inches / 60 cm	16–28	2.8–4.9
Lightweight concrete block ~8 inches / 20 cm	10–30	1.8–5.3
Insulating concrete form 12 inches / 30 cm	26–44	4.6–7.7

Scientists calculated these values using data from full-scale wall thermal-resistance tests performed at Oak Ridge National Laboratory. C-W R or clear-wall R-value describes the thermal resistance of the entire wall section, including structural material. The first four examples here are stud walls insulated with fiberglass batts. The last two examples show a range R/RSI values that include a thermal-mass factor.

The insulation's R/RSI value is the most a common metric for estimating the thermal resistance of a building assembly. Recent research has measured the thermal resistance of whole building assemblies through a device called a *guarded hot box.* These research results indicate a complex relationship among the insulation and structural materials in the tested assemblies. When experimenters include corners and joints, the whole-building thermal resistance is even less predictable.

The combination of insulation and other building materials creates the actual R-value (RSI) in any building assembly, which also depends upon these factors.

- Type and density of the insulation
- Thermal bridging in the assembly
- The insulation's installation quality
- Air leakage and convection
- Water's presence inside the insulated assembly
- Mass of the insulated assembly
- Outdoor temperature

See "Heat Transmission" on page 24.

Insulation type and density **—** Some common insulating materials have an ideal density, where the R-value per inch or RSI per centimeter is at its maximum. Compressing fiberglass, mineral wool, or cellulose to an ideal density maximizes their R-values per inch or RSI per centimeter. Compressing the fibrous insulation beyond the optimal level decreases the thermal resistance from its maximum.

Foam insulation exhibits the same density effect. Foam insulation with a higher density usually has greater R-value per inch or RSI per centimeter than foam with a lower density. Density is also a practical consideration for foam insulation. Greater density provides greater strength — for installation on roofs and underneath foundations, for example.

Approximate

Table 5-2: Insulation R and RSI Values per Inch or per Centimeter

Insulation Type	R/inch (US)	RSI/cm (metric)
Fiberglass batts, blown, board	2.6–4.2	0.18–0.30
Cellulose blown	3.2–3.6	0.22–0.25
Mineral wool batts, blown, board	2.6–4.4	0.18–0.31
Vermiculite or perlite	2.1–2.4	0.14–0.17
Expanded polystyrene (white)	3.6–4.4	0.25–0.31
Extruded polystyrene (blue/pink)	5.0	0.35
Polyisocyanurate board	5.6–7.6	0.40–0.52

R / RSI values vary by insulation type, density, and the quality of installation.

Thermal bridging **—** Thermal bridging is rapid heat transfer through thermally conductive structural materials like steel, aluminum, masonry, and wood,. The thermal bridge carries heat around the insulation. Thermal bridges are important for two reasons.

1. They create cold (or hot) places in building assemblies that cause condensation and comfort problems.
2. They affect the total heat flow through the building enclosure especially in energy-efficient buildings.

Strategies for reducing thermal bridging include minimizing structural materials and applying insulated sheathing to building surfaces as a *thermal break*. A thermal break is an insulating material that reduces heat transmission through a thermal bridge.

See "Heat transmission" on page 20.

Metal framing (and to a lesser degree, masonry structures) present a challenge to insulate adequately, especially in severe climates. Steel-framed assemblies have a low whole-wall R-value (RSI). Without insulated sheathing, the steel structures invite condensation, corrosion, and wetting of porous building materials, during both air conditioning and heating.

Air leakage and convection — When building cavities have airtight surfaces on all sides, the insulation achieves its nominal R or RSI value, listed by the manufacturer. If air from inside or outside the building leaks into an insulated cavity, this air leakage reduces the manufacturer's listed thermal resistance by 15–50%.

Insulation's installed coverage and density are important, especially in cold and very hot climates. Air flows quickly through loose fibrous insulation, slower through dense-packed fibrous insulation, and slowest through foam insulation. However the locations and surface area of the air leaks are even more important than the insulation's coverage and density.

Air convection inside insulated building cavities encourages heat transmission. Air convects heat off one surface and transports the heat to the adjacent surface — between drywall and batt insulation, for example. Convecting air can also find its way through channels around the insulation's edge gaps — between insulation, framing lumber, drywall and exterior sheathing for example. Edge gaps of only 4% of the insulated surface area can produce up to a 30% loss in effective R-value for R-19 (RSI-3.3) ceiling insulation. See "Finding Air Leaks" on page 99.

Moisture and insulation — Moisture and insulation have a simple relationship. The most important factors in this relationship are the average relative humidity (Rh) of the local climate and the vulnerability of the insulation to water saturation. Wet insulation contributes to wet building materials, insect and microbe infestation, corroded metal, and peeling paint. With cellulose, the insulation may also slump, leaving uninsulated gaps.

Most moisture problems in insulation relate to persistent Rh more than 85%. Organic insulation and open cell foam insulation are the most vulnerable to moisture problems because they can absorb more than 100% water by weight. Organic insulation includes cellulose, denim, and sheep's wool.

With fiberglass, mineral wool, and closed-cell foam, the R/RSI values don't change much with the small changes in absorbed moisture over the range of relative humidity from 50% to 100% Rh. Inorganic materials, like fiberglass and mineral wool, can hold a maximum of about 12% moisture by weight, and the moisture doesn't penetrate their fibers. Foam board and closed cell spray foam absorb a range of 8% to 16% moisture by weight. Thus foam insulation is relatively resistant to moisture problems. The glass fibers and dense closed-cell foams hold nearly the same thermal resistance even with persistent Rh over 90%.

Any porous material in a building assembly can absorb water vapor and water if a vapor barrier occupies one side of the cavity and the vapor pressure from the other side is strong enough. For example, XPS could double its density if it insulates the flat roof over a swimming pool in Alaska if a rubber roof covers the XPS. Water vapor penetrates most materials if exposed to enough vapor pressure.

Wind driven rain and flooding can completely saturate materials. However, most insulation moisture problems originate from the long-term high Rh of the local climate and the moisture vulnerability of the insulation. See "Vapor Retarders and Vapor Barriers" on page 159 and "Controlling Water Vapor" on page 51.

Figure 5-2: *Effect of Relative Humidity on 6 Types of Insulation*

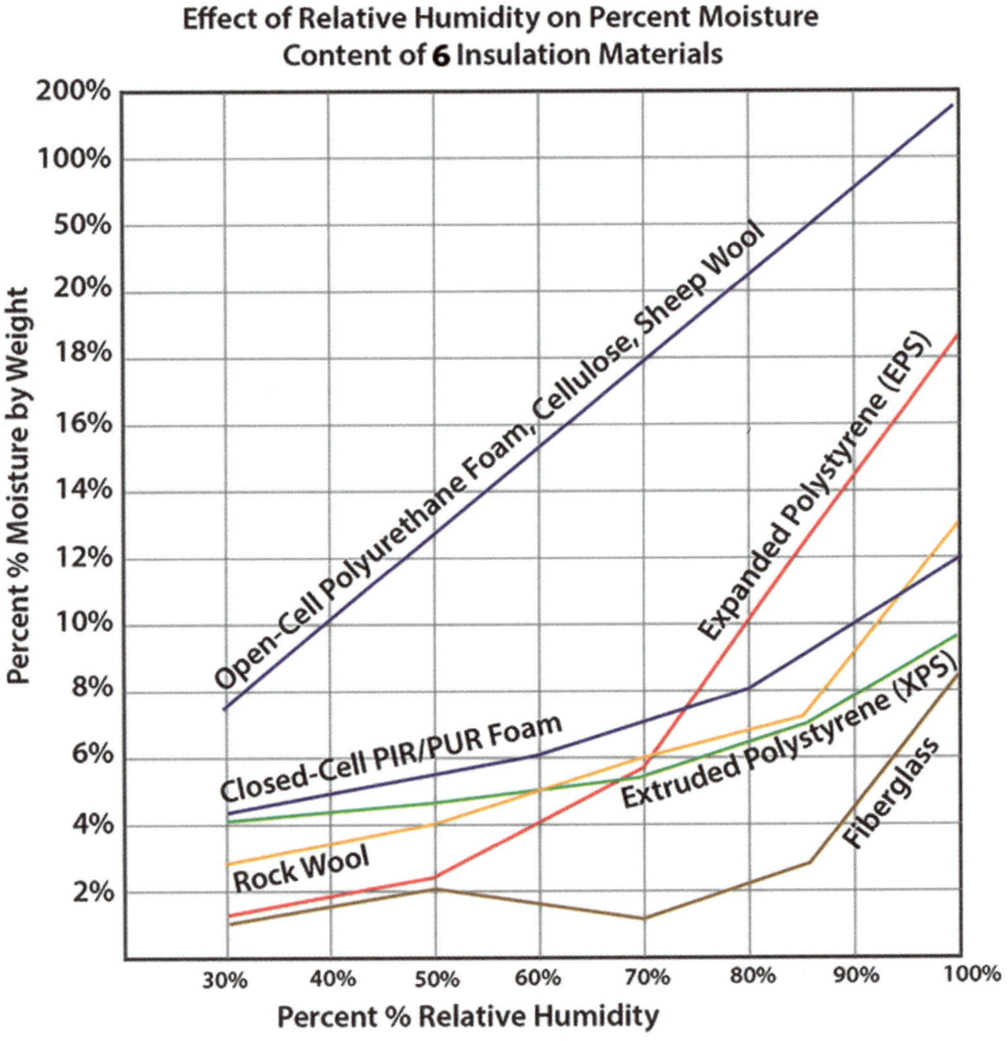

Y. Wang, S. Zhang, D. Wang et al., Experimental study on the influence of temperature and humidity on the thermal conductivity of building insulation materials, Energy and Built Environment, https://doi.org/10.1016/j.enbenv.2022.02.008

Thermal-mass effect — The mass of building components, particularly masonry walls, affects the heat transmission through those walls. In sunny climates with large daily temperature variations, massive building assemblies absorb surplus heat from both the indoors and outdoors during each day, slowing heat transmission in both directions. The mass delays heat transmission and averages the indoor and outdoor temperatures. This delay and averaging moderates the effect of the outdoor-temperature extremes.

The thermal-mass factor is a multiplier to the calculated R/RSI that increases the R/RSI, to account for the thermal-mass effect. This thermal-mass factor increases, if the R/RSI of the massive wall increases. The higher the massive wall's R/RSI the greater the mass factor. The mass factor varies from 1 to 2.6, according to tests performed at Oak Ridge National Laboratory (ORNL) on full-scale massive walls.

These tests show that masonry walls, insulated on the exterior, have higher thermal-mass factor and thus higher thermal resistance and than walls insulated on the interior. Walls insulated on the exterior have a maximum mass factor of 2.6, while walls insulated on the interior have a maximum mass factor of 1.5.

The mass factors vary according to both calculated R/RSI and climate. Of the six American cities simulated in the ORNL tests, dry hot Phoenix benefited most from insulated mass, and Minneapolis, with its mild humid climate, benefited least. Warmer climates benefit more than cooler climates from the mass effect. Cold climates show very little benefit from thermal mass.

See "Insulation R and RSI Values per Inch or per Centimeter" on page 138.

Outdoor temperature — In general, insulation demonstrates higher thermal resistance (R/RSI) at low outdoor temperatures compared to high outdoor temperatures. Very cold temperatures tend to increase insulation's thermal resistance. Warm outdoor temperatures tend to reduce thermal resistance.

R and RSI Values of Building Assemblies

Building scientists use five different types of R or RSI values to describe the different ways of measuring or calculating the thermal resistance of hollow building assemblies. These are the nominal insulation R/RSI, the center-of-cavity R/RSI, the clear-wall R/RSI, the whole-wall R/RSI, and the complete-assembly R/RSI. These R/RSI values apply to ceilings, roofs, floors, and foundations in addition to walls.

Nominal insulation R/RSI — A number that the insulation manufacturer provides, based on testing, provides the R/RSI of the insulation product only. Many people confuse this number with the R/RSI of the building assembly.

Figure 5-3: *Thermal Resistance of Insulation with Variable Outdoor Temperatures*

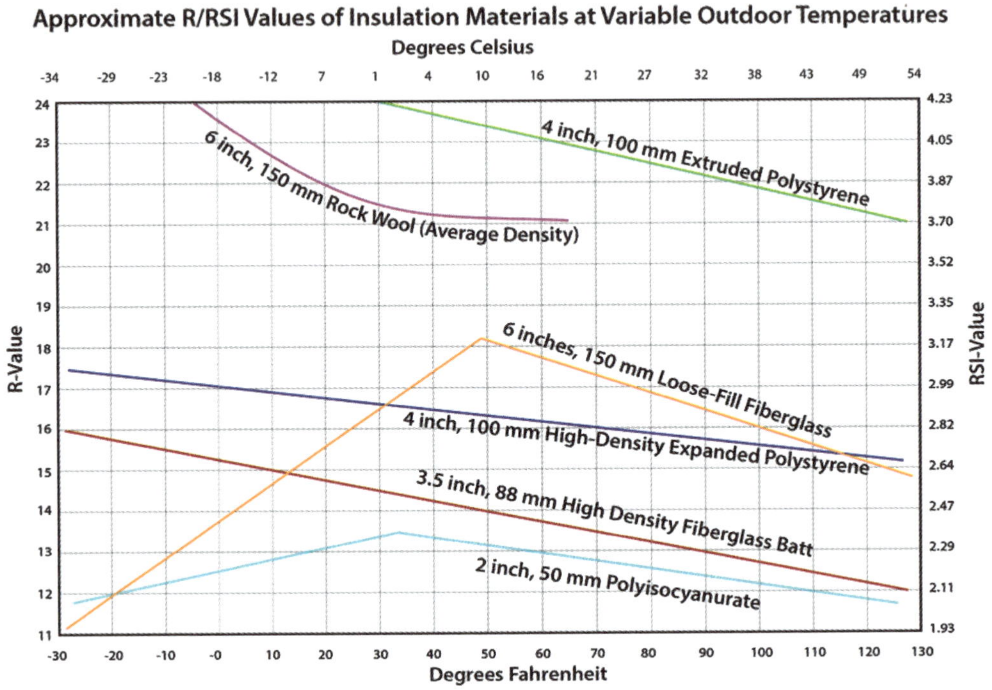

Center-of-wall R/RSI — A calculation or measurement of the center of the insulated building cavities that doesn't include the structural materials.

Clear-wall R/RSI — An area-weighted average or measurement of an insulated building assembly including both its insulated cavities and its framing material, but not including windows, doors, corners or connections to other building assemblies.

Whole-wall R/RSI — An area-weighted average or measurement, including the insulated cavities, the structural material, the corners, the intersections with other assemblies, and other complexities. Windows and doors aren't included.

Complete-assembly R/RSI — The whole-wall R/RSI, including windows, doors, and other sub-assemblies in the building enclosure.

Insulation Types

Gases conduct heat slower than solids or liquids. Insulation materials are mineral or organic materials that trap air or other gases inside small spaces. Mineral insulations include: mineral fibers, glass fibers, vermiculite, and perlite. Organic insulating materials include plastic foams and cellulose. *Organic* means that a material is composed of mostly carbon and hydrogen, similar to plants and animals.

Figure 5-4: Insulated Exterior Walls

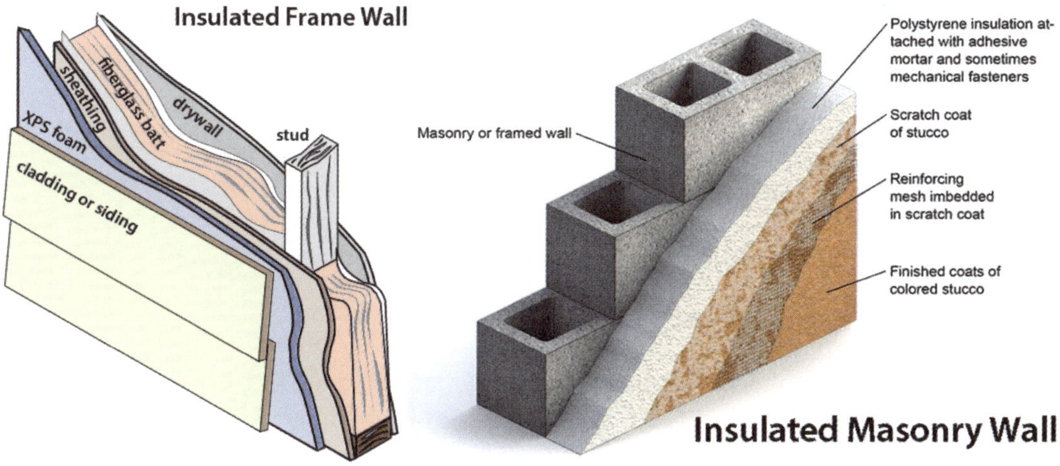

See "Insulation Characteristics" on page 136.

Insulation comes in various product types.

- Flexible materials, such as batts and blankets
- Rigid materials, such as foamboard and fiberboard
- Sprayed-on materials, such as polyurethane spray foam
- Loose-fill insulation, such as cellulose, fiberglass, and mineral wool

Loose-fill insulation has particular importance to energy retrofits because of its ability to fill spaces inside closed cavities, such as walls. Installers spray insulations to retrofit masonry walls, especially those with irregular surfaces, such as stone walls.

Insulation board is popular for retrofit projects where it attaches to framing or flat building surfaces. Most often, this insulation board is covered by new interior or exterior finish materials, such as siding, roofing, or drywall.

Fiberglass and mineral wool are the most popular insulating materials. Manufactures make fiberglass and mineral wool in batts, blankets, loose fill, and rigid boards. Batts are narrow blankets sized to fit between wall studs, floor joists, rafters, and ceiling joists.

Cellulose insulation is popular for residential buildings. Manufacturers make cellulose as a loose-fill insulation from wood fiber or recycled paper.

Energy specialists should understand each insulation material's maximum safe temperature, toxicity, fire resistance, and moisture characteristics. Here are some examples.

- High temperatures and sunlight damage foam insulation.
- Fiberglass irritates skin and lungs.
- Cellulose absorbs water in humid conditions.
- Kraft paper batt facing is flammable.
- And foam insulation produces toxic smoke when burned.

Fiberglass and Mineral Wool

Fiberglass and mineral wool are common insulations made from melting glass, rock, and slag and stretching the molten material into fibers. Manufacturers bind fibers together in batts, blankets, and boards, adhered together with organic resins. Insulators blow loose-fill fibers into open cavities like attics. Fiberglass and mineral wool absorb very little water and they're rot-resistant. Fiberglass and mineral wool are noncombustible, but batt facings and adhesives are combustible.

Mineral wool is denser and more moisture resistant than fiberglass. Mineral-wool batts are more rigid and self-supporting, than fiberglass batts. Mineral-wool batts also cost more than fiberglass batts.

Builders install batts in building cavities during construction. Batts fit between structural supports, with standard distances between them. As a retrofit, workers install batts in ceiling cavities and floor cavities. Blankets have a variety of special uses, such as metal-building insulation, duct insulation, tank insulation, and sound insulation. Installers attach boards to walls, roofs, and foundations, to insulate and to minimize thermal bridging through a building's structural materials.

Note: Fiberglass is North America's most popular insulation material. Mineral wool is Europe's most popular insulation material. The insulation industry classifies slag wool along with Rockwool™ and Stone Wool™ (product names) as *mineral wool.* Some people also refer to fiberglass as mineral wool. In this book, we don't use the name "mineral wool" for fiberglass. This rule helps to avoid using the trademarked product names for generic mineral-wool insulation.

Fiberglass and mineral-wool batts — Batts are flexible products made of compressed glass, rock, or slag fibers, held together by adhesive binders.

Fiberglass-batt manufacturers often attach a facing of kraft paper, foil-kraft paper, or vinyl during manufacturing as an air and moisture barrier and to facilitate fastening. These facings aren't usually helpful to installation. The facing hides the surface of the batt and creates difficulty in fitting the batt to completely fill its cavity.

Figure 5-5: Fibrous Batts and Boards

The main advantage of fiberglass batts is their low cost compared to mineral wool. The R-value (RSI) of fiberglass varies significantly with density as shown in "Insulation R and RSI Values per Inch or per Centimeter" on page 138. High-density batts have a higher R-value per inch or RSI per centimeter than low-density batts. We recommend that installers select fiberglass batts with a sufficient density to give a 3.5 inch (87 mm) batt with an R-value of 15 or RSI of 2.6.

The thermal performance of batts depends on careful installation. To provide maximum R-value (RSI), a batt should completely contact all its surrounding cavity surfaces. Workers should cut mineral-wool batts exactly to length. Workers should cut fiberglass batts slightly longer than the cavity height because, if too short, the small spaces at top and bottom cause air convection.

Installers cut dense mineral-wool batts with a large serrated knives. These dense batts are easier to cut because of their higher density, compared to fiberglass batts. The challenge to cutting fiberglass batts is making a square, accurate cut. The best installers cut fiberglass batts with an insulation knife or an electric meat-carving knife. The long knife avoids compressing the batt while cutting it. A good installer can use the cost advantage of fiberglass to nearly equal the quality or a good mineral-wool installation, but this isn't very likely. The workability of the mineral-wool batts usually results in a better installation compared to fiberglass. The performance advantage of mineral-wool batts is usually worth the extra cost compared to fiberglass batts.

Figure 5-6: Rigid Fibrous Insulation

Rigid mineral wool and fiberglass — Rigid fiberglass and mineral-wool board often insulate flat building surfaces such as: foundation walls, above-grade walls, and roofs. Mineral wool is denser, stronger, and more expensive than fiberglass. Mineral-wool boards have densities of 6-to-8 pounds per cubic foot or 96-to-128 kilograms per cubic meter. Semi-rigid and rigid fiberglass boards are 3-to-6 pounds per cubic foot or 48-to-96 kilograms per cubic meter.

Rigid fiberglass and mineral-wool board insulation excel at repelling and draining water, making them ideal for wet-climate applications. Installers can fasten fiberglass and mineral-wool board to wall surfaces with adhesive, furring strips, or screws with large washers.

Rigid fiberglass and mineral-wool board have these advantages over foam boards.

- Rigid fiberglass and mineral-wool boards have a better fire resistance than any kind of foam board.
- Mineral-wool and fiberglass are water-vapor permeable rather than a Class II vapor retarders like foam boards. For many building assemblies, this may be an advantage depending on the climate and type of building assembly. Vulnerable wood-frame walls avoid wetting of structural sheathing because the mineral-wool and fiberglass allow water vapor to move through the wall assembly readily in both directions.
- Mineral wool and fiberglass adsorb less water by weight and they dry faster, compared to foam board under very humid conditions.
- Mineral wool and fiberglass have a higher percentage of recycled materials, compared to foam.
- Mineral wool and fiberglass have durability advantages, including better resistance to ultraviolet radiation, heat, and fire compared to foam board.

Blown fiberglass and mineral wool — Manufacturers make two types of blown fiberglass: chopped-up batt waste and virgin short fibers. The batt waste has longer fibers and a binder, so its R-value per inch or RSI per cm is lower than virgin fibers, which have thinner and shorter fibers. The shorter thinner fibers create smaller and more numerous air spaces.

Manufacturers pack fiberglass and mineral wool for blowing into compressed bags. The compressed fiberglass or mineral wool requires a blowing machine with an powerful agitator. The agitator separates the compressed material into small pieces that travel fluidly through the blower and its hose without plugging the hose.

It's easy to over-aerate fiberglass in attics, leading to low densities (<0.5 lb/cf or <8 kg/m^3) and excessive air-permeability. Cellulose and mineral wool insulation are superior to fiberglass at resisting convection as attic insulation. Blown mineral wool is denser than fiberglass — averaging almost 2 lb/ft^3 or 32 kg/m^3. At this greater density mineral wool allows less air circulation within it, compared to fiberglass.

Figure 5-7: *Roofs with Ventilation above the Roof Deck*

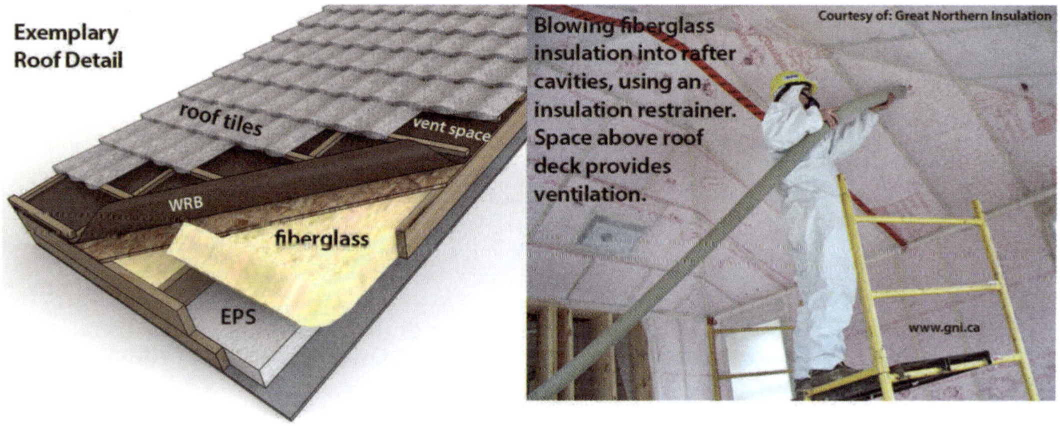

A quality blowing machine can achieve an adequate density for blown fiberglass. Such a blowing machine possesses the power to tear the fiberglass bale into small pieces and to then move it in a dense mass through the blower hose. Fiberglass in attics should be at least 0.90 lb/ft^3 or 14 kg/m^3 density. At that attic density, expect 3 to 6 percent settling.

Blown fiberglass or mineral wool give better R-values (RSIs) in closed cavities than fiberglass batts or cellulose. The density of blown fiberglass insulation should be greater than 1 lb/ft^3 or 16 kg/m^3 in closed floor and ceiling cavities. Fiberglass blown into walls should be greater than 2 lb/ft^3 or 32 kg/m^3.

To test the density of blown insulation, calculate the volume of a test cavity. Multiply the desired density times the volume to get the number of pounds or kilograms of insulation that should fit into that cavity. Look for the insulation's bag weight on its label, so you can compare the number of pounds or kilograms you install to the weight you calculated to sufficiently fill the test cavity or cavities.

Blown fiberglass is the best type of blown insulation for closed cavities that are structurally weak, such as the roofs and walls of metal-skinned mobile homes. Blown fiberglass installs inside roof and floor cavities at a lower density than cellulose or mineral wool blowing insulation. This lower density means that blown fiberglass puts less pressure on the structurally weak cavity, to reduce the possibility of damage.

Blown mineral wool and fiberglass both have better fire resistance compared to cellulose, which is recognized as a fire risk. Unlike cellulose, blown fiberglass and mineral wool are nearly non-combustible.

Figure 5-8: *Cross Section of Roof Insulated above the Roof Deck*

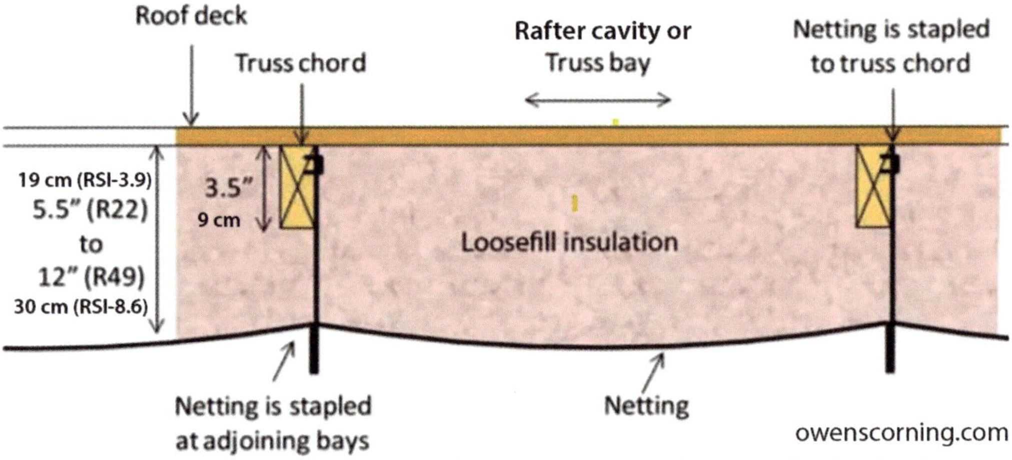

Insulators install netting as an insulation restrainer to increase the depth of insulation that they can install under the roof deck. Owens Corning product.

Blown Cellulose

Blown cellulose is newspaper, cardboard, cotton, straw, sawdust, hemp and corncob, often treated with fire retardants and pesticides. Newspaper is the most common form of cellulose used today.

Cellulose blows quickly and easily, achieving a high density in walls when installed with a fill-tube. Cellulose from newspaper contains small fibers that pack into cracks and crevices of closed building cavities, resisting airflow through these cavities. Insulators used this characteristic of cellulose to air seal existing buildings.

Blown-cellulose attic insulation should achieve a density of around 1.4 lb/ft^3or 22 kg/m^3. At this density, expect at least 15% settling. Wall insulation should be at least 3.5 lb/ft^3 or 56 kg/m^3 to prevent settling.

Figure 5-9: *Insulation-blowing Equipment*

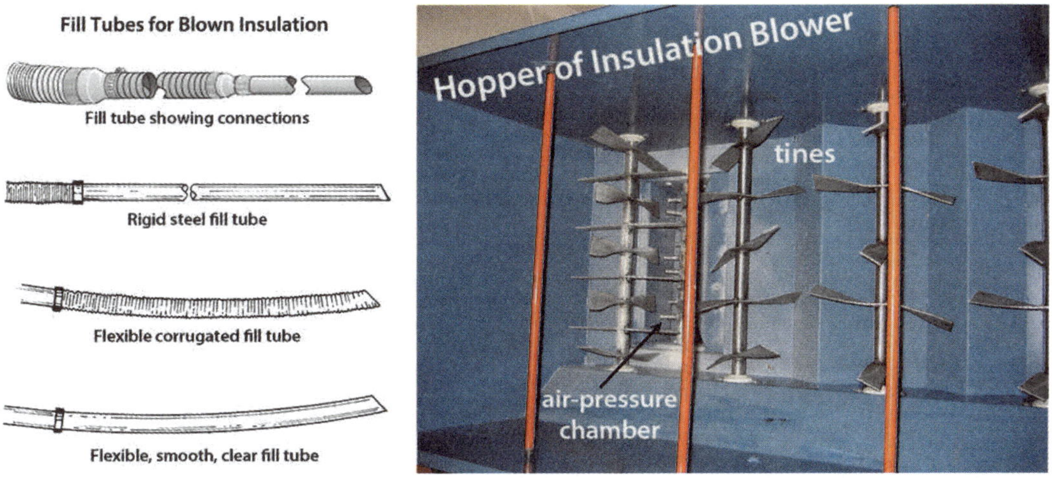

The major disadvantage of cellulose is that it can absorb more water than fiberglass or mineral wool. This high water *absorption* becomes a problem if water leaks from the outdoors because cellulose wicks water into itself and stores it. Water adsorption is also a problem in humid climates with consistently high relative humidity (Rh). In fact, cellulose can absorb over 100% of its weight in water if subjected to constantly high Rh — over 90%. Fog, ground-water saturation, occasional flooding, and summer discomfort characterize high-humidity climates.

Also, water can carry fire retardants out of the cellulose. The solution of water and standard fire retardants can corrode metal building materials. If insulation must touch bare metal building materials, use cellulose treated with a non-corrosive fire retardant.

Vermiculite and Perlite

Manufacturers make vermiculite and perlite by heating rock pellets until they pop, like popcorn. Vermiculite pellets are gray or brown and $^{3}/_{16}$ inch or 5 mm in diameter. Vermiculite has an R-value of R-2.4 per inch or RSI-0.17 per cm, and it is relatively expensive. For these reasons, vermiculite isn't a popular insulating material.

Installers usually pour vermiculite and perlite into cavities. Vermiculite and perlite aren't likely to leave voids by damming up inside walls like blown fibrous insulations. Vermiculite and perlite work well in vertical cavities with rough surfaces and obstacles that might dam blown insulation — concrete-block cores, for example.

Masons sometimes mix vermiculite or perlite with cement and aggregate to produce a lightweight and less heat-conductive mortar or concrete.

Assume that vermiculite contains asbestos, unless a laboratory test proves the material to be asbestos-free. Workers must not disturb asbestos, unless they are licensed to handle asbestos. Perlite doesn't contain asbestos.

Rigid Foam Panels

Rigid insulation panels have a wide variety of applications. Polyisocyanurate (PIR) and polystyrene boards function as interior and exterior insulated sheathing for residential and commercial construction.

Foam insulation reduces heat transmittance and thermal bridging through structural materials like wood, steel, and masonry. The foam can function as both surface insulation and thermal break.

Cover plastic foam insulation with a cladding that resists ultraviolet radiation, mechanical damage, and moisture when you install foam outdoors above ground. Protective coverings include a variety of plastic, metal, and masonry materials.

Fire and rigid foam — Rigid foam is combustible, and it releases toxic gases when it burns. When used on the building's interior facing the living space, you must cover foam panels with a *thermal barrier* of half-inch (13 mm) drywall. When used to insulate roofs from the attic or crawl-space walls in the crawl space, building codes may require installers to cover foam panels with an *ignition barrier* such as drywall or galvanized steel.

Moisture and rigid foam — Water-vapor permeance is important to ensure the drying potential of new or renovated building assemblies. Most rigid foam boards are Class II or Class III vapor retarders. Permeance decreases with increasing thickness as a function of the permeance value for one inch or one centimeter divided by the number of inches or centimeters of the material's thickness.

The choice between less permeable foam and more permeable foam depends on climate and the building assembly it insulates. The goal of a correct choice of insulation is to avoid condensation near the exterior or interior of the building's thermal enclosure.

In hot, humid climates, you may want to minimize the flow of water vapor from the hot humid outdoors to the cool dry indoors. Also it's best to avoid less permeable foam and other impermeable materials on the building enclosure's interior in hot humid regions because of the potential for condensation in building cavities.

If installed on the inside of a an enclosure, foil-faced PIR can experience condensation on its surface because it's a Class I vapor retarder (vapor barrier). Foil-faced-PIR can block outdoors-to-indoors vapor diffusion when installed outdoors in a hot humid climate. This blockage could stop moisture diffusion and prevent condensation during air conditioning. However, if the climate also experiences long periods of cold weather, the outdoor-facing PIR could cause condensation in the cavity during winter from warm humid indoor air diffusing into the cavity.

Rigid-foam products **—** Builders commonly use foam-insulation panels for exterior building insulation. Technicians fasten foam panels to roof assemblies and cover them with rubber membranes or other roofing materials. Plastic foam insulation is available in a variety of thicknesses with plywood, drywall, or other materials bonded to it. These products can be useful to retrofit masonry buildings.

Foam board's rigidity sometimes improves structural strength to building assemblies, particularly when combined with wood sheathing. For example, polystyrene sandwiched between plywood or composite wood sheathing serves as structural wall and roof systems in energy-efficient new buildings. Polyurethane, installed inside hollow exterior doors, adds to their rigidity.

Compressive strength is important for load-bearing, for example under footings, under concrete slabs, and over roofs where people walk. A product density of 2-to-3 pounds per cubic foot or 32-to-48 kilograms per cubic meter are common for foam boards that require structural strength.

Types of Rigid-foam Boards

Exterior insulation avoids disturbing the building's interior. Exterior insulation is expensive and requires careful flashing and other water-protection details to avoid water damage.

Common uses for rigid foam boards include these.

- As an exterior thermal break for wood-framed and steel-framed walls
- Above-grade and below-grade masonry-wall insulation in new and retrofit applications
- New and retrofit roof insulation, especially to reduce roof solar gain and to keep roof cavities warm and dry in winter
- Under footings and concrete slabs for new buildings

Figure 5-10: *Rigid Foam Insulation*

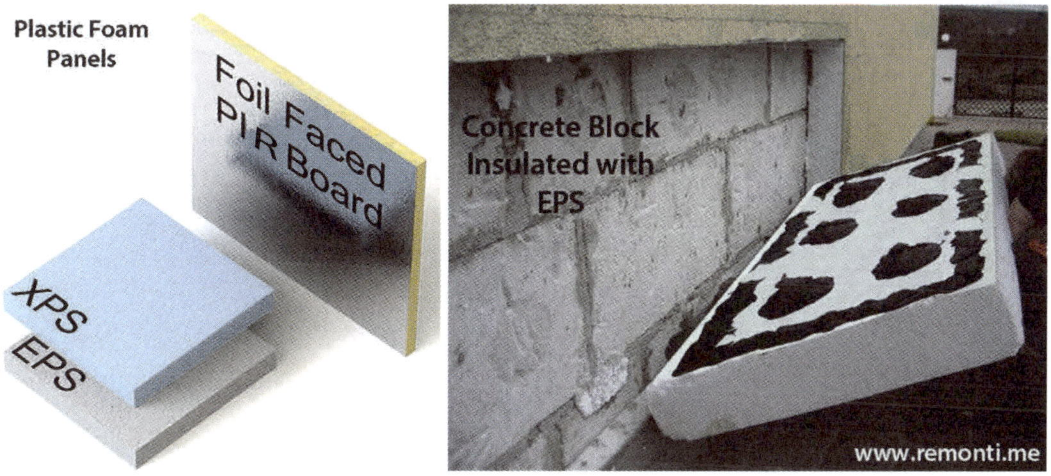

Expanded polystyrene (EPS) — Manufacturers make EPS in many small factories throughout the world by expanding polystyrene granules in a mold. Builders and manufacturers use expanded polystyrene (EPS) to insulate roofs, walls, foundations, exterior doors, and garage doors.

The EPS panels vary significantly in density depending on their use. Higher density reduces EPS's water-vapor permeance and increases its compressive strength. Thermal resistance varies with density between 1 and 2.5 pounds per cubic foot or 16-to-40 kilograms per cubic meter are typical for EPS boards. EPS has an R-value of an average of R-3.6 per inch or 0.25 RSI per centimeter.

Don't install EPS in direct contact with dark-colored roofing or wall cladding. EPS begins to soften and may shrink at temperatures greater than 165°F or 75°C. Ultraviolet radiation also damages EPS, so install a UV-resistant covering as soon as possible after installation.

EIFS insulation — Special masonry coatings with acrylic adhesive can adhere to polystyrene insulation after installers attach EPS panels to the building's exterior walls. Insulators install these stucco-like coatings, known as exterior insulation-and-finish systems (EIFS) in multiple layers. A fibrous fabric, applied with the first coat, reinforces this latex stucco. EIFS is one of the few practical cladding options for exterior insulation of solid masonry walls.

Figure 5-11: *Polystyrene Insulation: Expanded and Extruded*

Extruded polystyrene — Only a few manufacturers make extruded polystyrene. Extruded polystyrene (XPS) is more expensive than expanded polystyrene (EPS). Extruded polystyrene has better compressive strength than EPS. XPS has an R-value of about R-5 per inch or 0.35 RSI per centimeter.

XPS has a lower water-absorption rate compared to EPS. Many experts believe that this characteristic makes XPS a better choice for damp environments such as roof insulation and foundation insulation. Other experts worry that XPS doesn't dry as fast as EPS, in moist environments. XPS can absorb and retain moisture if a vapor barrier blocks its drying potential — for example, if covered by rubber roofing in a cold climate.

Don't install XPS in direct contact with dark-colored roofing or cladding. XPS starts to soften and can possibly shrink at temperatures of 165°F/75°C or less. Ultraviolet radiation also damages XPS, so install cladding as soon as possible after installation.

Figure 5-12: *Roof Products that Include a Roof Deck or the Roofing Material*

Polyisocyanurate **—** Polyisocyanurate (PIR) panels provide excellent service when space is limited, because its thermal resistance is R-6 per inch or 0.42 RSI per centimeter of thickness. PIR is usually faced with aluminum foil or a glass fiber, to protect its fragile surface.

Foil-faced PIR is a vapor barrier (Class I vapor retarder), and fiber-faced PIR is a Class II vapor retarder. Consider the effect of PIR's vapor permeance before you use it in a building assembly.

If one or both of its surfaces face an air space, the foil could function as a radiant barrier, although this type of installation isn't common. Polyisocyanurate's R-value (RSI) deteriorates slightly over time as air, with its lower thermal resistance, replaces PIR's pentane inter-cellular gas.

Figure 5-13: Foamboard Insulation Examples for Cold Climates

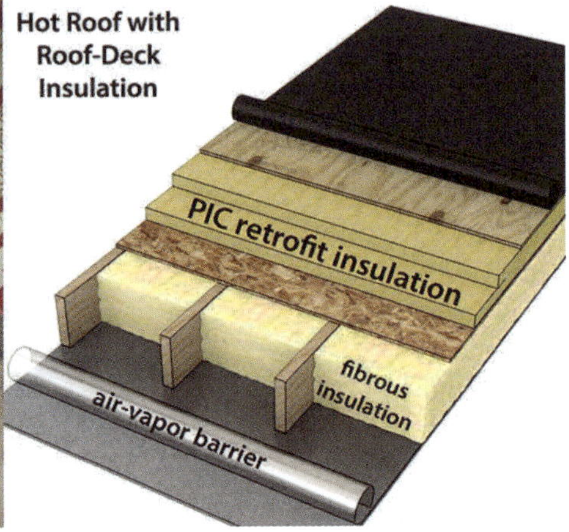

Foam Insulation: Sprayed and Injected

Manufacturers formulate several types of foam insulation to spray on surfaces or inject into cavities. During installation, polyurethane resin combines with a blowing agent, such as hydro-chlorofluorocarbon gas (HCFC) or carbon dioxide (CO_2). The blowing agent causes the resin to expand into a foam, encapsulating tiny gas bubbles. The most common types of spray foam (SPF) are high-density or closed-cell and low-density or open-cell polyurethane.

Figure 5-14: Spray Insulation Products and an Example of Spray-roof Insulation

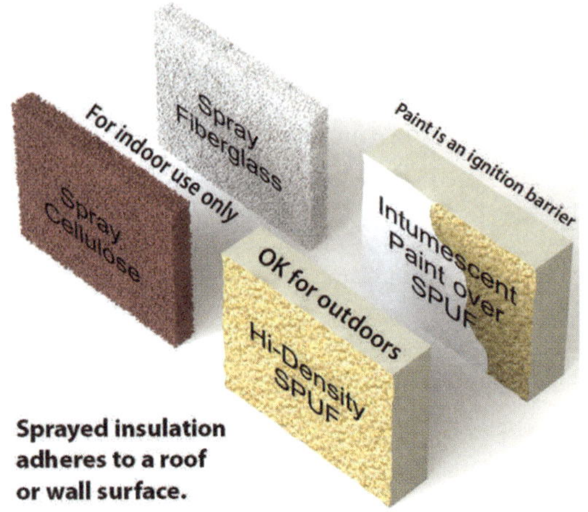

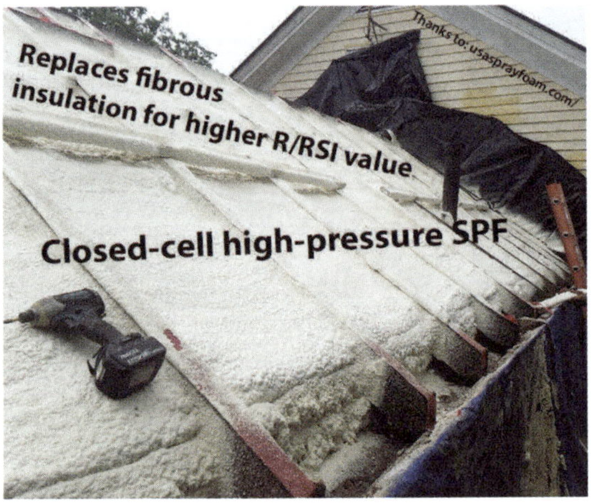

High-density polyurethane — High-density sprayed polyurethane insulation has excellent thermal resistance, air-sealing capacity, adhesion, and structural strength. Workers install high-density SPF using two methods.

1. *High-pressure method:* truck mounted equipment sprays foam at a high pressure and high volume for large jobs. Workers wear full-face, fresh-air respirators with coveralls and toxin-resistant gloves.

2. *Low-pressure method:* Workers spray foam from two-component kits in smaller amounts. Workers wear organic-vapor respirators with coveralls and special gloves for this method.

HCFC gas that fills the foam's cells accounts for polyurethane's superior thermal resistance of R-6 per inch or 0.42 RSI per centimeter of thickness. High-density polyurethane foam insulates roofs, masonry walls, metal buildings, and many manufactured building products. High-density foam's initially high R-value per inch (RSI/cm) decreases over time as air replaces the hydrocarbon gas.

Low-density polyurethane — Low-density polyurethane foam can be either sprayed or injected. Low-density polyurethane is water-soluble and has a dry density of 0.5-to-1.0 lb/ft^3 or 8-to-16 kg/m^3. Its R-value is between 3.7-to-4.7 per inch or RSI between 0.26-and-0.33 per centimeter of thickness.

Figure 5-15: *High-Density Spray Foam (SPF)*

AirKrete® — AirKrete® is a proprietary sprayed-injected foam, composed of expanded Portland cement and sand. AirKrete is the only non-combustible foam in common usage.

AirKrete excels at filling cavities of framed buildings and hollow-masonry-block buildings. It also has less toxicity concerns compared to spray polyurethane foam.

AirKrete blocks are structural building blocks with the same range or R-value as the sprayed-injected foam, which is 3.7-to-6.0 per inch or 0.26-to-0.42 per cm, depending on the mass factor. See "Thermal-mass effect" on page 141.

Figure 5-16: Retrofits with Injectable Foam

Facings and Barriers

This section discusses barriers and facings that protect insulation materials. The most important types of barriers are water-resistive barriers, water barriers, air barriers, vapor retarders, and fire barriers. These barriers are also called *control layers*. These barriers also prevent water and water vapor from entering a building and increasing its relative humidity, which affects summer comfort.

Insulation Facings

Manufacturers fasten various facings to insulating materials during manufacturing. Fiberglass batt or blanket insulation is available with a variety of facings. Some common insulation facings are: kraft paper, aluminum-foil-kraft-paper laminate, and flexible vinyl.

Facings protect the insulation's surface, provide an air barrier and/or vapor retarder. They also facilitate fastening and help to hold fragile insulation together. All of the common facings attached to fiberglass batts are air barriers and most are also vapor barriers or vapor retarders. However, their effectiveness as air barriers depends on nearly flawless installation. Besides being an air and vapor barrier, aluminum foil is also a radiant barrier, minimizing radiant heat flow when it faces an air gap.

Most facings are combustible. Kraft paper by itself or attached to other materials, such as aluminum foil or polyethylene, is flammable — meaning it burns if a flame touches it. Vinyl facings are also flammable.

Figure 5-17: *Types of Environmental Barriers*

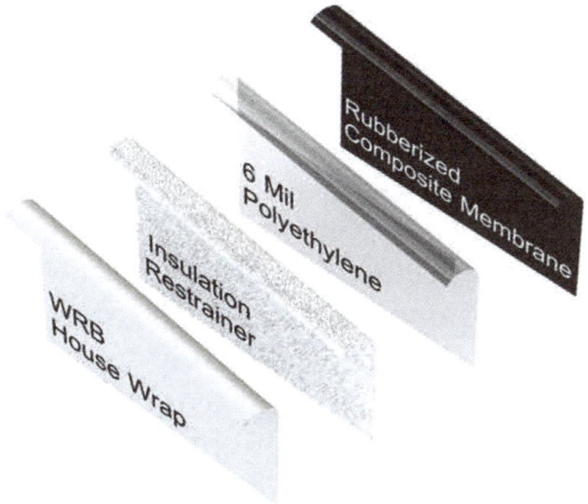

Drainage planes and rain screens — A water-resistive barrier (WRB) may border a narrow space, called a *drainage plane*, where water can drain to the bottom of the wall and then to outdoors. When a planned space exists between the exterior cladding and the water-resistive barrier, this assembly is called a *rain screen*. A rain screen should have openings at its bottom for drainage.

When the rain screen is open to the outdoor-air pressures at the top and bottom of the wall, this is called a *pressure-equalized rain screen*. This assembly drains and ventilates the space, and it also reduces the pressure of wind and wind-driven rain across the wall assembly.

Masons install masonry-wall cladding such as brick veneer and stone veneer with drainage planes. The best method puts a space between the structural sheathing and masonry cladding wide enough to drain the wall and also to allow for the accidental mortar droppings into that space.

With stucco and thin masonry veneers, masons may install a drainage mat and weep screed. A drainage mat creates a very narrow space for water to drain down behind the thin masonry veneer. A *weep screed* is a long metal or plastic bracket with weep holes to drain water out the bottom of a masonry veneer.

Water Barriers

There are four main types of water barriers or *water-control layers*.

1. The *water-resistive barrier* stops water penetration into walls and roofs. Manufacturers offer WRBs in a variety of water-vapor permeabilities.
2. In cooperation with the water-resistive barrier, *flashing* protects vulnerable areas in foundations, walls, and roofs by moving water away from the building.

3. *Waterproofing* stops water and water vapor from penetrating foundation walls.
4. *Capillary barriers* stop ground water from rising into the building.

See "Foundation Water Leakage: Prevention and Treatment" on page 48.

Air Barriers

Retrofit air sealing is one of the most effective and economical energy retrofits for reducing air conditioning costs in electricity consumption. Air barriers stop hot and humid outdoor air from entering buildings. Air barriers also stop air-conditioned air from leaving buildings. Air barriers reduce both air leakage and air convection in building assemblies. Air barriers also reduce wind-driven air intrusion around and through insulation. This air intrusion reduces the insulation's thermal resistance, when air leaks around or through insulation.

See "Approximate Air Leakage per Unit of Surface Area at 50 Pascals Pressure" on page 86.

Figure 5-18: *Water Barriers, Drainage Plains, and Flashing*

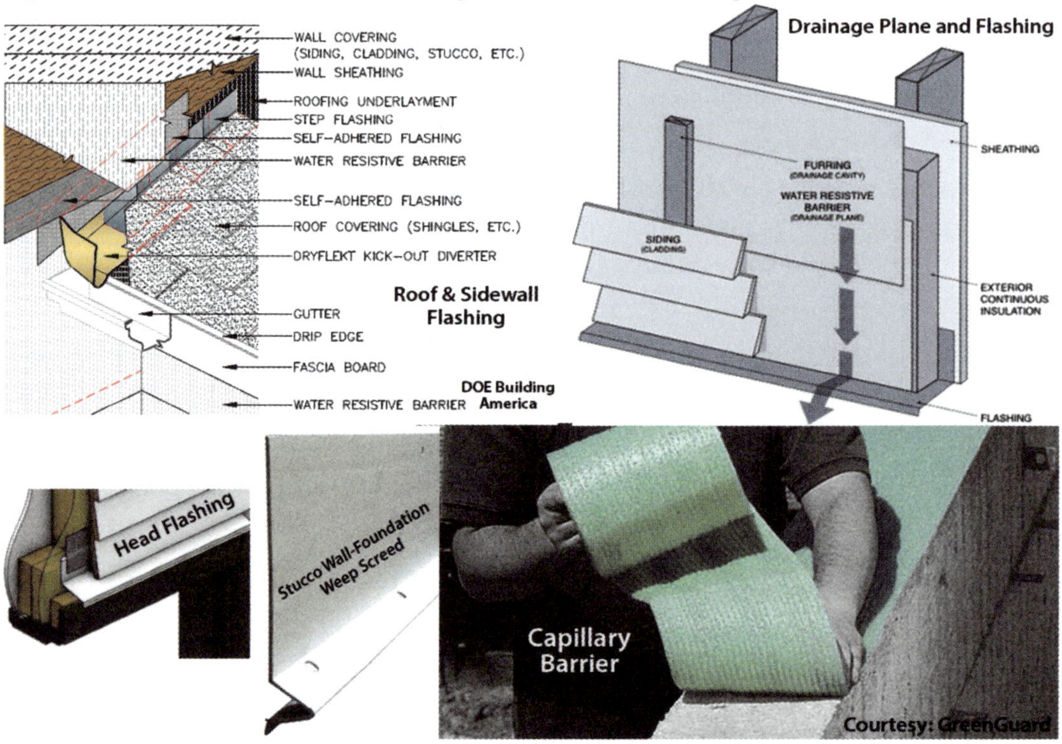

An increasing number of energy-efficient new buildings use interior drywall and/or the exterior sheathing as their air barriers. Drywall and wood sheathing materials like plywood and oriented strand board are good air barriers. We recommend structural materials instead of house wrap as air barriers. Although house wrap itself is airtight, house wrap is both difficult to seal and structurally weak. Plywood, OSB, and drywall are more effective, and they are strong enough to resist powerful wind gusts without damage, unlike house wrap.

In existing buildings, you can consider any airtight interior or exterior surface as an air barrier. Plaster and stucco are good air barriers because they are continuous and airtight. Carefully installed drywall is an effective air barrier. Drywall or plaster walls in existing homes become effective air barriers when someone seals their joints with adjacent materials. Concrete is an air barrier, but brick, concrete block, and other masonry units aren't usually air barriers.

Retrofitting an existing building requires caulk or foam around windows, doors, electric switches, electric receptacles, the floor-wall junction, and the wall-ceiling junction. However, the airtightness of a retrofit job that seals all the above areas is never as good as a new building, built with a designed air barrier and tested before the builders install the finish materials.

Figure 5-19: *Wall Vapor Retarders*

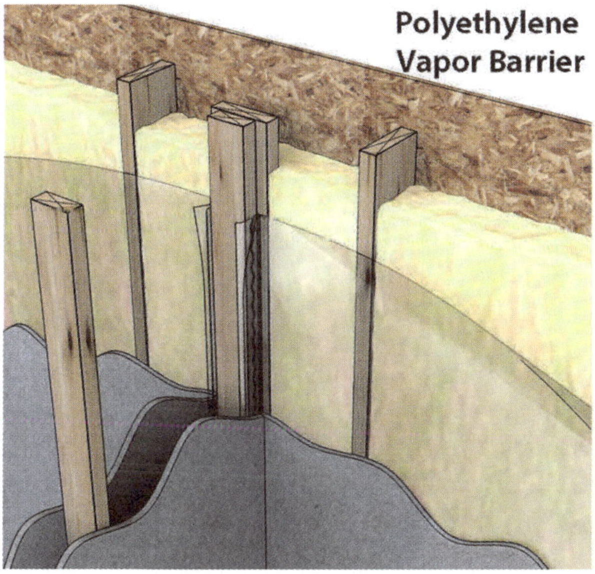

Vapor Retarders and Vapor Barriers

A vapor retarder is a material that resists vapor diffusion through the building enclosure. A *vapor barrier* is a very effective vapor retarder, or a Class I vapor retarder. A *Water-vapor-control layer* is an alternative term for a vapor retarder.

During cold weather, large indoor-outdoor temperature differences lead to high vapor pressure between cold, dry outdoor air, and warm, moist indoor air. The most common application of vapor barriers is on the interior surfaces of exterior walls in cold-climate buildings to prevent condensation in building assemblies.

Vapor pressure moves water vapor from outdoors toward indoors during humid summer weather, while an air-conditioner is cooling and drying indoor air. Builders sometimes install vapor barriers on exterior walls in hot humid climates. Foil-faced foam wall and roof sheathing works well for this purpose. However, if the vapor pressure is strong in both summer and winter, an exterior vapor barrier might trap moisture and cause condensation in building cavities. In this case, a class III vapor retarder would be the best choice to control water-vapor diffusion moving both in both directions.

Vapor retarders are semi-permeable to water vapor and include: wood composite panels, EPS, XPS, and fiber-faced PIR.

Table 5-3: *Classes of Vapor Retarders (or Vapor control layers)*

Vapor Retarder	Perms USA	Perms Metric
Class I vapor retarder (vapor barrier)	<0.1	<0.15
Class II vapor retarder	0.1–1.0	0.15–1.5
Class III vapor retarder	1.0–10	1.5–15
Class IIII vapor retarder	>10	>15
1 USA perm = 1 grain per square foot per hour per inch of Mercury pressure. 1 metric perm = 1 gram per square meter per day per millimeter of Mercury pressure.		

See "Controlling Water Vapor" on page 51.

Fire Barriers

The building codes focus on preventing the spread of fire inside buildings and between buildings. *Fire barrier* is a word with a specific meaning in building codes: **A building assembly that has been tested and certified to withstand and contain a fire for a specified time period.**

- A *thermal barrier* is a material or assembly that protects the materials behind it from reaching a temperature of 250°F or 107°C during a fire. One-half-inch or 13-millimeter paper-faced drywall, a common thermal barrier, provides 15 minutes of fire protection. Fiberglass-faced drywall provides a 1-hour thermal barrier.
- An *ignition barrier* is a material used with foam insulation to prevent the foam from igniting. The code specifies a number of materials that can function as ignition barriers including drywall, plywood, galvanized steel, and intumescent paint.

- A *fire partition* is a fire barrier that prevents the spread of fire between rooms on one level of a building.
- A *firewall* is a structural fire barrier between buildings that should remain standing during and after a fire. No building renovation or energy retrofit should damage or penetrate a firewall or fire partition.
- A *fire stop* is a passive fire barrier that seals openings and joints in a fire-rated building assembly.

Never damage any of these fire barriers during energy retrofits. Air sealers should repair these barriers during their air-sealing work, if necessary.

Insulation Restrainers

Installers stretch *insulation restrainers* across cavities to hold blown fibrous insulation in the open cavities. The insulation restrainer also allows the installer to pack the insulation to the required density.

Insulation restrainers are typically strong, flexible materials like plastic netting, plastic-sheet vapor retarders, and strong open-weave fabrics.

Figure 5-20: *Retrofit Blown Fibrous Insulation*

Retrofitting Insulation

The biggest challenge to installing retrofit insulation is to access the building cavities. Insulators access building cavities by drilling holes or by removing building materials. These activities can create dust, which is a respiratory hazard for workers and residents. Contractors must follow health-and-safety rules about protecting workers and customers from dust, and especially lead dust.

Like many contractors whose work can be dirty, hot, and unpleasant, insulation companies have problems with employee training, motivation, and turnover. These human challenges can affect installation quality, which is essential to the insulation's performance. For this reason, it's important that energy specialists who specify, supervise, and install insulation be familiar with insulation principles, characteristics, effective installation methods, and safety procedures.

Where to Insulate

Choices of where to insulate depend on existing insulation, insulation cost, the air barrier's location (if present), access to building assemblies, and a number of site-specific, practical considerations.

Which is more important to insulate, the walls or the ceiling? The answer depends on the existing R-value (RSI) in each area. Select the lowest-R-value choice of locations if there isn't enough money to insulate both. The unit-price difference in insulation for the two areas is also an important decision factor.

Figure 5-21: *Baffles Promote Attic Ventilation*

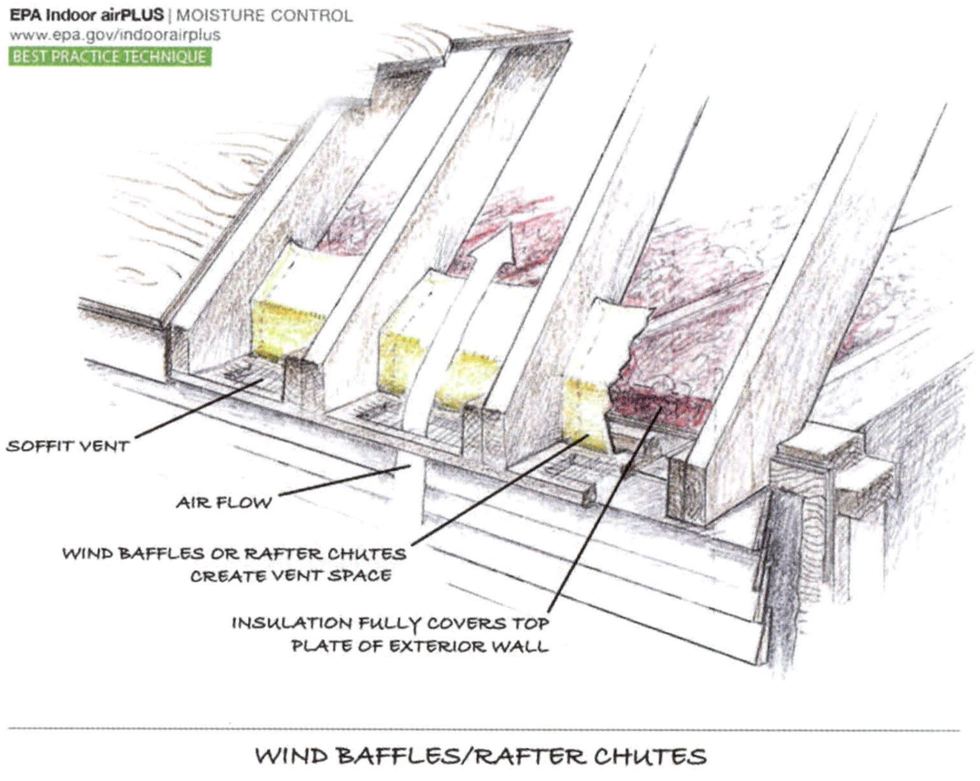

Blowing Loose-fill Insulation

Insulation blowing machines break up the compressed insulation and move the insulation into a sealed chamber at the bottom of the machine's hopper. Air pressure, created by a compressor or blower, blows the insulation out of this chamber and through a hose to its destination.

The two common types of insulation blowing machines are electric-powered and gasoline-engine-powered. Most professional insulation companies use gas-powered machines. Gas-powered machines blow insulation more quickly and hold more bags in their hoppers, compared to electric-powered machines.

Smaller electric machines do a good job of blowing cellulose insulation. However, blowing fiberglass or mineral wool is difficult for smaller electric machines. Machines may blow slowly if they aren't designed for blowing fiberglass or mineral wool, or if the machine is poorly maintained.

Controls on the blowing machine adjust the air pressure and insulation delivery rate. The operator varies air pressure and delivery rate depending on whether workers are insulating an open cavity or closed cavity. If blowing into an open attic, the material delivery rate should be high. Installers strive to install insulation at near its settled density and to avoid fluffing the insulation to a low density. In closed cavities, the machine's delivery rate and pressure should fill the cavity quickly and steadily without bulging interior or exterior surfaces and without clogging the hose.

Hoses, tubes, and fittings — Hoses and fittings are important to blow insulation efficiently. Connections between blower hoses, sections of hose, and fill tubes should be as airtight as possible to prevent pressure loss.

Use the biggest tube diameter that fits comfortably into a cavity. In closed cavities (walls, floors, and ceilings: closed on both sides), blown insulation won't reliably travel more than 18 inches (46 cm) from the end of a tube. If installers assume the insulation will travel farther, this causes empty spaces and low-density insulation. A fill-tube, inserted into the farthest corners of the cavity, fills the cavity evenly, without leaving spaces.

When you fill closed building cavities, reduce the diameter of the hose from 3 or 4 inches (75 or 100 mm) down to $1^{1}/_{2}$ or 2 inches (38 or 50 mm). Make these hose reductions no closer than 6 feet (1.8 meters) apart. Use metal or rubber reducer fittings to reduce the hose's diameter an inch (25 mm) at a time to avoid plugging the hose.

A flexible fill tube is considerably more rigid than a standard blower hose. Its semi-rigidity allows the installer to push the fill tube 6 feet (1.8 meters) or more into a closed cavity without bending or crimping the tube. But the tube is still flexible enough to move around corners and obstacles.

Use rigid fill tubes for filling roofs and floor cavities that are inaccessible from indoors. Metal electrical conduit, copper drain pipe, steel muffler pipe, or plastic pipe all work well to fill these large cavities.

Insulation blowing safety — Remember that blowing insulation creates dust that can irritate or damage lungs. Always wear a respirator when you blow insulation. Installers should protect their skin with gloves, long sleeves, and coveralls, and protect their eyes with goggles or safety glasses.

Static electricity sometimes builds up in fill tubes and shocks installers. Fill tubes made of plastic pipe are most likely to shock. Fill tubes made of lightweight electrical conduit, muffler pipe, and copper drain pipe are more static-free than plastic.

Figure 5-22: *Blown Fibrous Wall Insulation*

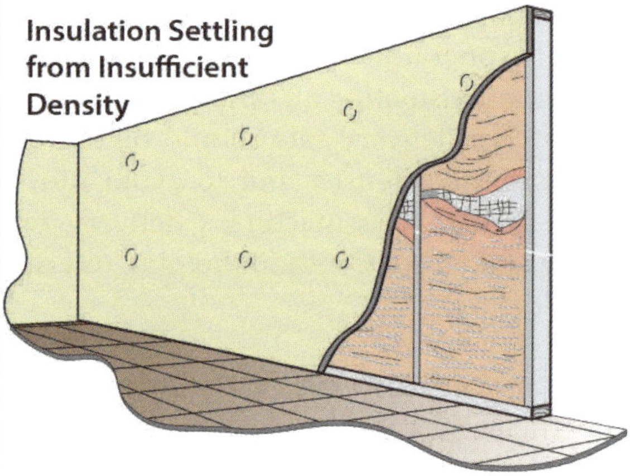

Figure 5-23: *Fibrous Retrofit Ceiling-and-Roof Insulation*

Ground the plastic pipe with a grounding wire or spray small amounts of water into the hopper of the blowing machine, if you need to reduce or eliminate static electricity.

Sprayed and Injected Insulation

Sprayed insulation is a logical choice to insulate rough surfaces or to install in places where it's difficult to fasten insulation batts or boards. Insulators spray both fibrous and foam insulation onto surfaces and into cavities.

Insulators inject foam insulation into closed building cavities through holes — usually as a retrofit option.

Figure 5-24: *Retrofit Wall Insulation*

When sprayed and injected insulation fills the entire cavity, it prevents convection and air leakage better than batts or loose fill insulation.

Caution: Most liquid foam formulations release toxic gases. Installers need fresh-air respirators or organic-vapor respirators along with toxin-resistant gloves and overalls. Consult the product's safety data sheet (SDS) for instructions about personal protective equipment (PPE) before installing spray foam. The building's residents may suffer sickness or lung damage from spray foam, unless they to leave temporarily or unless the installers provide continuous and effective ventilation.

Sprayed-and-injected foam insulation — Various manufacturers make sprayed and injectable foam insulation in at least 5 chemical formulations.

1. *High-pressure, high-density two-part polyurethane foam:* Installers spray this very common formulation from truck-mounted equipment. This type of foam is structurally strong and moisture resistant. This type of SPF is the most popular spray foam worldwide.
2. *Low-pressure, high-density, two-part polyurethane foam:* Installers spray this formulation from disposable or recyclable bottles and cans. Each unit has two bottles. Two hoses meet at the nozzle and mix their contents during installation. This type of foam seals air leaks, fills gaps, and insulates surfaces. It is almost as dense, strong, and moisture resistant as the high-pressure foam.
3. *Low-pressure, high-density, one-part injectable polyurethane foam:* Installers inject these formulations into holes and gaps from single disposable or recycla-

ble containers. This popular type of foam seals air leaks and reduces thermal bridging.

4. *Low-pressure, low-density, water-soluble two-part spray polyurethane foam:* Installers spray this very common formulation from truck-mounted equipment. This cavity-filling foam isn't strong or moisture resistant, but is low-cost, compared to other formulations.
5. *Low-pressure, low-density, water-soluble, injectable foam with various formulations:* Installers inject or spray polyurethane or AirKrete® foam into closed wall cavities with a fill tube. This type of foam fills cavities of masonry block and wood-frame building cavities.

Spray foam is a challenge to install and requires careful mixing, temperature control, equipment maintenance, and installer training. Installers may accidentally create bubbles that lead to poor thermal resistance. Bubbles happen when the sprayed mixture isn't perfect, the temperature isn't correct, or sprayed layers are too thick.

Polyurethane is available in a variety of formulations that install at 0.5 to 3.0 lb/ft^3 density (8–48 kg/m^3). Installers spray all densities of polyurethane, but they only inject the low-density foam (<1 lb/ft^{3}or <16 kg/m^3). Installers inject low-density polyurethane foam through multiple holes in each cavity using a nozzle that penetrates the cavity.

Sprayed cellulose, fiberglass, and mineral wool — Installers spray cellulose, fiberglass, and mineral wool insulation using an insulation blower, equipped with a water-pump, hose and spray nozzle. The fibrous insulation moves through a blower hose and meets a sprayed water-glue mixture at a nozzle. The sprayed liquid dampens the insulation and helps propel it toward the surface. The water-glue mixture binds the loose fill fibrous insulation together and adheres it to the surface.

Installers spray fiberglass and cellulose on metal buildings and masonry walls. Sprayed fibrous insulation provides good fire resistance, thermal-bridge reduction, and noise reduction. However, sprayed fibrous insulation allows air to penetrate and move within its fibers.

The dried sprayed insulation has a fragile surface, so don't specify it for places where people may damage it. In humid locations, cellulose sprayed insulation and low-density SPF may absorb a lot of water, reducing their thermal resistance and weakening their adhesion.

Installers spray fibrous insulation into wall cavities in new buildings. To avoid dampness and moisture problems in wall cavities, the builder must allow the sprayed fibrous insulation adequate time and plentiful ventilation to dry before covering it.

A proprietary variation of damp fiberglass insulation is the blow-in batt system (BIBS™), which uses less water during installation, compared to sprayed fiberglass. Installers blow the BIBS™ fiberglass insulation into open wall cavities through a stretched-net restrainer. The net, stapled to the wall studs, holds the BIBS insulation in place.

Figure 5-25: *Insulation Fasteners for Masonry*

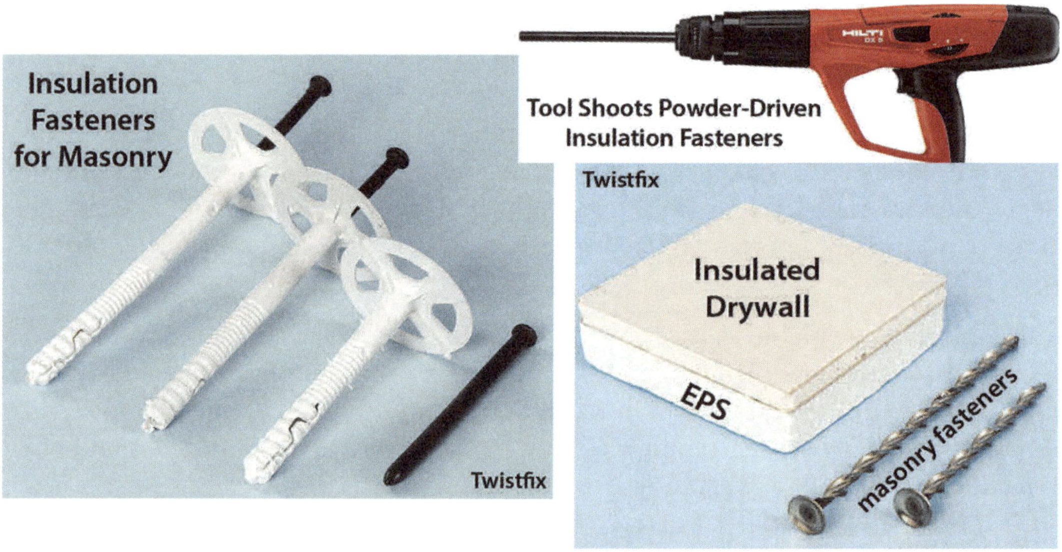

Masonry-wall Retrofit Insulation

Masonry-wall insulation is often the most difficult type of insulation retrofit, especially in wet climates. The first decision is whether to insulate on the exterior or the interior of a wall.

Advantages of insulating on the exterior include these.

- Exterior installation gives better coverage and continuity of the air barrier, water-resistive barrier, and insulation compared to interior installation.
- Exterior installation avoids disturbing or displacing residents.
- Exterior installation creates less moisture risk in wet climates.

Advantages of insulating on the interior include these.

- Installers work indoors, out of the weather.
- Materials and labor costs are usually less than exterior insulation.
- Interior insulation is usually easier to install than exterior insulation.

Caution: The building's age is a decision factor, especially for brick buildings. The mortar in very old brick buildings or masonry-block buildings may no longer adhere well to the masonry blocks or may crumble with abrasion or pressure. The wall could collapse under too much pressure. In this case, the structural risk may prevent any retrofit insulation. See "Masonry Air-sealing and Repair" on page 116.

Exterior masonry-wall insulation — Exterior insulation is more continuous airtight, moisture-resistant, and thermally resistive compared to interior insulation. Exterior masonry insulation is usually the more expensive method because of these requirements.

- Engineering-and-installation expertise
- Rigid foam or rigid mineral wool insulation
- New cladding or siding
- New windows or window-finish details
- Expertly designed-and-installed flashing
- Liquid-applied water-vapor-open water-resistive barrier (Class III vapor retarder)

The three most common exterior masonry-insulation strategies are these.

1. *Insulated traditional stucco* requires expert masonry installation. Expert masons apply this cladding in a variety of ways, depending on their experience and the manufacturers' specifications for the cladding components. The stucco is approximately ¾ inch or 18 mm thick. This heavy-stucco assembly may require its own structural support at the wall's bottom.
2. *Exterior insulation and finish systems (EIFS)*: Foam board or mineral-wool board with proprietary synthetic stucco, approximately ¼ inch or 6 mm thick.
3. *Foam board or mineral-wool board* attached to the masonry exterior using structural furring strips that also provide a fastening method for weather-resistant exterior cladding, such as steel siding.

See "Insulated Masonry Building Assemblies" on page 176.

Interior masonry-wall insulation — Interior masonry insulation is usually less expensive and more practical than exterior insulation. However, interior insulation isn't as thermally continuous as exterior insulation because of transitions at floors, ceilings, interior walls, parapet walls, and other obstacles that make air-sealing-and-insulation continuity difficult.

The two most common interior insulation strategies are these.

1. Attach rigid foam or mineral wool to the inside masonry surface using furring strips on the insulation's exterior. Then install drywall or some other interior finish materials to cover the furring strips and insulation.
2. Build a frame wall with a space between it and the masonry wall. Install foam board, mineral-wool board, or spray foam to create a thermal break between the masonry wall and the frame wall. Fill the cavities with dense-packed min-

eral wool or fiberglass for economy. Or fill the cavity with high-density spray foam, if money isn't as important as structural strength and adhesion.

Figure 5-26: *Exterior Insulation and Finish System (EIFS)*

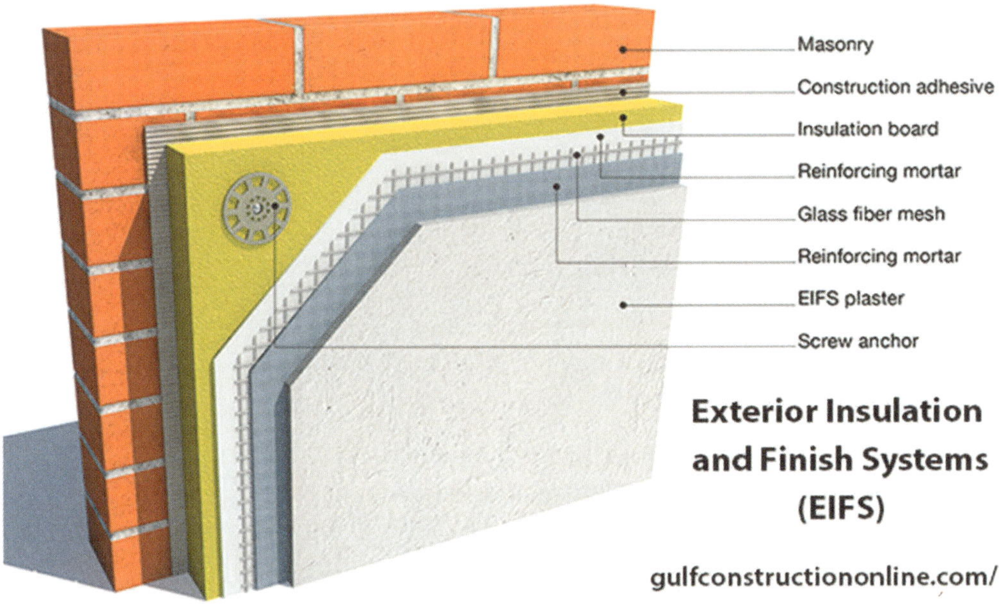

Figure 5-27: *Spray-foam Masonry Applications*

Caution: With porous masonry, like brick, interior insulation causes the brick to store more water than before insulation. Because of the new interior insulation, the masonry becomes colder than before, during cold weather. The colder temperature condenses more water vapor inside the masonry pores and inhibits drying. More water storage in the porous masonry may cause freezing in cold weather, which damages the masonry material. This problem occurs most commonly in cold and wet climates such as Montreal, Canada; Juneau, Alaska; and Bergen, Norway and less commonly in warmer and drier climates. See "Masonry Air-sealing and Repair" on page 116.

Figure 5-28: *Masonry-wall Foam Insulation*

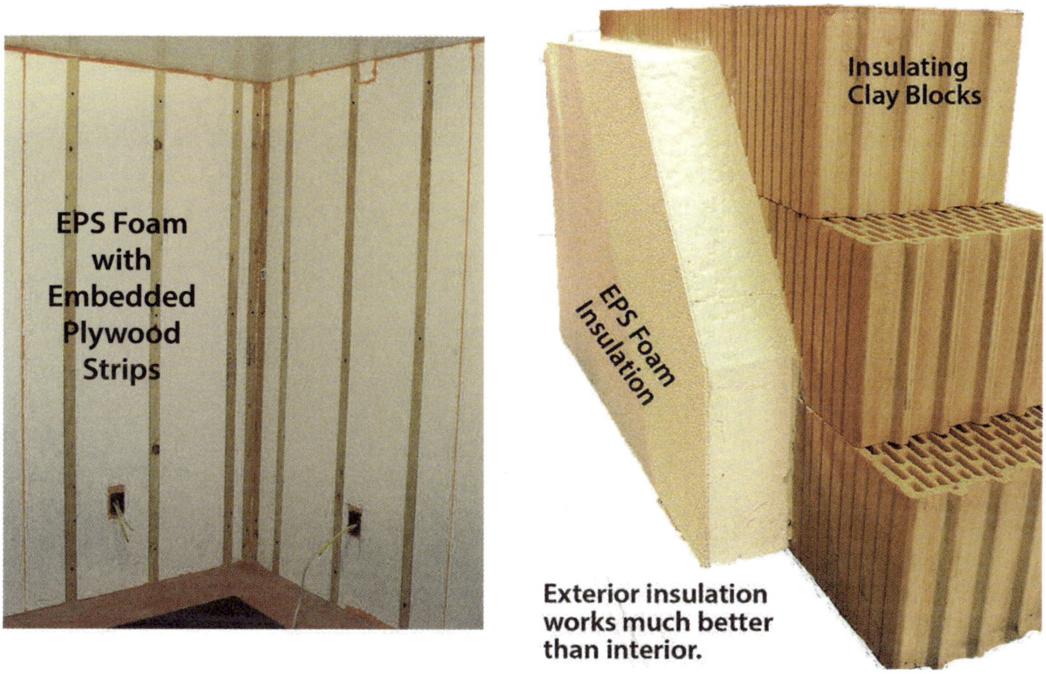

Insulating Unreinforced Masonry Walls

Unreinforced means that the builders used no metal reinforcement. There are four types of unreinforced masonry walls.

1. *Traditional brick walls* with header bricks that hold two layers of stretcher bricks together. Larger buildings may have three or more brick layers instead of two.
2. *Cavity-brick walls* with a layer of brick on either side of a cavity. Metal or wood ties hold the two walls together.
3. *Wood-frame brick-veneer walls* with a single layer of brick veneer that attaches to a wood-frame wall with metal or wood ties. A cavity between the brick or masonry block and the wood-frame wall measures 2 inches or 50 mm deep or more.
4. *Other types of masonry building blocks* that are old and weathered. Mortar may be soft and crumbling. Hundreds of thousands of buildings like this have collapsed in recent earthquakes.

All four of these masonry assemblies may have structural problems depending on the condition of the masonry units and mortar joints. Proceed with caution and expert guidance, if and when you insulate old masonry walls. Walls may require exterior jacketing with reinforced stucco before installation of interior or exterior insulation. These walls may require reinforced-stucco jacketing anyway, if we expect the walls and the people they shelter to survive.

Caution: Mortar can turn to dust over many decades. Cavity brick walls can be frighteningly fragile, and small movements can topple 100-year-old brick veneer. Consider consulting a structural engineer or masonry expert before making any modification to an unreinforced-masonry building, especially in places that may experience earthquakes. See "Masonry Air-sealing and Repair" on page 116.

Figure 5-29: *Cavity-brick Walls: Insulating their Empty Space*

Thermal Enclosures for New Construction or Major Retrofit

Builders use many innovative products and methods for integrating high thermal resistance into new homes and new multifamily buildings. Improved framing techniques substantially improve the overall thermal resistance of modern wood-framed buildings by minimizing structural lumber. Structural insulated panels and high-R, high-mass walls minimize cooling-energy consumption, using innovative materials and construction methods. These modern construction methods and materials produce buildings that are very airtight and require mechanical ventilation for acceptable indoor-air quality.

Superinsulated Wood Buildings

Wood has a low thermal resistance compared to insulation. Minimizing framing lumber allows insulation to occupy a greater percentage of a building enclosure's surface area and volume.

The simplest improved wood-framing methods include these.

- Spacing studs, rafters, and joists on wider centers, compared to conventional framing.
- Using more insulation and less framing in corners, interior wall intersections, and headers.
- Using insulated sheathing behind siding.
- Superinsulation framing techniques create deeper insulated cavities. The most common of these framing techniques uses double stud walls or specialized composite framing studs.
- Raised-heel trusses and scissor trusses allow installers to blow the full depth of attic insulation above the exterior-wall top plate. by providing higher roof cavities, compared to the shallower cavities, such as cathedral roofs.

Structural insulated panels (SIPS) — Interlocking foam-filled building panels have created some very energy-efficient buildings because they eliminate most thermal bridging and air convection in the walls and roofs. Wall-and-roof panels are available in a variety of thicknesses. Structural insulated panels require special skills, techniques, and tools, not common to other construction methods.

Figure 5-30: *Superinsulated Frame Buildings*

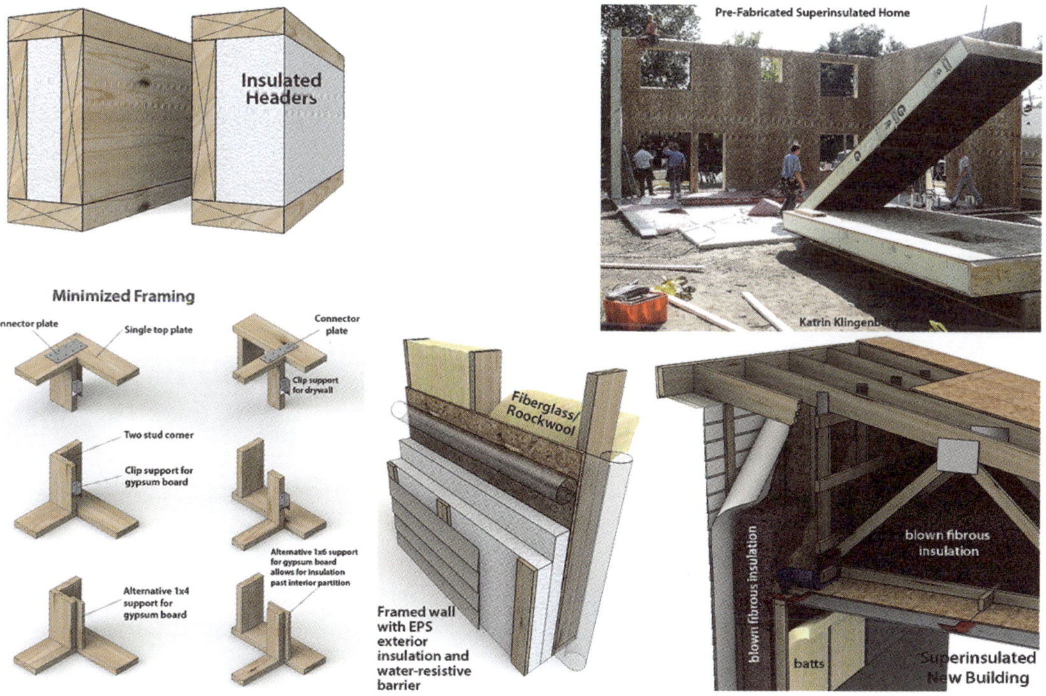

New-Building Insulated Masonry Assemblies

High-mass wall systems have good thermal resistance and very little thermal bridging, air leakage, or convection when builders use the right materials. Insulated masonry walls have thermal mass factors ranging from 1.2 to 2.6, depending on climate. A sunny climate with high daily temperature variations is ideal to benefit from thermal mass. Builders use many types of masonry blocks with insulated cores. Combined with surface-installed board insulation, they provide an energy-efficient and easy-to-build wall assembly.

Figure 5-31: *Structural-insulated Panels (SIPS)*

Site-installed core insulation for conventional masonry blocks includes: low-density polyurethane foam, vermiculite, polystyrene beads, and AirKrete™.

Aerated, autoclaved concrete blocks show promise for high-R, high-mass walls. Other lightweight masonry blocks use sawdust, polystyrene beads, foam-board inserts, and other insulating strategies as part of their manufacture. These insulated blocks can have R/RSI-values as high as R-36 (RSI-6.3) in sunny warm climates depending on their thickness.

Figure 5-32: Insulating Building Block

Tilt-up concrete-wall construction is also a viable option in some regions. Local workers can pour concrete in a horizontal slab on top of insulating material, stack the modular units, transport them to the site, and then tilt the wall up into place. Builders may use similar insulated concrete panels for floors and roofs.

See "R and RSI Values of Building Assemblies" on page 141 for more information on effective R-values of high-mass walls.

Insulated concrete forms — Insulated concrete forms (ICFs) have revolutionized insulation for foundation walls, above-grade walls, floors, and roofs. Many proprietary insulated concrete forms are available. ICFs provide some of the most energy-efficient and disaster-proof buildings in the world.

Figure 5-33: *Insulated Masonry Building Assemblies*

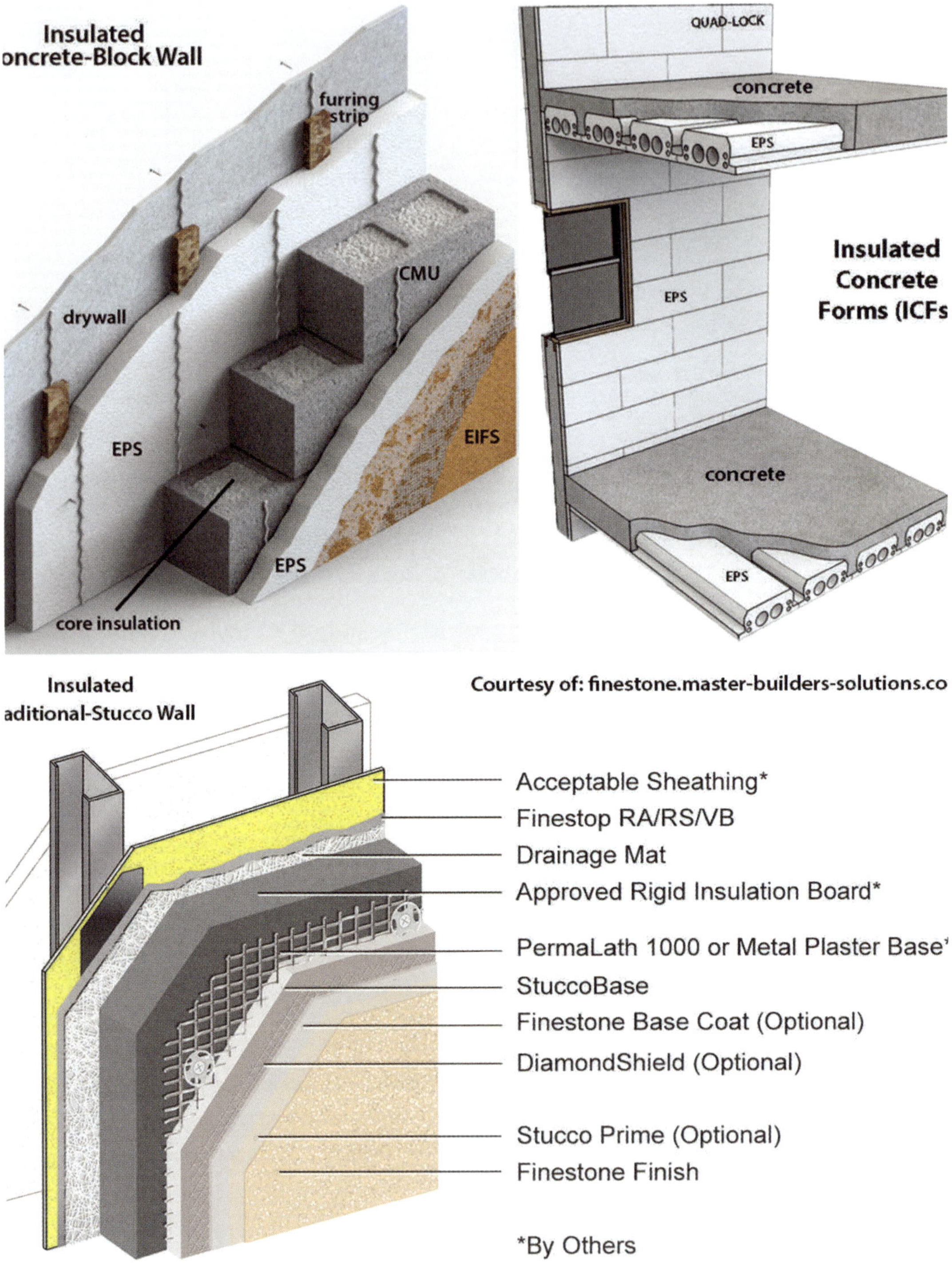

See "Exterior masonry-wall insulation" on page 169.

6 Cooling Devices and Systems

Before the development of electricity, cooling systems used the wind, differences in air temperature, and water evaporation to provide indoor cooling. With the right outdoor temperature, air can flow between indoors and outdoors to remove heat. Or air can just circulate indoors to provide a wind-chill affect.

Modern homes and multifamily buildings now use electricity for cooling in a variety of ways. This chapter discusses both the electric and non-electric devices that provide comfort during hot weather, including these.

- Centuries-old systems like *badgirs*, *earth tubes*, *wind catchers*, and evaporative coolers that use little or no electricity.
- Small portable fans that circulate air within rooms.
- Larger portable fans that suck cool night outdoor air in and push warm indoor air out, to remove heat from buildings.
- When installed in a ceiling or wall, powerful whole-house fans push warm indoor air out, which then sucks cooler outdoor air in through windows.
- Air conditioners (ACs) remove a building's heat using refrigeration.
- Air conditioners, equipped with economizers, use only their blower to move cool outdoor air into the building at night and move warm indoor air out of the building, when the temperature and humidity are favorable.
- In dry climates, evaporative coolers reduce outdoor-air temperature and blow that cool outdoor air into a building to provide hot-weather comfort.
- Dehumidifiers remove excess water vapor from indoor air to reduce indoor relative humidity, which improves summer comfort.

Inspect the Building First

Keep an open mind about the main causes of hot-weather discomfort and high energy costs. People often blame cooling equipment for comfort problems. However, the main causes of the discomfort come more often from flaws in the building enclosure. Consider both the enclosure and the equipment when seeking a cause.

Improve the building enclosure — Before evaluating the HVAC equipment, inspect and evaluate the building enclosure, especially in the least comfortable rooms. The enclosure problems may include air leaks, insufficient insulation, unblocked solar radiation, and windows with poor thermal resistance. Always evaluate the building enclosure before assuming that the HVAC system is the main problem.

If most of the HVAC system is outside the thermal enclosure, up to 25% of the system's cooling energy may escape. You might reclaim this 25% by enclosing the air handler and ducts within the thermal enclosure by insulating and air sealing the building assemblies surrounding the equipment.

Air Circulation Cools Your Body

Air circulation creates a wind chill indoors, which cools residents directly. Smaller air-circulating fans are less expensive to operate compared to larger fans, evaporative coolers, air conditioners, and dehumidifiers. When residents use air circulation effectively, they can save energy and achieve affordable comfort.

In moderate summer weather, fans keep residents cool without air-conditioners. Circulating fans also save energy by improving comfort while air-conditioners, evaporative coolers, or dehumidifiers operate. Circulating fans may provide acceptable comfort at a higher thermostat setpoint compared to no air circulation.

Outside the first world, ceiling fans provide adequate comfort for heat-tolerant residents, even with high outdoor temperature and high relative humidity. Ceiling fans, floor fans, and table fans cool your skin, and not the air around you. Circulating air convects heat away from your skin and helps your sweat evaporate.

Moving air is more comfortable than calm air by up to 8°F (4.4°C) of perceived comfort. Air speeds of up to 350 feet per minute (1.75 meters per second) are tolerable. The faster the air speed, the greater the cooling effect.

See "Energy, Comfort, and Climate" on page 25.

Figure 6-1: *Night Ventilation with and without a Fan*

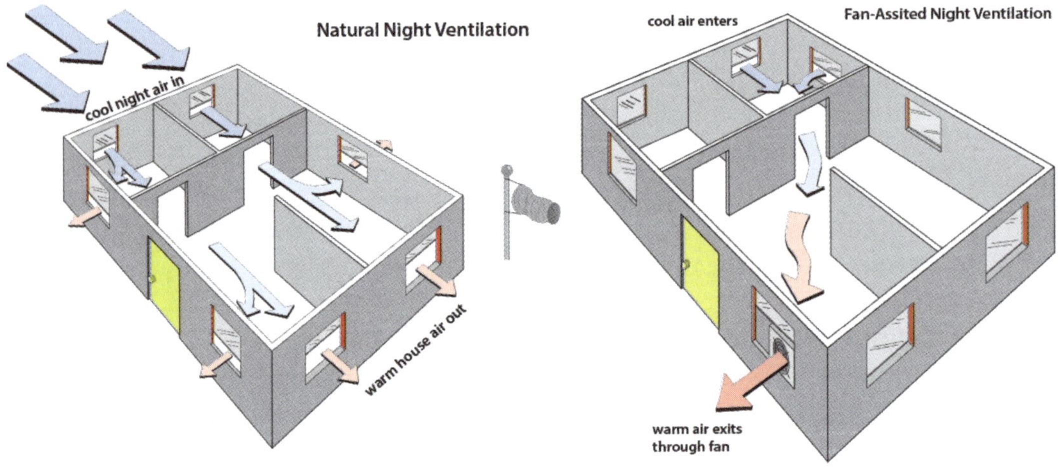

Ventilation Cools Buildings

Residents of homes and multifamily buildings can use natural ventilation for cooling if they know the wind intensity, wind direction, outdoor temperature, and outdoor relative humidity. Clever residents open windows at night to flush accumulated heat out, and then close and shade windows in the morning to minimize daytime solar heat gain.

Ventilation can remove accumulated heat from buildings by circulating cool outdoor air through rooms. Ventilation works best in regions with hot days, cool nights, and low relative humidity. Ventilation also works well with rapid air circulation in humid tropical climates, where the outdoor temperature isn't too uncomfortable for local people. Nighttime ventilation removes solar-heat and internal heat gains that accumulate during the day.

The effectiveness of ventilation for cooling depends on these factors.

- The temperature and humidity of outdoor air during the nighttime hours.
- The ability to ventilate, during night and early morning, for 4 hours with a approximately 10°F (5.6°C) temperature difference (ΔT) or 6-8 hours with a 5°F (2.8°C) ΔT.
- The amount of shade around the building and access to a cool outdoor zone, from which to draw outdoor air.
- Residents willing to turn fans on and off at the appropriate times and to endure the white noise that the fans produce.
- Residents willing to open and close windows and blinds at the appropriate times.

Figure 6-2: *Cooling Solutions Depend on Outdoor Relative Humidity / Dew Point*

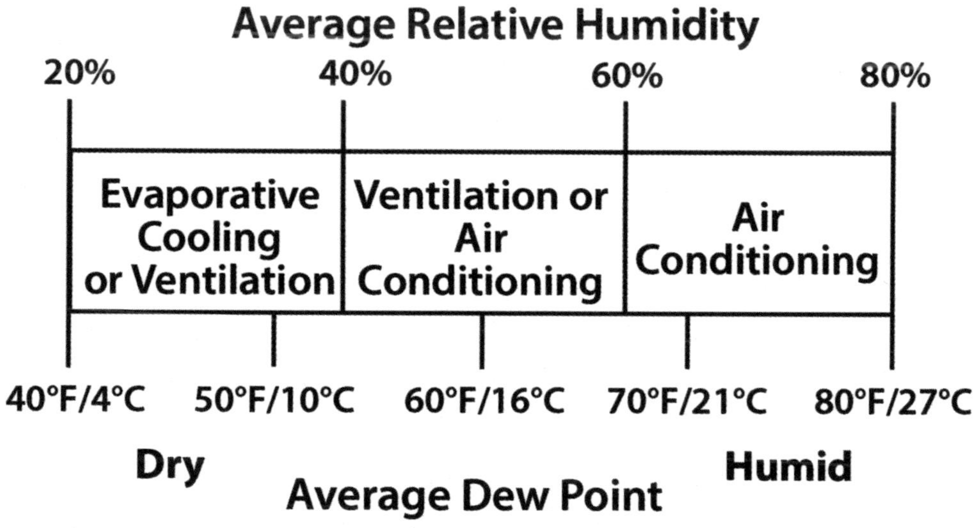

Fan-powered Ventilation for Cooling

Fan-powered ventilation, when combined with shading and other low-cost cooling measures, can provide adequate cooling during many warm spring, summer, and fall days. These night-ventilation fans must be powerful to move enough air to strip accumulated heat from the buildings indoor surfaces.

Powerful window fans and whole-house fans can provide effective cooling by night ventilation. Use floor, table, and ceiling fans, at the same time, to create a wind chill indoors where residents sleep. Small fans are much less effective as whole-house ventilators compared to these large powerful fans.

Whole-house Fans

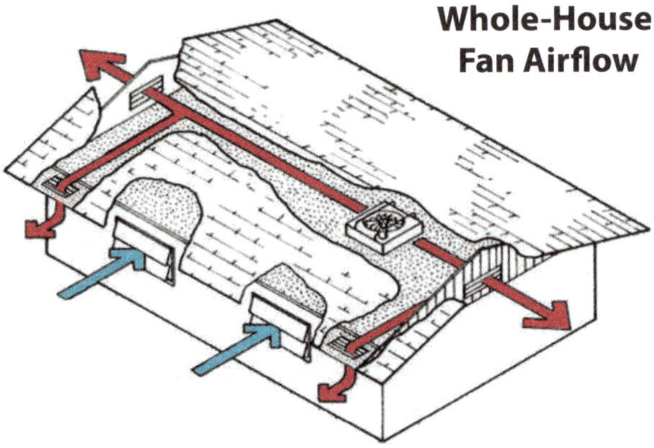

Figure 6-3: *Fans for Night Ventilation and Air Circulation*

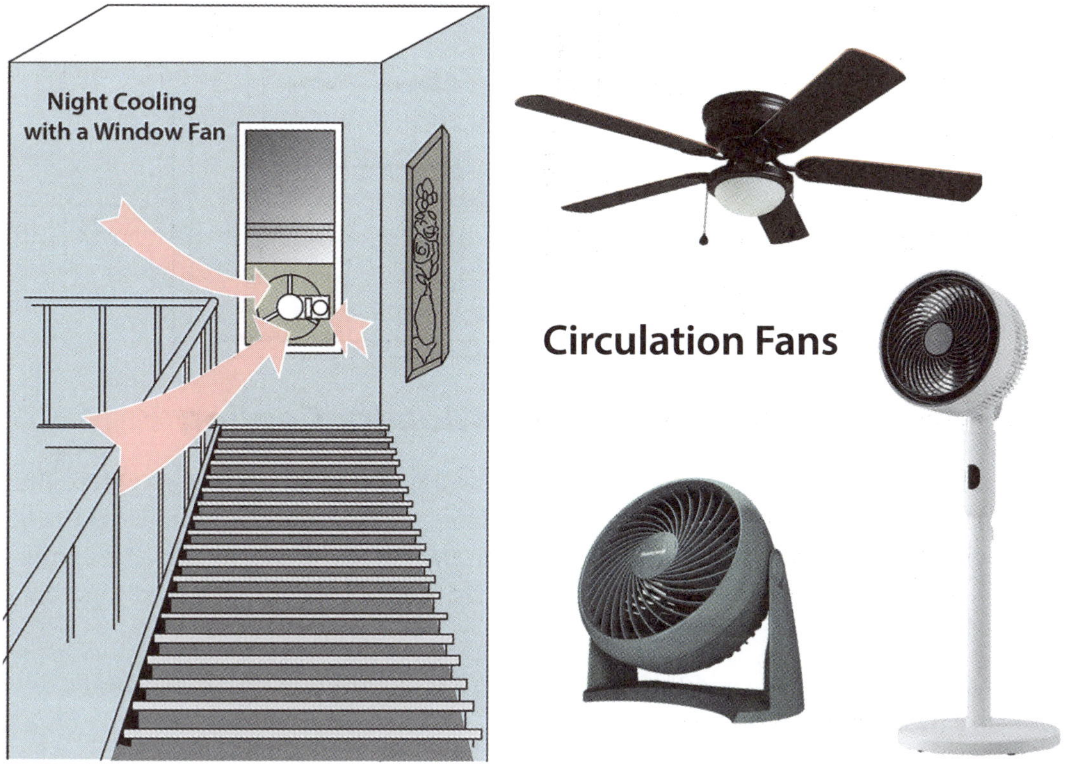

Caution: **Powerful fans can create strong house pressures that could cause backdrafting or flame roll-out in combustion water heaters. Provide adequate window openings during ventilation to avoid excessive house pressure.**

Attic-and-Roof Ventilation

If a hot roof radiates solar heat to the ceiling beneath it, that solar heat then radiates from the ceiling to the rooms below. Solar heat moves by radiation. Attic ventilation doesn't resist radiation heat flow unless the ventilation air flows through the ventilating space between the roofing material and the roof deck.

A ventilated space between the roofing and roof deck removes much of the solar heat from the roofing. The roof deck requires ventilation to keep it dry in any climate. If the ventilation air flows between the roofing and the roof deck, this is ideal for reducing summer solar gain through the roof.

Wind-powered, solar-powered, or electric attic ventilators don't save much electricity in summer because radiation moves most of the heat from the roof deck through the attic to the ceiling below. In many cases, electric roof ventilators run all the time and waste a lot of electricity.

Figure 6-4: *Centuries-old Alternatives to Air Conditioning*

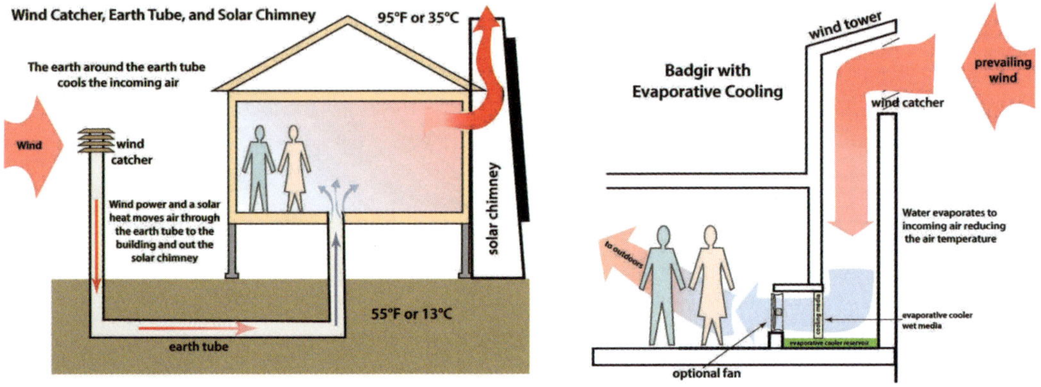

Wind-and-solar-powered Ventilation for Cooling

Wind-powered and solar-powered ventilators work well in dry climates where the relevant design and building skills are available. Developing countries possess these skills because they have been using non-electric cooling systems for centuries. Five types of ventilation devices, working by themselves or combined together, are good examples.

1. Wind catchers, also called wind towers, use wind along with air-temperature differences to move air.
2. Solar chimneys use air-temperature differences to move air.
3. Wind-powered exhaust fans create suction to remove hot air.
4. Solar-powered exhaust fans create suction to remove hot air.
5. Evaporative cooler media evaporates water into water vapor, absorbed by incoming air, which reduces the air temperature.

Figure 6-5: *How Badgirs Move Air and Cool Buildings*

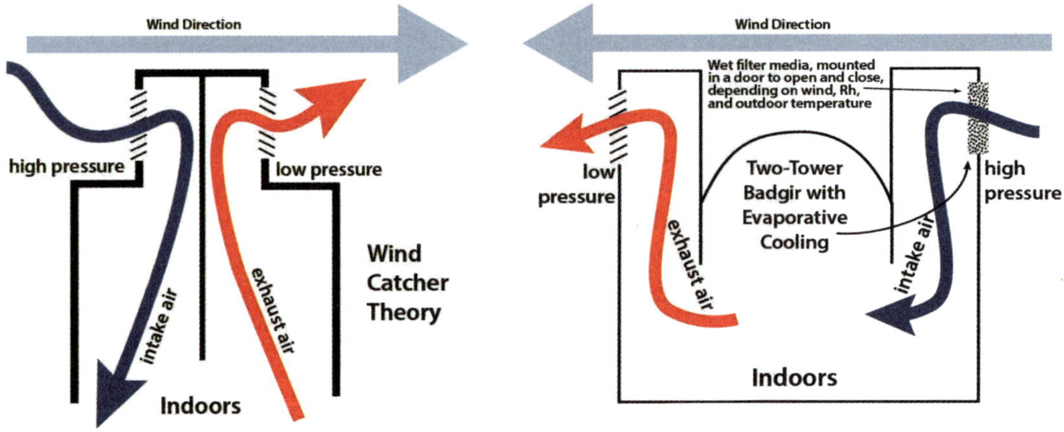

Non-electric Cooling with Ventilation & Evaporation

Most ancient and modern dry-climate non-electric cooling systems use wind catchers. The oldest of these systems is called the *badgir*. Badgirs are the oldest known ventilation-cooling systems, and you find them in many dry-climate regions.

Evaporative cooling is a good strategy for dry climates because you can add water vapor to air to reduce the air temperature, without creating excessive relative humidity or moisture problems. For centuries, these systems have made some of the hottest places on earth livable. For example, Ahvaz, Iran has experienced temperatures as high as 130°F or 55°C and wouldn't be livable without its many badgirs.

Figure 6-6: *Non-electric Devices Cool Buildings with Ventilation*

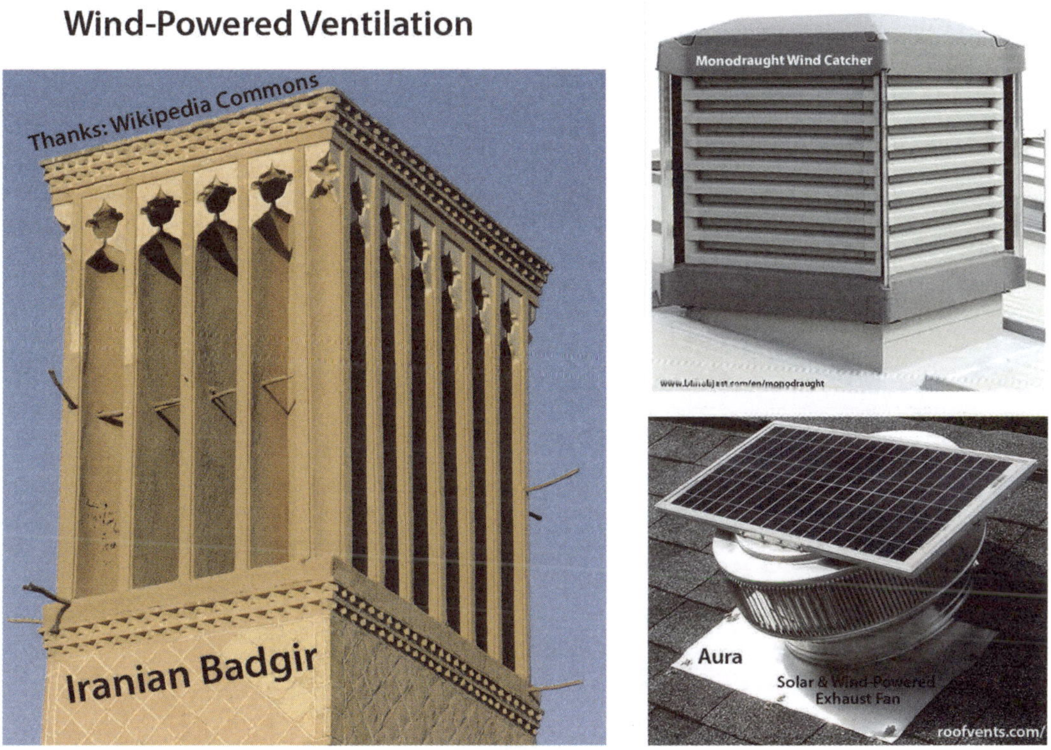

Cooling with earth tubes — An earth tube is an earth-to-air heat exchanger. Earth tubes often couple with wind catchers or solar chimneys, which drive air into and out of residential and commercial buildings. Wind towers and solar chimneys use wind or solar heat to move air. People have used in earth tubes and solar chimneys in hot arid regions for centuries, beginning in the middle east. "Centuries-old Alternatives to Air Conditioning" on page 182 shows you how earth tubes work.

Non-electric Evaporative Cooling

Ancient engineers and builders designed and installed evaporative-cooling systems at least 700 years ago and possibly as long ago as 2000 years. Most people in hot, arid countries can't afford air conditioning. Wind-powered evaporative cooling is common in hot dry places, such as Iran, Egypt, and Syria. Wind-powered evaporative cooling still provides comfort in millions of dry-climate residential buildings around the world.

These ancient systems display a wide variety of designs, many of which are site-built, one-of-a-kind systems. The photos and drawings in this section introduce the ways these systems work. These innovative cooling systems may well replace many residential air-conditioning systems if future electricity costs and electricity shortages become more common. Non-electric systems are simple and require less maintenance, compared to electrical cooling systems.

Figure 6-7: *Evaporative Cooling with Little or No Electricity*

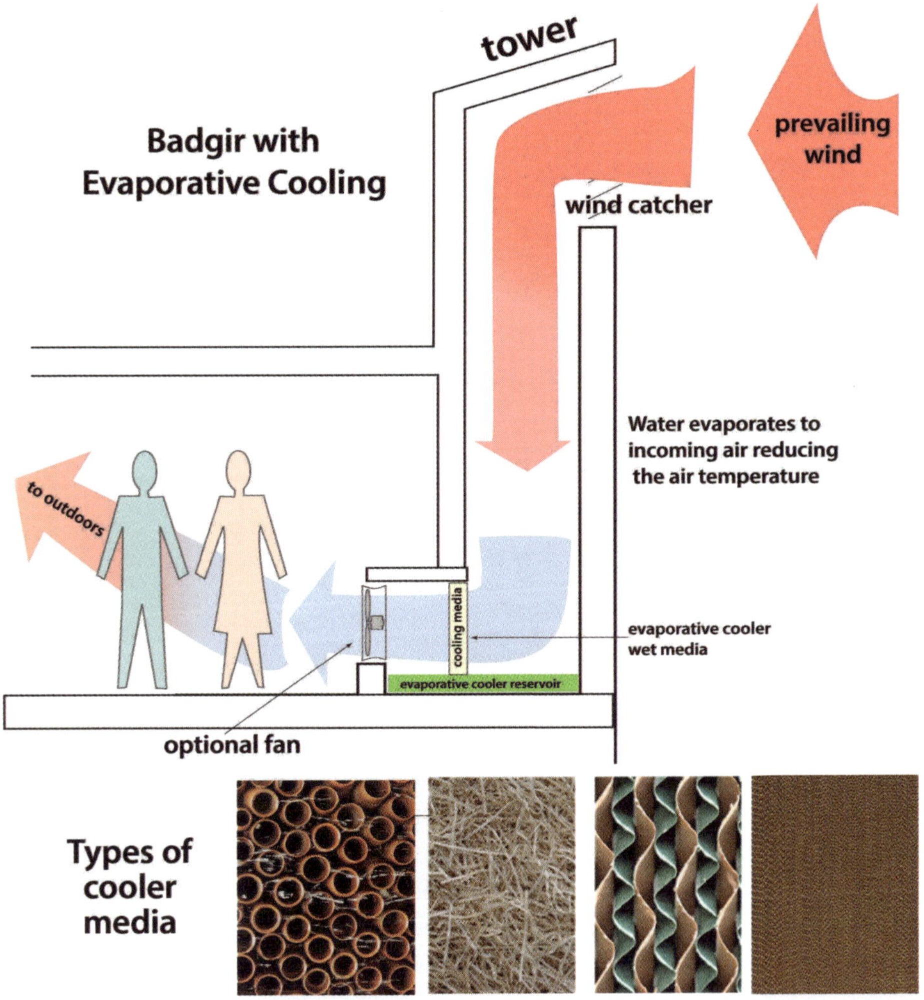

Electric Evaporative Coolers

Evaporative coolers work well in climates where the summer outdoor relative humidity is less than 50%. Evaporative coolers use different principles from air conditioners. Evaporative coolers reduce the temperature of outdoor air without removing heat from the air. Evaporative coolers also humidify the air, but not so much that the increased humidity reduces comfort. See "Yearly-Average-Relative-Humidity Map" on page 292 for information about dry-versus-humid climates.

A large blower sucks outdoor air in through the evaporative cooler's louvered cabinet and absorbent pads (cooler media). A water pump delivers water to absorbent pads. A large fan (blower) pushes outdoor air through the wet pads. This air movement evaporates water, which reduces the incoming air's temperature. Then that relatively low-temperature outdoor air enters the building and pushes higher-temperature indoor air out through open windows.

Evaporative coolers cost less than half as much to install as central air conditioners. Evaporative coolers are 2 to 3 times more energy-efficient in providing summer comfort than the most efficient air conditioners.

Building owners install evaporative coolers on roofs, through windows, through walls, or on the ground outdoors. The evaporative cooler blows air directly into a room or hallway, or it flows through ducts that carry the cool air to rooms.

Evaporative-cooler performance factors

The evaporative cooler performance directly relates to the cooler's ability to evaporate water at a specific outdoor relative humidity. The performance increases as the outdoor air becomes hotter and thus dryer. The morning's outdoor air is usually the day's highest relative humidity. As the day progresses the outdoor temperature rises and its relative humidity falls. The hotter the day, the dryer the outdoor air becomes, and the evaporative cooler's cooling capacity increases.

Airflow: recommended amount — Air changes per hour (ACH) is the number of times the air within an indoor space exchanges with outdoor air each hour. Evaporative coolers in the hottest climates need more air changes than those in cooler climates, as shown in the next illustration.

Figure 6-8: Suggested Air Changes per Hour for Evaporative Coolers

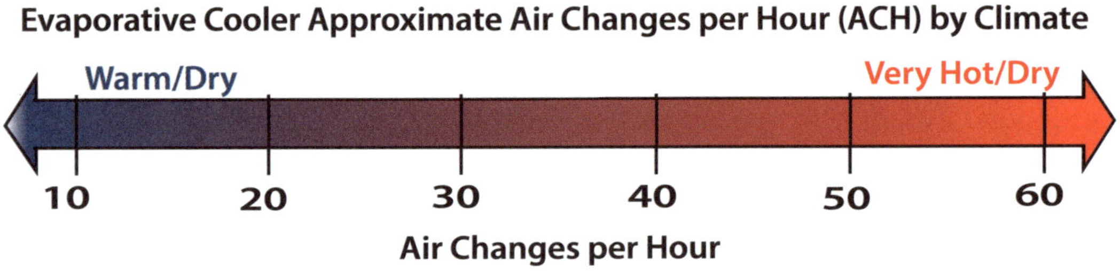

Air changes per hour = airflow per hour ÷ space volume

Airflow per hour expressed in cubic feet per hour or cubic meters per hour
Space volume expressed in cubic feet or cubic meters

Evaporative-cooler Operation

Evaporative coolers must operate with open windows or special vents called *up-ducts*. Up-ducts are square openings in the ceiling that let hot air out of the building into a ventilated attic. The cooler's air pressure opens and closes the up-ducts. If you turn off a cooler's water pump, the cooler then functions as a whole-house fan at night. To be effective, the blower must deliver cooler outdoor air, compared to indoor air. Many cooler controls give you this option.

Figure 6-9: Cooler-grille Cover and Cooler Control

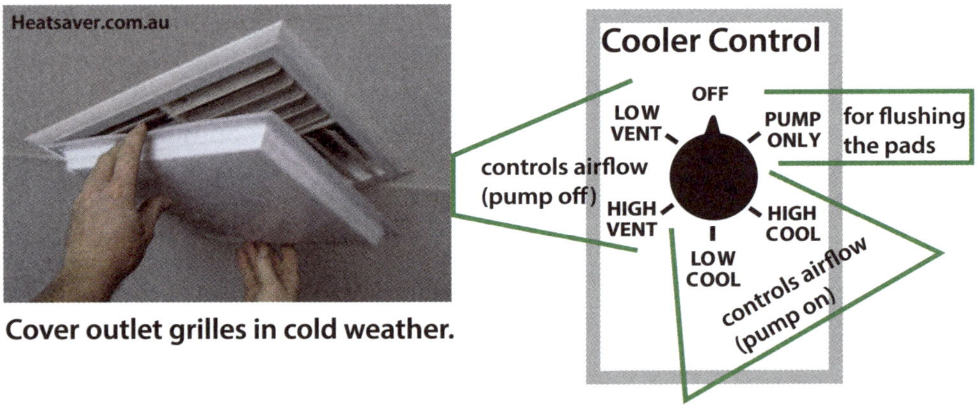

Cover outlet grilles in cold weather.

Open windows in occupied rooms, and close windows in unoccupied rooms to deliver cool air where people need it. Open your windows to provide approximately 1 to 2 square feet of opening for each 1,000 cubic feet per minute (CFM) of cooler airflow, or one square meter of opening for each 160 cubic meters per hour (m^3/hr) of airflow. The cooler's data tag should give you its airflow rate. Experiment with which windows to open and how wide to open them.

If you open the windows too wide, hot outdoor air enters. If the windows aren't open far enough, indoor humidity increases and that reduces comfort. Experiment to find the right windows to open and how wide to open them.

When the cooler has been idle for a few days, let the pump run before starting the fan to wash dust out of the pads. Modern cooler controls allow for separate water-pump and blower operation. The best controls are programmable thermostats, designed for evaporative coolers.

Figure 6-10: *Evaporative Coolers and their Operation*

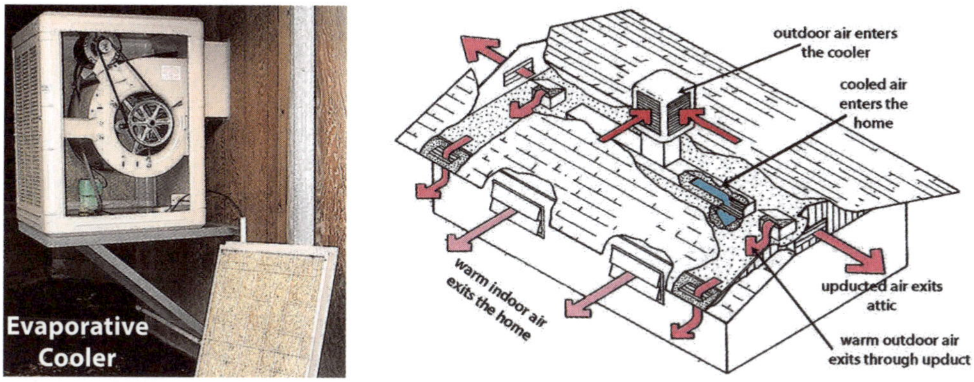

Routine Maintenance

Cooler pads, made of aspen-wood fibers, cellulose, plastic, or terra cotta collect dirt from the outdoor air. Dirt in the cooler pads washes into the reservoir. Most evaporative coolers have a bleed tube or a separate pump called a *dump pump*. The dump pump drains the reservoir water during cooler operation to remove dirty water.

Figure 6-11: *Cooler Up-duct, Overflow Tube, and Drain*

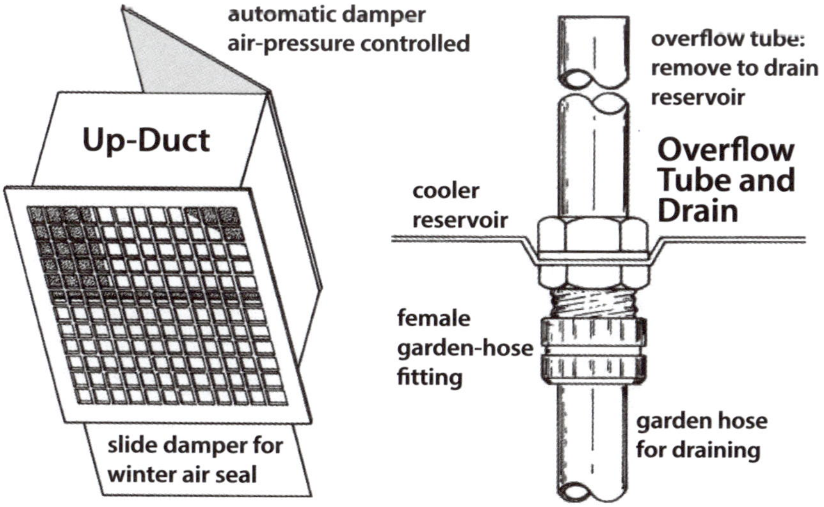

A cooler needs manual cleaning at least once a year. The longer the cooler runs, the more dirt it collects, unless the dump-pump system works well. Be sure to disconnect the electricity to the unit before cleaning it manually.

Here is a list of annual maintenance work for evaporative coolers.

- ✓ Soak pads in soapy water to remove dirt. Replace the pads when they become non-absorbent, thin, or full of scale and dirt.
- ✓ Drain and clean the reservoir.
- ✓ Test the operation of the drip tube or dump pump, whichever the cooler uses to maintain water quality.
- ✓ Clean the blower's fan blades.
- ✓ Clean the louvers in the cooler cabinet and at the air outlet into the building.
- ✓ Oil the blower motor, and check its belt tension.

Two-Stage Evaporative Coolers

Two-stage evaporative coolers are a newer, more efficient cooler which use a pre-cooler, more effective pads, and more efficient motors. The pre-cooler cools the incoming air without adding any moisture to the air. This pre-cooling provides an extra 3° to 9°F (1.7°C to 5.0°C) temperature drop, so the air feels drier and more like it has been refrigerated. These systems are gaining greater acceptance, especially in the southwest, as customers become more aware that refrigerants may be destroying the ozone layer of the earth's atmosphere.

Two-stage coolers provide comfort much more efficiently than air conditioners. Two-stage evaporative coolers aren't rated with SEERs like air conditioners, but if they were, the SEER would be well over 40.

See "Air-conditioner Efficiency" on page 194.

Figure 6-12: *Two-stage Evaporative Cooler*

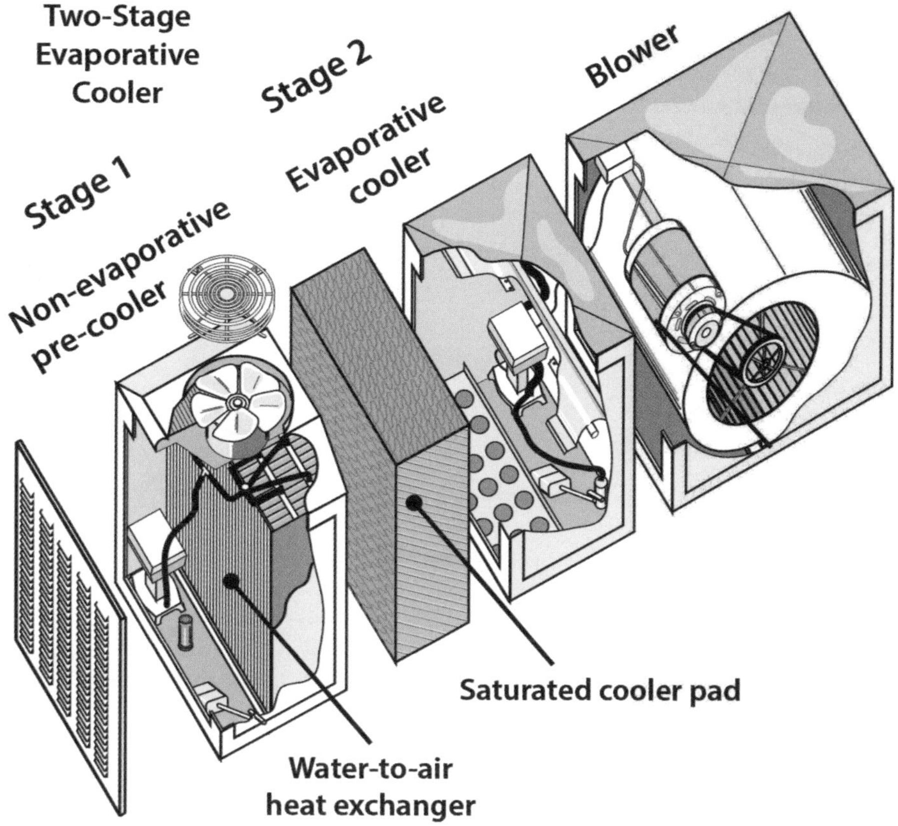

Air Conditioners

Air conditioners use the same principles as refrigerators. An air conditioner cools a building with a cold coil, called the *evaporator*. A hot outdoor coil, called the *condenser*, releases the collected heat outdoors. The evaporator and condenser coils are serpentine copper pipes surrounded by aluminum fins. **Heat pumps are just reversible air conditioners that provide both heating and cooling. Most of what you read in this section describes both air conditioners and heat pumps. Instead of outdoor air, geothermal heat pumps use the ground or a water source to transfer heat from indoors to outdoors or from outdoors to indoors, depending on the season.**

A fluid, called a *refrigerant*, collects heat at the evaporator coil and releases it at the condenser coil. The compressor forces the refrigerant through the coils where the refrigerant evaporates in the evaporator and condenses in the condenser. The air conditioner's fans move indoor air through the cold evaporator, which strips heat from the indoor air. The hot condenser heats cooler outdoor air to transfer collected indoor heat to outdoors. **For heat pumps, we call these two coils: the**

indoor coil and the outdoor coil because these coils change functions between the heating season and the cooling season.

Figure 6-13: *Ducted Central Air Conditioners or Heat Pumps*

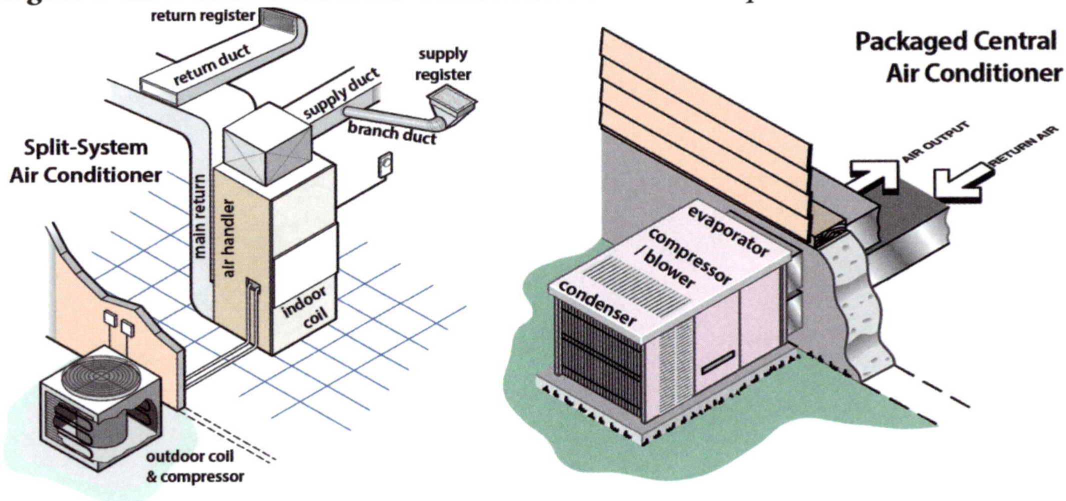

Types of Air Conditioners (AC)

Air conditioners are either ducted or ductless. Ducted air conditioners may provide ventilation air in addition to cooling. A correctly sized and installed ducted air-conditioning system provides consistent room-air temperatures throughout the building. This is the main advantage of cooling with ducted air conditioners.

The ductless air conditioners are more efficient than ducted ones because they avoid duct-energy losses, which vary from 10% to 40% of the energy contained in the ducted air. Residence might notice indoor-temperature variations because much of the building is remote from the fan coils that distribute cool air. Ductless systems work best in airtight and well-insulated buildings because an energy-efficient building heats the cool air slower than an inefficient building, giving the air time to circulate.

Central-ducted air conditioners — Ducted air-conditioning systems have supply and return ducts that connect to a central air handler. An air handler is a metal box with a blower along with other heating and cooling equipment. There are two types of ducted central systems: *split systems* and *packaged systems.*

Split-systems have their evaporator coil inside an their indoor air handler. The condenser coil is outdoors with its own fan. *Packaged air conditioners* have both coils in an outdoor air handler. These two ducted systems are the most common air conditioners in the USA and many other countries.

__Figure 6-14:__ Packaged Central Air Conditioners or Heat Pumps

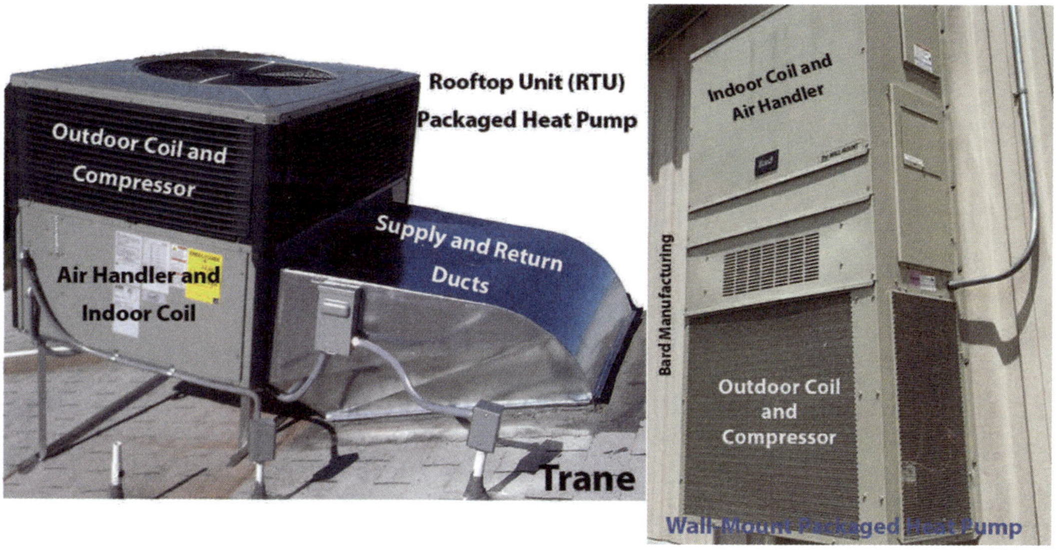

Room air conditioners — Room air conditioners are common throughout the world. They work well for homes, apartments, and commercial buildings. Room air conditioners install in a window or a rectangular hole through a wall. Their condenser faces outdoors and their evaporator faces indoors. Room air conditioners save energy by cooling individual rooms instead of cooling the whole building like ducted AC systems do. A packaged terminal air conditioner (PTAC) is a type of room air conditioner.

Ductless mini-split air conditioners — Ductless mini-splits are the most efficient common type of air conditioner. Mini-splits cool rooms with an indoor-evaporator fan-coils. The fan-coil is a small air handler with an evaporator, a fan, and a drain to carry away condensed water that drips from the evaporator. An outdoor condenser rejects the collected heat. Ductless mini-splits combine features of room air conditioners and ducted split-system air conditioners. They circulate refrigerant from indoors to outdoors and back again. Each indoor evaporator has its own fan so ductless mini-splits don't need ducts.

Designers and contractors install ductless mini-splits because they are easy to retrofit into buildings without ducts, they're easier to install, and they're more efficient than ducted systems. Mini-splits can save energy compared to central air conditioners because they can cool occupied rooms instead of the whole building. To cool only occupied rooms isn't common practice with ducted central air conditioners.

Figure 6-15: *Room and Ductless Heat Pumps or Air Conditioners*

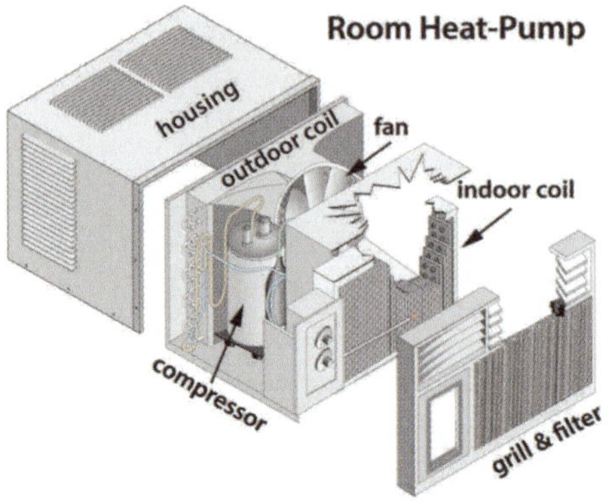

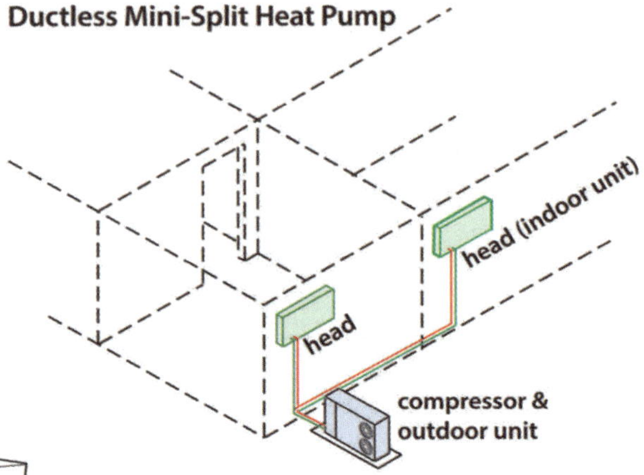

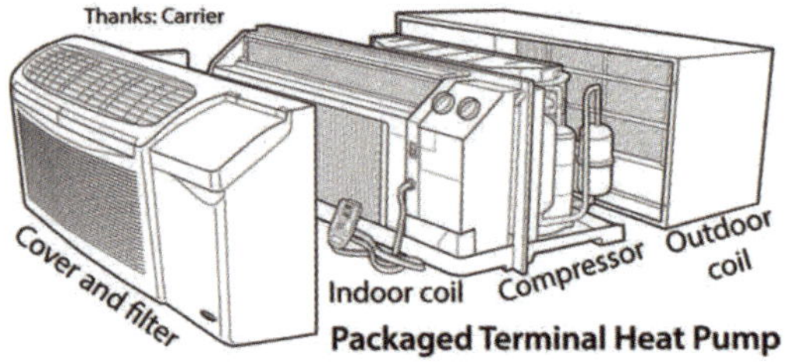

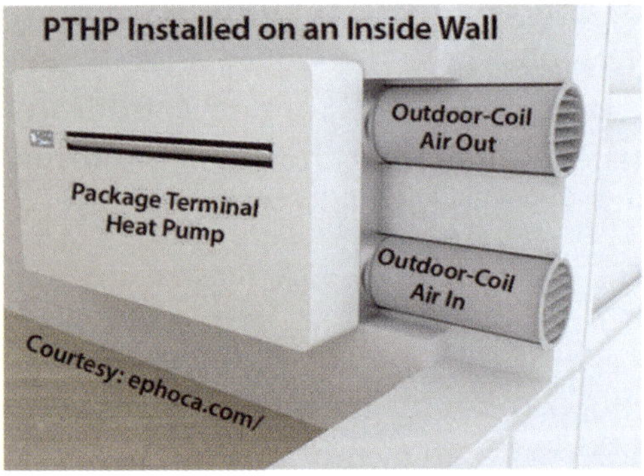

Ductless Heat Pumps

The room heat pump and the packaged terminal heat pump (PTHP) are ductless units popular for hotel rooms and apartments. They are more energy efficient than almost any kind of central heating and cooling system. The ductless mini-split heat pump outperforms these two room heat pumps. Mini-splits can move more air through their outdoor coils and thus move more heat, compared to PTHPs and room units.

Ducted mini-split air conditioners — Ducted mini-splits combine the features of ductless mini-splits and ducted central air conditioners. These systems use small air handlers located inside the thermal enclosure of a home or multifamily building. Ducted mini-splits avoid long duct lengths with numerous joints that lose a lot of cooling energy. Ducted mini-splits work well to replace a large, poorly installed central-ducted air-conditioning system. Or, designers specify them for new energy-efficient buildings.

Ducted mini-split air conditioners — Ducted mini-splits combine the features of ductless mini-splits and ducted central air conditioners. These systems use small air handlers located inside the thermal enclosure of a home or multifamily building. Ducted mini-splits avoid long duct lengths with numerous joints that lose a lot of cooling energy. Ducted mini-splits work well to replace a large, poorly installed central-ducted air-conditioning system. Or, designers specify them for new energy-efficient buildings.

Figure 6-16: *Ductless and Ducted Mini-Split Heat Pumps or Air Conditioners*

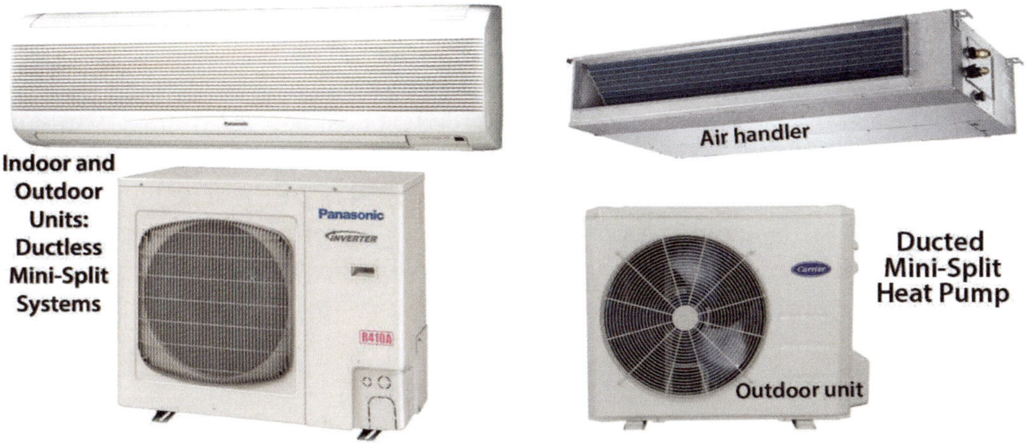

Variable refrigerant flow (VRF) — VRF heat pumps are split systems that direct warm or cool refrigerant to fan coils in rooms of large buildings as needed. VRF systems can even heat one part of a large building and cool another. Water heating is another possibility for VRF systems.

Their disadvantages include the possibility of leaks in the many refrigerant-line connections. Also, the skill and care required for successful design, installation, and control is considerably greater, compared to simpler heat pumps.

Figure 6-17: *VRF System for Multifamily Buildings' Heat Pumps or Air Conditioners*

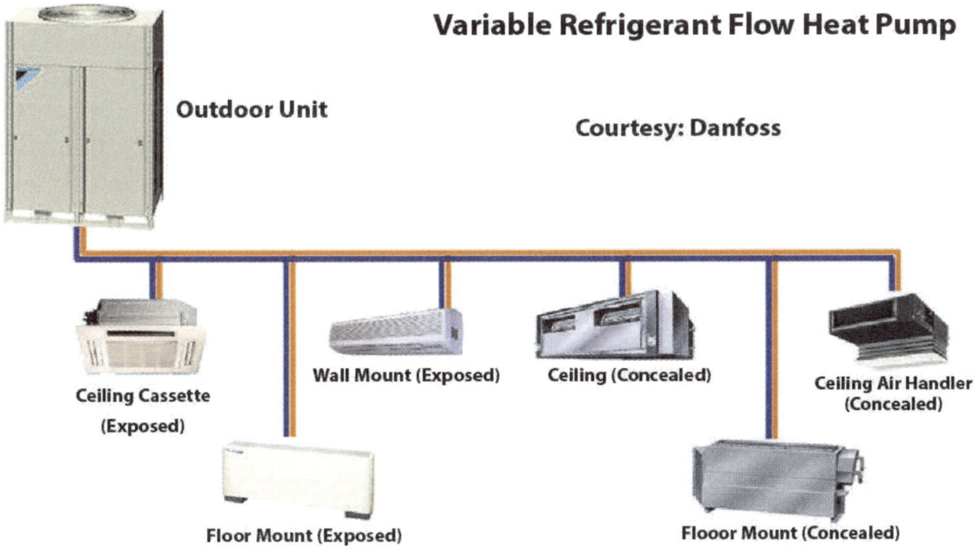

Air-conditioner Efficiency

The energy ratings of air conditioners quantify how many BTUs per hour or watts of heat the unit removes for each watt of electricity the AC uses. For central air conditioners, we call the efficiency rating the *Seasonal Energy Efficiency Ratio*, or SEER. For room air conditioners, we call it the *Energy Efficiency Ratio*, or EER.

Air conditioners with higher EER or SEER cost more than less efficient models. However, the energy savings pay for the initial extra cost several times during the air conditioner's lifespan. The energy-guide label lists EER or SEER or an energy class. In the USA the energy-guide label shows the EER or SEER. The buyer should know how that rating compares between competing products. In Europe, the energy-guide label provides an energy rating, which varies from A+++ to D, which simplifies the AC product comparison.

Imperial (IP)

$$SEER\ or\ EER = \frac{BTUs\ per\ hour\ heat\ removed}{Watts\ of\ electric\ power\ drawn}$$

Metric (SI)

$$SEER, EER, or\ COP = \frac{Watts\ of\ heat\ removed}{Watts\ of\ electric\ power\ drawn}$$

Buy the most efficient air conditioner you can afford, especially if you live in a hot climate, or if your air-conditioning costs are high.

The SEER for central air conditioners depends on when they were manufactured. A wide selection of modern air conditioners have SEERs up to 22 (5.8 SI) or EERs up to 12 (3.5 SI). Before 1979, the SEERs of central air conditioners ranged from 4.5 to 8.0 (1.3 to 2.3 SI). If you replace a 1970's central air conditioner with an SEER of 8 (2.3 SI) with a new AC unit that has an SEER of 16 (4.7 SI), the new AC saves 50% of your air-conditioning costs compared to your old unit.

New Energy-efficient Air Conditioners

The efficiency of today's air conditioners is one-and-a-half to two times greater than air conditioners made in the mid-1970s. This efficiency improvement results from technical advances such as these.

- Two-speed compressors
- Inverter-driven variable-speed compressors

***Table 6-1:** Approximate SEER Ratings for Cooling Devices*

Cooling Device	kW	SEER (USA)	SEER (SI)
Central air conditioner (SEER)	2–50	14–21	4.1–6.2
Room air conditioner or PTAC (EER)	0.5–10	10 –12	2.9–3.5
Ductless mini-split (SEER)	2–50	15 – 23	4.4–6.7
Evaporative Cooler (SEER estimate*)	0.4–5	30–70	8.8–21

*Evaporative coolers don't have SEER ratings. The last row gives a conservative estimate of their performance in a hot dry climate.

- Variable-speed or multi-speed blowers
- Aluminum fins spaced closer together and perforated to improve heat transfer to air
- Improved electric motor design

Inverter-driven variable-speed compressors and variable-speed blowers are particularly effective in humid climates to employ the AC's power to provide maximum dehumidification.

***Figure 6-18:** Comfort and Efficiency Benefits of Variable-Speed Compressors*

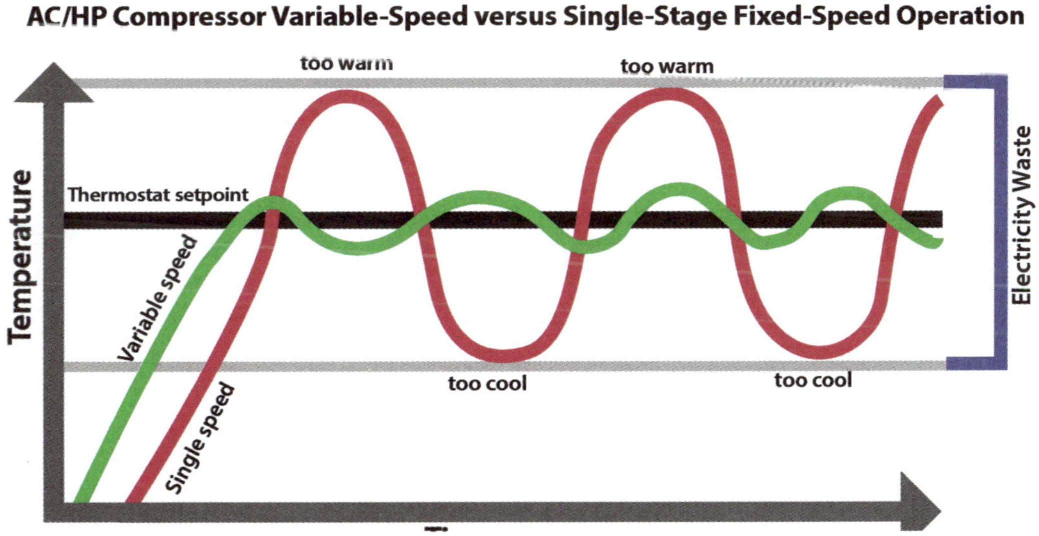

Cooling Strategy for Humid Climates

Air conditioners are the most powerful dehumidifiers now used in buildings. ACs use a lot of electricity to remove excess moisture in an effort to provide sufficient comfort. Moisture sources, especially wet ground, humid outdoor air, and rain contribute to disappointing summer comfort and high cooling costs. Both air conditioners and dehumidifiers struggle to remove enough water vapor when water vapor constantly emerges from these sources of humidity.

The ideal cooling strategy for humid climates is different compared to the ideal cooling strategy for dry climates. In humid climates, the building's air conditioner may provide acceptable dehumidification, or it may not. If the AC doesn't run for long enough and if its evaporator coil isn't cold enough, the AC can't remove enough water vapor from indoor air. Short-cycle operation and insufficiently cold evaporators may allow indoor relative humidity above 60% Rh, which we consider unacceptable for comfort, health, and building durability. See "Yearly-Average-Relative-Humidity Map" on page 292 for information about average Rh in your local climate.

In humid climates, an AC must run long enough with a cold-enough evaporator to provide sufficient comfort and dehumidification. If a humid climate isn't very hot, a building may need both AC and a dehumidifier. Some humid-and temperate-climate buildings may only need the dehumidifier, especially if the building is energy efficient.

See "Controlling Water Vapor" on page 51.

Sizing and Selecting Air Conditioners

Size is a very important consideration to achieve comfort and minimize energy cost when selecting new air-conditioning equipment. Oversized air conditioners cycle more than correctly sized ones. Oversizing is a problem you will find consistently if you maintain or service air conditioners.

With each cycle, the air conditioner heats the condenser, cools the evaporator, and cools the ducts before it begins to cool the building. The air conditioner wastes this start-up energy when it cycles off. Thus, more cycles waste more energy.

Contractors normally oversize air conditioners by 30% to 100% to ensure that they are big enough to cool the building. However, correctly sized air-conditioning systems are more efficient and more effective at removing moisture than oversized systems.

Sizing air conditioners — When you select an air conditioner, consider AC size, moisture removal, and energy efficiency. The amount of heat gain during the hottest and most humid weather determines the air conditioner's size or *capacity*. HVAC contractors use computer software to calculate cooling capacity, based on standard methods.

In the USA, this capacity is measured in BTUs per hour, or "tons" of cooling. Each ton equals 12,000 BTUs per hour (1 ton = 12,000 BTUs/hour). In Europe, the capacity is more commonly expressed in kilowatts (kW). One ton of cooling capacity = 3.5 kW.

Controlling relative humidity with AC — Moisture removal capacity of the system — expressed as the *sensible-heat ratio (SHR),* is an important consideration to select a central AC system. The SHR rates the air conditioner on its ability to remove moisture. The SHR is a decimal number between 0.5 and 1.0. The lower the SHR, the more moisture that the unit removes from the indoor air. The SHR depends on the AC unit's evaporator-coil temperature, its compressor speed, and its blower airflow. A single air conditioner can have a high SHR or a low SHR according to the blower airflow and other settings, selected by the installer.

Air conditioners provide cooling comfort by reducing air temperature and removing water vapor from the air. The air conditioner must operate for long enough and at a cold enough evaporator-coil temperature to remove the necessary moisture. The moisture condenses on the cold coil, drips into a pan, and flows away through a pipe and drain, connected to the pan.

Oversized air conditioners: cycle frequently, remove less moisture, and waste energy. Long cycles are good for both energy efficiency and moisture control. However, an air conditioner that runs constantly isn't necessarily the correct size. It could be oversized, and straining to overcome problems with maintenance, repair, or adjustment. Or much less likely, the AC could be undersized.

Building owners in dry and moderate climates should choose a high SHR (>0.77) because the AC needs to remove less moisture compared to humid climates. Air conditioners with high SHR are more efficient. But, air-conditioning systems for humid climates normally require a lower SHR between 0.67 and 0.77 to provide acceptable comfort. Indoor relative humidity less than 50% discourages biological pests that cause respiratory problems. This represents another benefit to the AC's moisture removal in addition to reliable comfort.

Correct size and equipment selection are very important for new higher efficiency air conditioners. These new energy-efficient AC units must also have the correct SHR to provide adequate moisture-removal for a humid climate.

Figure 6-19: *Relationships among Airflow, Efficiency, and Moisture Removal*

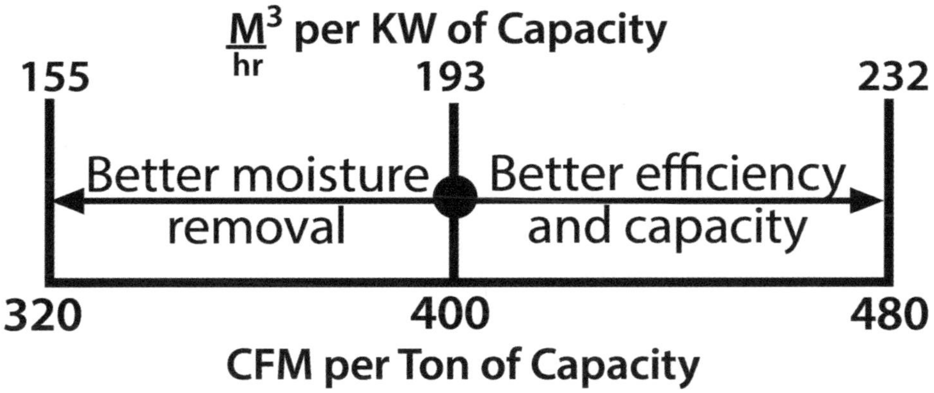

Figure 6-20: Hygrometers Show the Effectiveness of ACs and Dehumidifiers

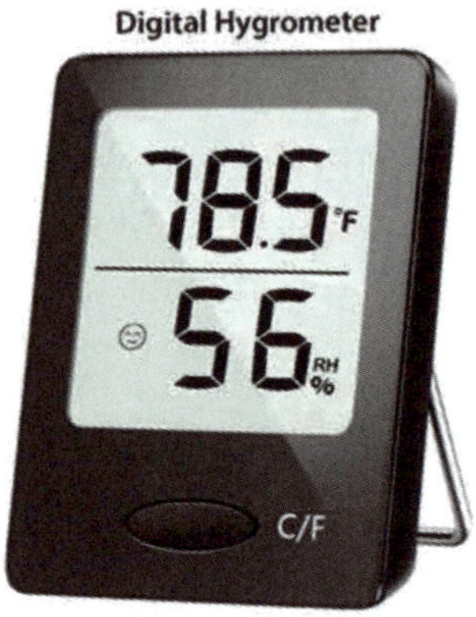

How air conditioners fail to dehumidify — The air conditioner may fail to dehumidify the building for these reasons.

- Thermostats control air conditioners by temperature, not Rh.
- Fish tanks, plants, cooking, and showering create too much humidity for the AC to remove.
- The installer may have chosen the wrong settings.
- The air conditioner is oversized.
- AC evaporator runs too warm, or the airflow is too high, or both.
- Humid air leaks through the building enclosure, introducing too much water vapor.
- Mild outdoor temperature during spring, fall, and nighttime doesn't activate the AC thermostat.
- Air sealing, insulation, exemplary windows, and shading combine to make indoor temperature and sensible-cooling load too low to activate the AC.

Building owners should do whatever they can to improve drainage. Vapor retarders and other moisture barriers can minimize the amount of water and water vapor in contact with the building.

Consider adjustments to air conditioner to run longer with a colder evaporator coil. Consider also a dehumidifier as part of your cooling strategy if you live in a humid climate and your indoor relative humidity reaches 60% or more.

Your AC can provide optimal efficiency or it can provide optimal dehumidification, but it can't usually provide both. Evaporators, designed for energy-efficiency, may not run cold enough to condense sufficient water vapor. An 85% SHR AC probably won't dehumidify sufficiently in Miami or in Singapore.

AC versus dehumidifier to control Rh — Modern air-conditioning systems have the capacity to remove more moisture than residential dehumidifiers because the AC systems have more powerful compressors and blowers. However, a building won't benefit from an acceptable Rh without a correctly sized air conditioner set to the correct sensible-heat ratio (SHR).

For optimal efficiency, the advanced AC needs a warmer evaporator coil and higher airflow per ton or kilowatt. For optimal dehumidification, the system needs a colder evaporator coil and lower airflow per ton or kW. Thus, in humid climates, the AC service tech may need to adjust both the compressor speed and the blower speed to allow the evaporator coil to remain below the indoor-air dew point, while the AC runs as continuously as possible for both moisture control and efficiency.

A skilled HVAC technician can strike an acceptable balance between SHR and optimal efficiency. This balance can be easy with an energy-efficient building enclosure and a modern AC with variable-speed compressor and a variable-speed blower.

However, if an insulated, airtight building has a cool roof, sun-blocking windows, and effective shading, comfort may demand a dehumidifier, especially for the spring and fall when the AC doesn't run long enough to adequately dehumidify.

Air-conditioner Installation

Many air conditioners fail to achieve their rated efficiency or moisture control because of installation problems. The most common problems are insufficient airflow, incorrect refrigerant charge, duct air leakage, and incorrect settings. These problems result from insufficient planning, too small installation space, and inadequate installer training. If you select a low-bid contractor instead of the most qualified one, you may experience installation problems.

Technicians should follow the manufacturer's instructions for installation, maintenance, and service. In a building's planning, consult the manufacturer's installation instructions for the physical sizes and clearances of equipment. Ducts need special consideration, during the building phase, because they are large, and they connect to all the main rooms. Duct installation should occupy first priority for space among a building's utilities. The HVAC contractor should work with building contractors, carpenters, plumbers and electricians to ensure that these workers don't obstruct the pathways, needed for ducts.

Replacing an air-conditioning system requires planning. Consider these five (5) improvements to ensure the new AC's optimal performance.

1. Insulate, seal air leaks, and provide shade to the building enclosure.
2. Test ducts for air leakage, and seal the leaks.
3. Insulate supply ducts in unconditioned spaces.

4. Select a smaller AC unit, compared to existing.
5. Check the new AC system for minimal duct leakage, correct airflow, and correct refrigerant charge.

If you accomplish the above list, your previously undersized ducts are probably quite adequate for your new AC system's lower airflow. The new system with its equipment improvements may be twice as efficient as the old one. Together with the building-enclosure improvements, you can maintain comfort and minimize energy cost into the future, even with our warming climate.

Thermostatic Control of Air Conditioners

If you reduce the thermostat setpoint to rush a comfortable temperature after setting the temperature higher, this doesn't make the air conditioner cool a building any faster. Instead it wastes energy. Residents who change the thermostat frequently to satisfy their different comfort demands, cause AC systems to operate erratically and inefficiently.

The thermostat's location may cause problems in controlling cooling systems. Install thermostats away from direct sunlight. A thermostat located on a warm outside wall may cause the air conditioner to over-cool and operate inefficiently.

Residents who use programmable heating/cooling thermostats or smart thermostats may save money, if they have predictable schedules of occupancy and vacancy.

Thermostats may not provide good comfort control in humid climates. Manufacturers sell air-conditioning controls that respond to both temperature and humidity. These controls adjust variable-speed blowers to provide better cooling and humidity-control than single-speed or multi-speed blowers.

Rooftop Units (Air Handlers)

Rooftop units (RTUs) are air handlers located on roofs, but you find these same units on ground slabs, or platforms outdoors. RTUs may contain one or more of the following.

- A combustion furnace
- All the components of an air conditioner (RTUs are packaged or unitary air conditioners)
- All the components of an heat pump (RTUs are may be packaged or unitary heat pumps)
- An economizer for free cooling and ventilation.

***Figure 6-21:** Rooftop Packaged Air Conditioner*

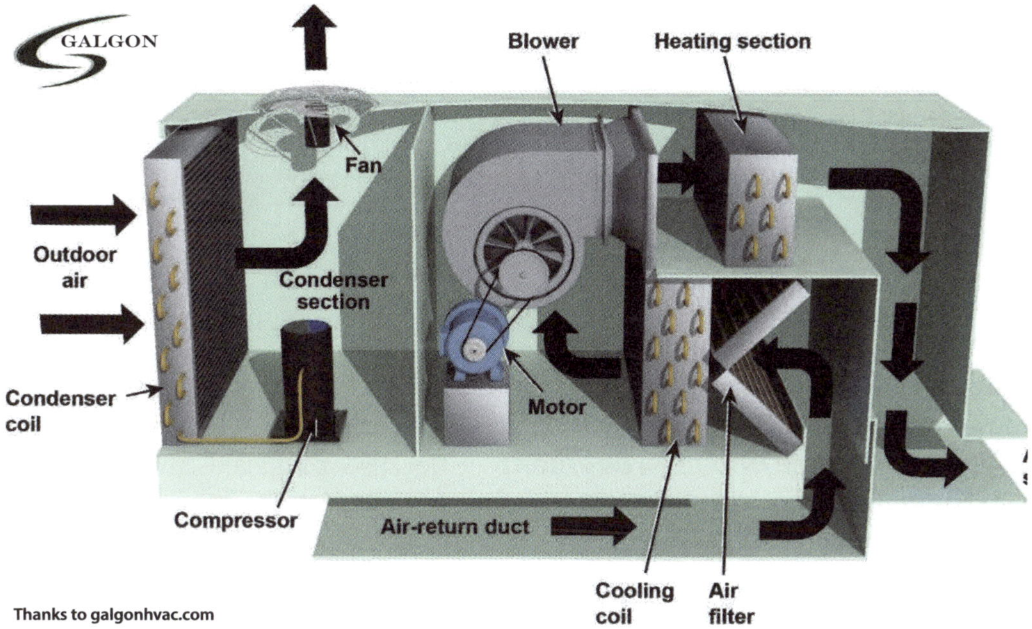

Economizers for large AC systems — Economizers typically operate at night when the outdoor air is cooler than the indoor air in a process called free cooling. Economizers mix enough outdoor air into the indoor air in order to meet the economizer-thermostat setpoint, without using the compressor. **Dry climates and not humid climates benefit most from an economizers' free cooling.**

The value of economizer for free cooling depends on climate, building type, system type, and control settings. A controller in the economizer measures the temperature and humidity of the outdoor air and indoor air. When the temperature-and-humidity conditions are favorable, the control switches the air conditioning compressor off and cools the building with outdoor air instead. The economizer uses far less cooling energy compared to air conditioning.

Economizers' energy savings are highest in dry climates and lowest in humid climates. In the USA, ASHRAE Standard 90 for multifamily buildings requires economizers in dry climates for air-conditioning systems, larger than 5 tons or 19 kW.

While saving energy at night, the fresh outdoor air that economizers exchange with indoor air can also function as ventilation. Thus the ventilation system (if present) can run for fewer hours and avoid operating during the day's heat and peak electrical load. See also "Yearly-Average-Relative-Humidity Map" on page 292.

Figure 6-22: Economizers and USA Climates where They Work Well

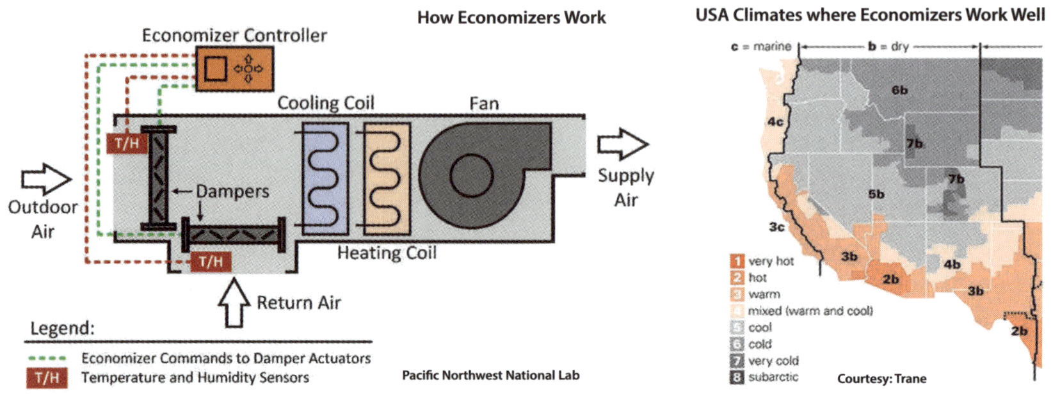

RTU maintenance and improvement — Because packaged air conditioners (RTUs) are outdoors, they are more likely to be neglected compared to systems with indoor air handlers. Fortunately though, the RTU's components are more accessible compared to indoor air handlers because there is usually more space to work on them.

Consider the following maintenance and improvements for RTUs.

- ✓ Clean or change filters, provide extra filters, and educate the building owner on filter maintenance.
- ✓ Clean the evaporator and condenser coils as specified on "Airflow Improvements at the Air Handler" on page 215.
- ✓ Test the RTU and its ducts for air leakage because many RTUs systems have excessive duct leakage. See "Troubleshooting Duct Leakage" on page 223.
- ✓ Test and adjust the economizer to maximize its benefit for both free cooling and ventilation. **This requires an elite HVAC-control technician.**
- ✓ Educate the building owner or operator on economizer function and control. Replace the thermostat, if necessary to control the economizer's free-cooling and ventilation functions. **Note: Economizers are difficult for many service techs to understand, and they require frequent maintenance. Thus many economizers function poorly.**

Geothermal Heat Pumps

Geothermal heat pumps use refrigeration to move heat between the ground or an outdoor water source to heat or cool the building. Geothermal heat pumps, like air-source heat pumps, work best in well-insulated and airtight buildings because they provide heated water at a relatively low temperature — 125°F or 52°C or less.

Geothermal heat pumps are more efficient than air-source heat pumps because of the ground's moderate and consistent temperatures. However, they're also more expensive than air-source heat pumps. Homeowners and multifamily owners select geothermal heat pumps because existing buildings have radiators or other hydronic heat emitters.

Figure 6-23: *Air-to-water and Water-to-water Heat-Pumps*

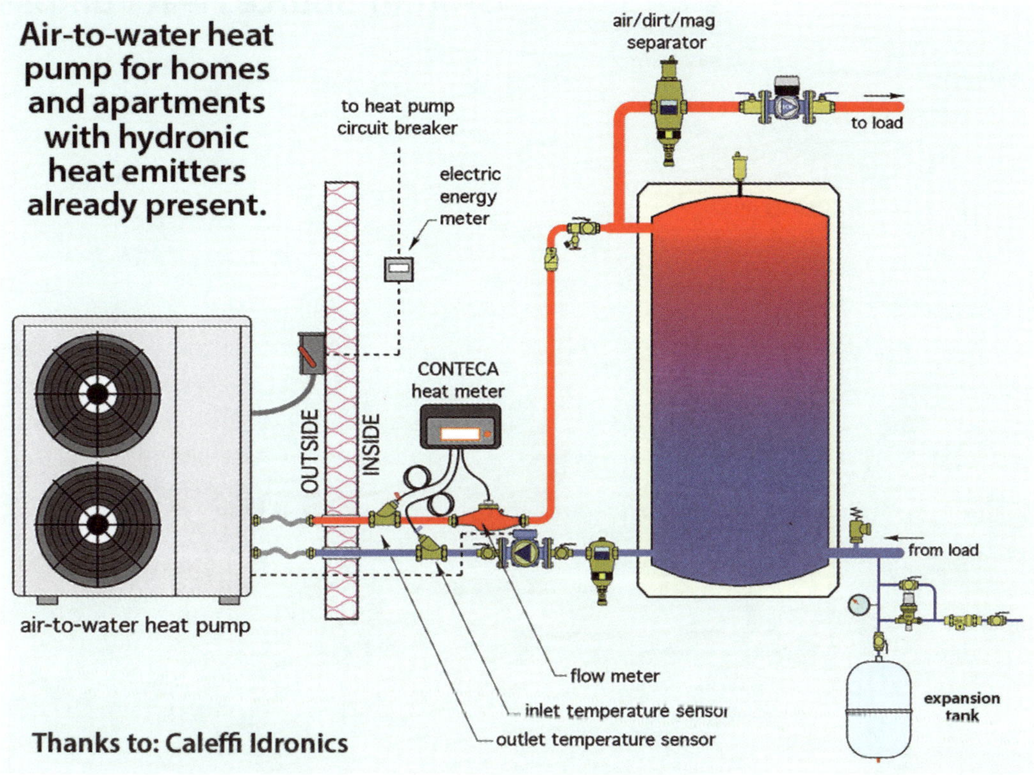

The two most common types of geothermal heat pumps are these.

1. *Closed-loop geothermal heat pumps* also called *ground-source heat pumps* or *ground-coupled heat pumps*. They collect the earth's heat with a ground coil. A pump circulates water through a buried, sealed coil of pipe, running between the ground and the heat pump. During the heating cycle, the heat pump collects heat from the circulating water and heats the building with it. The pump recirculates the cooled water back through the warmer ground, collects more heat, and then returns the water again to the heat pump. The heat pump reverses this process for the cooling cycle.
2. *Open-loop geothermal heat pumps* also called *water-source heat pumps* or *groundwater heat pumps*. They use water as their heat source. A pump moves water from a well or nearby water source and circulates the water through the heat pump. The heat pump collects heat by refrigerating the water from the water source. That refrigerated water then returns to the water source to collect more heat.

See "Professional Service and Commissioning" on page 208.

Figure 6-24: *Ground-source Heat-pump Collectors*

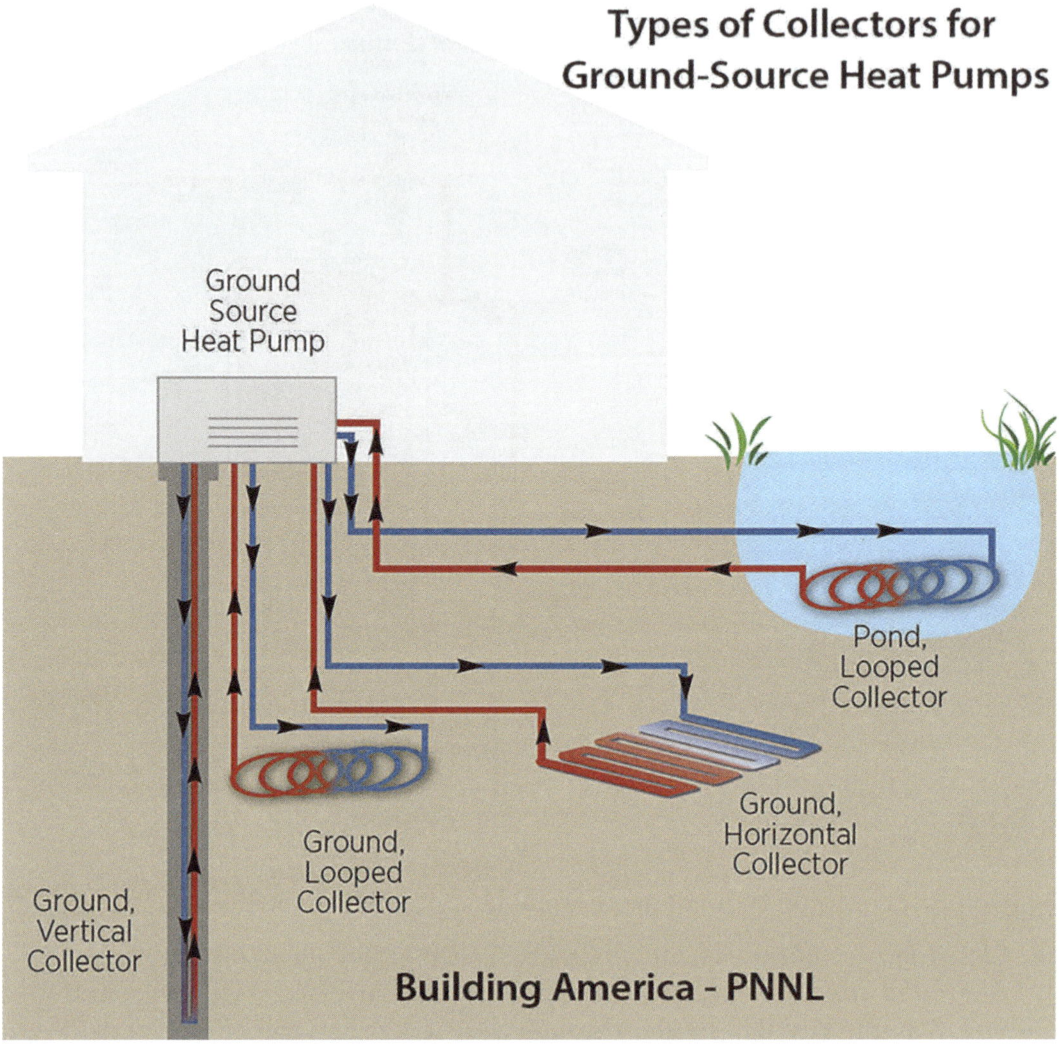

Airflow and Performance in Central Air Conditioners

The AC system's cooling capacity and efficiency depend on airflow. Central air conditioners are designed to cool the whole building. Closing registers in unused rooms reduces system airflow and seldom saves energy, except with large zoned systems with smart controls.

Blowers and ducts should provide about 400 cubic feet per minute (CFM) airflow for each ton of air-conditioning capacity. In SI units, this translates to about 200 cubic meters per hour (m^3/hr) or about 70 liters per second (l/s) of airflow per kilowatt (kW) of capacity. A HVAC contractor or designer may specify more airflow for drier climates to optimize efficiency, or less airflow for more humid climates to optimize moisture removal.

See "Air Handlers, Ducts, and Forced Air" on page 210.

Measuring airflow — Service technicians measure the airflow in the air-conditioning system in a variety of ways. They should bring the airflow up to a good standard before sealing the ducts because making duct-airflow improvements could be difficult in a recently sealed duct system.

A variety of diagnostic tools, designed for measuring HVAC system airflow, may provide accurate-enough measurements. If calibrated correctly, flow hoods can also give a reasonably accurate measurement of airflow through grilles, registers, and diffusers. Pressure-drop or temperature-difference across coils and air handlers can indicate whether or not the system airflow is sufficient.

See "Measuring Airflow in the System" on page 212.

Airflow problems and solutions — Obstructions in the supply-air ducts include dents, debris inside the ducts, and bent or dirty registers. Supply registers can have severe blockages if the vanes have been flattened by foot traffic or if dust has packs the registers' vanes.

Insulated flexduct is often kinked or crushed, causing insufficient airflow. Crushed or damaged duct board can also cause insufficient airflow. Fiberglass duct board has far less structural strength than galvanized-steel ducts. Fiberglass duct liner (installed inside metal ducts) may deteriorate over time from moisture and vibration and fall apart, partially blocking the duct.

Air from every supply register should have an unobstructed airway back to a return grille. Blockage of supply registers or return grilles can pressurize or depressurize parts of the building. This pressure causes poor air-conditioning performance and increased air leakage through the building enclosure. Retrofits include cutting off the bottoms of doors or installing grilles through walls or doors. Installing more return ducts or ducts between rooms are also options to improve air circulation.

Technicians increase the airflow by repairing or enlarging ducts, cleaning evaporator coils, and increasing blower speed. In some AC systems, enlarging ducts or replacing ducts might be the only remedy for low system airflow and the resulting comfort problems and high cooling costs.

Supply registers sometimes deliver too much or too little cool air to the rooms. Adjust the room airflow in one of these three ways.

1. Adjust the balancing dampers at the branch duct's takeoff, remove them to increase airflow, or install them to reduce airflow.

2. Adjust movable vanes in registers or remove them to increase airflow.
3. Seal off portions of the over-supplied registers.

Don't block registers altogether, because this reduces system airflow and cooling efficiency.

See "Duct Airflow Solutions" on page 215 and "Airflow and Performance in Central Air Conditioners" on page 204

Sealing Leaky Ducts

Testing ducts for air leakage, finding duct air leaks, and sealing ducts are all discussed elsewhere in this book.

See "Sealing Duct Leaks" on page 226 and "Troubleshooting Duct Leakage" on page 223.

Refrigerant Charge

Manufacturers charge room air conditioners and packaged air conditioners with refrigerant at the factory. Installation technicians charge split-system air conditioners and heat pumps during installation. Many ducted split systems have too much or too little refrigerant. Split systems that have the correct refrigerant charge and airflow usually perform very close to manufacturer's specifications for capacity and SEER. Manufacturers recommend that a technician measure airflow before checking charge because the charge measurements aren't accurate unless airflow is sufficient.

When the refrigerant charge is correct, the system operates at the specific refrigerant temperatures listed by the manufacturer. For satisfactory performance and efficiency, a single-family central AC system should be within one ounce (30 grams) of the correct charge as specified by the manufacturer.

Saturation temperature is the point between liquid and vapor that exists in both the evaporator and the condenser. Two commonly listed measurements are *superheat* and *subcooling*, which are the number of degrees F or C that the HVAC system heats or cools the refrigerant beyond its saturation temperature.

The vapor needs superheat to dry it in preparation for compression so that the wet refrigerant doesn't enter the compressor. The liquid from the condenser needs subcooling to insure that it doesn't flash into a gas before it reaches the expansion device and the evaporator.

See "The Refrigeration Cycle for Air Conditioning" on page 32 to review air-conditioner operating principles.

Evaluating Refrigerant Charge

Air-conditioning replacement or service includes refrigerant charge-checking. The efficiency of the air-conditioning system is directly related to the correct amount of refrigerant.

The best option for charge-checking is to measure key temperatures in a process known as *noninvasive refrigerant testing.* Every time that technicians connect their gauges and hoses, they create a risk of contamination, lost refrigerant, and the resulting problems and system failures.

Noninvasive refrigerant testing — Noninvasive testing provides the following benefits, compared to connecting gauges and hoses.

- Noninvasive testing saves time for airflow testing and thorough preventive maintenance.
- It saves refrigerants, which benefits the environment and saves money.
- It prevents refrigerant cross-contamination and moisture contamination.

Invasive refrigerant testing — Use a refrigerant manifold, gauges, and hoses, in these two situations.

1. When noninvasive testing isn't adequate to troubleshoot a refrigerant problem.
2. During installation, refrigerant charging, and major repairs.

HVAC technicians evaluate refrigerant charge by two invasive methods, depending on what type of expansion valve the air conditioner has.

1. If the expansion valve has a fixed orifice, they do a superheat test.
2. If the valve is a thermostatic expansion valve (TXV) or an electronic expansion valve, they do a subcooling test.

Figure 6-25: *Correcting Refrigerant Charge for AC Systems*

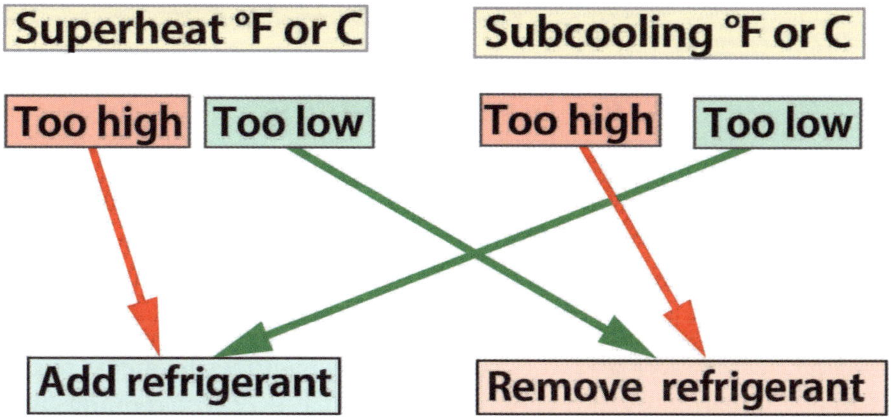

Maintenance and Service

Air-conditioner efficiency depends upon routine maintenance. The percent difference between the electricity consumption of a well-maintained AC system and a neglected system ranges from 10% to 35%.

A skilled technician or homeowner should do the following routine maintenance work.

- ✓ Clean or replace filters regularly (depending on operating time, every 1 to 6 months).
- ✓ Clean supply and return registers and straighten their vanes (dirt and damage are visible).
- ✓ Clean the blower's fan blades.
- ✓ Clean evaporator coil and condensate pan every 2 to 5 years.
- ✓ Clean condenser coils each year or two (dirt may or may not be visible).
- ✓ Remove debris from around condenser.
- ✓ Straighten bent fins in evaporator and condenser coils.

Figure 6-26: *Professional Service and Commissioning for Air Conditioners*

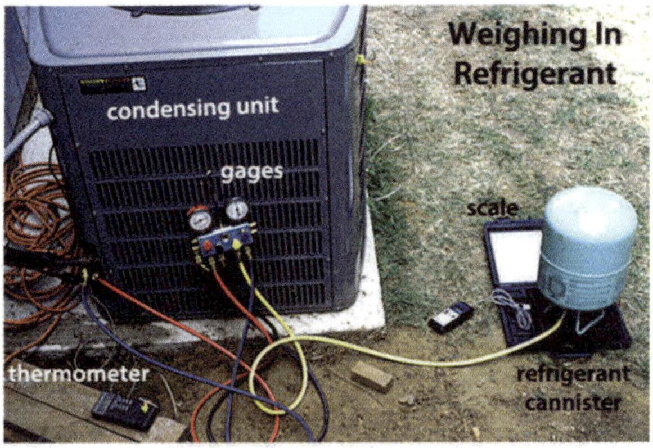

Professional Service and Commissioning

Only professionals should adjust and repair air conditioners. Even many professionals don't know how to service air-conditioning systems comprehensively. Thus you have no guarantee that basic operating requirements exist in an air-conditioning system. These requirements include: correct size, sufficient airflow, airtight ducts, and the correct refrigerant charge.

When a building owner hires a HVAC contractor or engineer to measure performance and make adjustments and repairs, this service is *commissioning*. A commissioning professional may do the following procedures to commission a central air conditioner.

- ✓ Examine ducts, filters, blower, and indoor coil for dirt and other obstructions.
- ✓ Verify correct electric control, especially that heating is locked out when the thermostat calls for cooling and vice versa.
- ✓ Diagnose and seal duct leaks.
- ✓ Verify adequate airflow by measurement.
- ✓ Verify correct refrigerant charge by non-invasive or invasive procedures, as required.
- ✓ Examine electric terminals — clean and tighten connections and apply non-conductive coating, if necessary.
- ✓ Oil motors and check belts for tightness and wear.
- ✓ Verify correct thermostat operation.
- ✓ Verify that drains and their traps move condensate to outdoors.

Figure 6-27: *AC Drains and Condensate Traps*

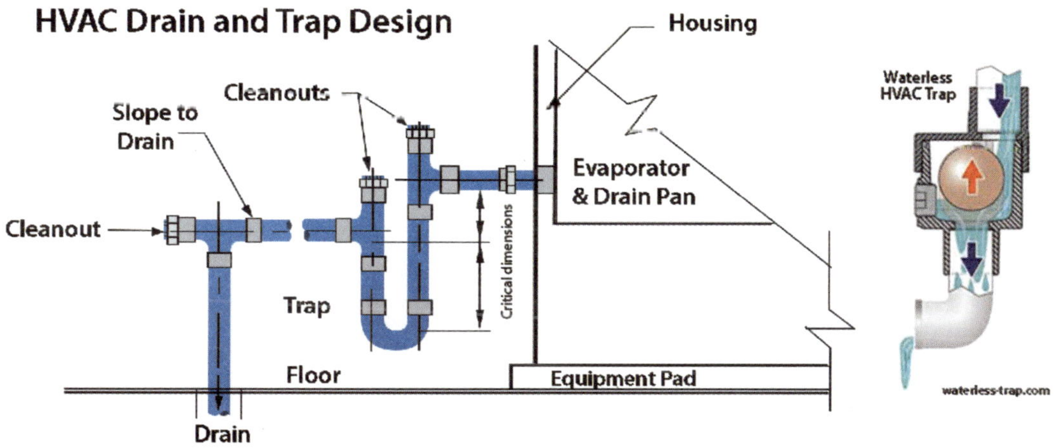

HVAC condensate drains and traps — HVAC condensate drains and traps are essential for the operation, durability, health, and safety of air conditioning and dehumidification systems. An HVAC trap works like a plumbing trap. The HVAC trap uses condensate water to block the drain from transporting conditioned air to outdoors and polluted outdoor air into the building.

A nonfunctional drain or trap can cause the following problems.

1. Condensate spills into attics, basements, roofs, and other places, possibly causing damage.

2. Polluted air leaks through the drain and trap, pulled by blower's negative pressure, if the trap is dry.
3. Dust and other pollutants enter a building because air leaks through a dry trap.

HVAC technicians should inspect, repair, replace, or clean HVAC condensate drains and traps as often as necessary. Waterless traps solve the problem of dry traps by using a ball the an air barrier. The air handler's pressure and the evaporator's accumulating condensate to move the ball back and forth.

Air Handlers, Ducts, and Forced Air

The goal of this section on forced-air systems and ducts is to explain how the systems work and to discuss how to inspect, evaluate, troubleshoot, and improve forced-air systems in existing buildings.

A forced-air system consists of an air handler (furnace, heat pump, air conditioner) with its heat exchanger and its ducts. Whole-building dehumidifiers and whole-building ventilators also use air handlers to move air.

Duct evaluation and improvement has a logical sequence of steps. The logic of this sequence is to prevent one procedure interfering with another procedure. Comfort and energy problems might disappear if we accomplish everything on this list.

1. Evaluate the efficiency of the building-enclosure, and make cost-effective improvements.
2. Correct air-handler equipment settings.
3. Optimize system airflow.
4. Correct duct-induced building pressures.
5. Seal duct air leakage.
6. Install duct insulation as needed in unconditioned spaces.
7. Evaluate the desired filter characteristics and install appropriate filters.

Figure 6-28: *Well-installed Air Handler and Ducts*

Forced-air Systems

Air, as a heating fluid, has a theoretical efficiency advantage over hot water and steam because air circulates at a lower temperature. Forced-air heating is an effective residential HVAC system because forced air allows heating, cooling, and ventilation through the same air handler and distribution system. Modern energy-efficient furnaces meat or exceed the highest efficiencies of any combustion heaters.

However, forced-air distribution systems often leak air, spread dust, and dry indoor air excessively during the heating season. Forced-air heating systems are also difficult to zone. Zoning means dividing a home or apartment building into zones that separate thermostats control. Zoned forced-air systems can save energy if experts design and install the duct system very carefully.

However, separating forced-air systems into zones isn't usually practical or economical, except in large ducted distribution systems. Residents may close off registers and closing doors in unused rooms, which usually reduces airflow through the air handler and ducts, reducing the system efficiency.

The forced-air distribution system consists of an air handler, furnace, indoor coil, blower, ducts, filters, and registers.

Forced-air distribution systems waste energy in the following ways:

- Air leaks into and out of ducts.
- Obstructed airways.
- Incorrect blower-speed causes incorrect airflow.
- Poorly adjusted controls make the blower's operating time too short.
- Uninsulated ducts in unconditioned areas lose excessive heat.

Cumulative savings from the operating and maintenance improvements, discussed in the next few sections, vary from 5% to 30% of fuel consumption. Savings depend on how much energy the distribution system currently wastes.

Figure 6-29: *Air-handler Configurations*

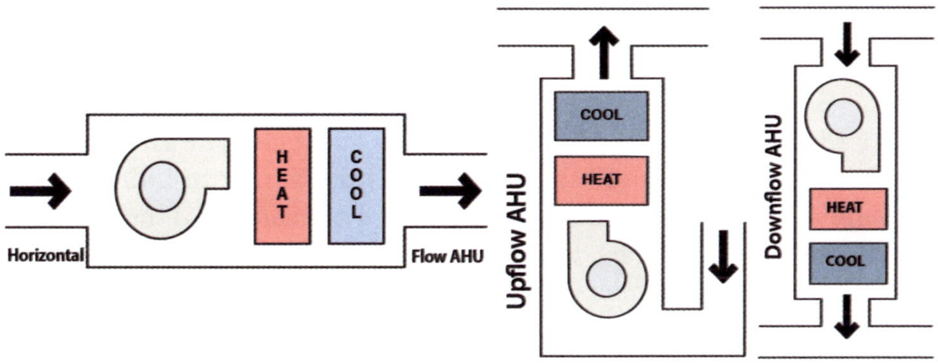

Measuring Airflow in the System

We diagnose system airflow first because we may need to make changes in the ducts to achieve an acceptable airflow. If we seal or insulate the ducts first, the sealant and the insulation make it difficult to improve the airflow.

Measuring total external static pressure (TESP) — Airflow is difficult to measure but is possible if you have the right tools. Technicians usually measure static pressure because it's easier to estimate, using inexpensive manometers and available manufacturers' data.

The blower creates the duct pressure, which we measure in inches of water column (IWC) or pascals (pa.). The return static pressure is negative and the supply static pressure is positive. Total external static pressure (TESP) is the sum of the absolute values of the supply and return static pressures.

Absolute value means that you ignore the positive or negative signs when adding supply static pressure and return static pressure together to get TESP. This addition represents the distance on a number line as shown in the next illustration.

Figure 6-30: *TESP Measurement and Number Line*

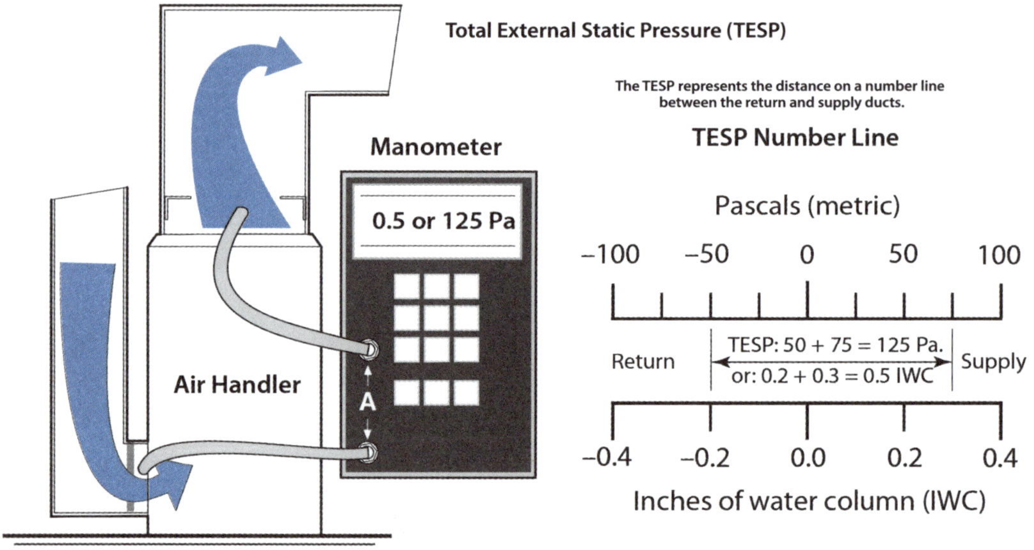

TESP gives a rough indicator of whether airflow is adequate. The greater the TESP, the lower the airflow. The supply and return static pressures can indicate whether the supply, the return, or both sides are restricted. For example, if the supply static pressure is 0.10 IWC (25 pascals) and the return static pressure is –0.5 IWC (-125 pascals), this indicates that undersized or restricted return air ducts cause the airflow problems. The TESP gives a rough estimate of airflow if the manufacturer's graph or table for static pressure versus airflow is available.

1. Measure pressure on each side of the air handler to obtain both supply and return static pressures separately. This test helps to locate the main problems related to either the supply or the return or both.
2. Find online manufacturer's literature for a table of TESP versus airflow for the blower or the air handler. Find airflow for the TESP measured above.
3. Make sure that the manufacturers model number on the literature is the same as the model number of the air handler. Then, choose the right blower speed to find the estimated total airflow.

Static pressure guidelines — Air handlers deliver their airflow at TESPs varying from 0.30 IWC (75 Pascals) to 1.0 IWC (250 Pascals) as found in small and medium-sized buildings. Manufacturer's recommended static pressure is usually a maximum 0.50 IWC (125 pascals) for standard air handlers. TESPs greater than these guidelines indicate inadequate airflow in standard residential forced-air systems.

The popularity of pleated filters, electrostatic filters, and high-static high-efficiency evaporator coils, prompted manufacturers to introduce premium air handlers that can deliver adequate airflow at a TESPs of greater than 0.50 IWC (125 pascals). Premium residential air handlers can provide adequate airflow with TESPs of up to 0.90 IWC (225 pascals) because of their variable-speed blowers. TESPs greater than 0.90 IWC indicate the possibility of inadequate airflow in these premium residential forced-air systems.

Figure 6-31: *Static-pressure Measurement and Static-pressure Budget*

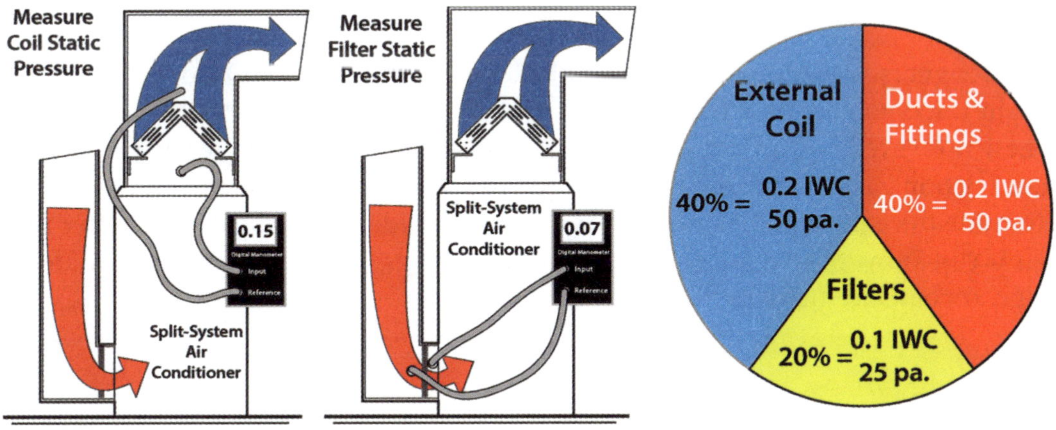

Air-conditioner blower performance — In the cooling mode, AC performance depends upon outdoor temperature, indoor relative humidity, in addition to airflow through the indoor evaporator coil. If the settings are correct, the 2-speed and variable speed compressors adapt better than single speed compressors to variations in outdoor temperature and indoor relative humidity.

Most AC blowers have two or more speeds or variable speed. The manufacturer publishes tables with airflow at various TESPs. You can find this information online by searching for the AC manufacturer and model number.

The blower speed depends upon which terminal the blower motor connects to. The ideal airflow is between 325 CFM and 475 CFM per ton or 45-66 m^3/hr per kW of capacity, depending on outdoor relative humidity. Dry-climate systems don't need to dehumidify the indoor air. Drier climates benefit from higher airflows and thus higher efficiencies. If you expect the system to dehumidify, the blower must run slower.

Table 6-2: *Total External Static Pressure (TESP) Versus System Airflow: Example*

TESP (IWC)	0.30 IWC 75 Pa	0.40 IWC 100 Pa	0.50 IWC 125 Pa	0.60 IWC 145 Pa	0.70 IWC 175 Pa	0.80 IWC 190 Pa
Airflow	995 CFM 1690 m^3/hr	945 CFM 1600 m^3/hr	895 CFM 1520 m^3/hr	840 CFM 1430 m^3/hr	760 CFM 1290 m^3/hr	670 CFM 1140 m^3/hr

Duct-system equivalent length — Each type of fitting, including elbows, takeoffs, reducers, diffusers, grilles and register boots has its own *equivalent length*. Equivalent length compares the resistance of the fitting to a length of round pipe or rectangular duct. For example, a standard register boot has a resistance, equivalent to 30 feet (9.2 meters) of straight duct of the same dimension.

Duct designers use the sum of all equivalent lengths of assembled fittings and straight ducts to calculate the *friction rate* for each room. The designers use the highest measured equivalent length from a room to the air handler as the maximum friction rate for the duct system. As a strategy to increase airflow, you can determine the equivalent lengths of rooms and consider reducing it with better fittings if you know the equivalent lengths of the fittings. Calculations related to duct design are beyond the scope of this book.

Figure 6-32: Forced-air Negative and Positive Static Pressure

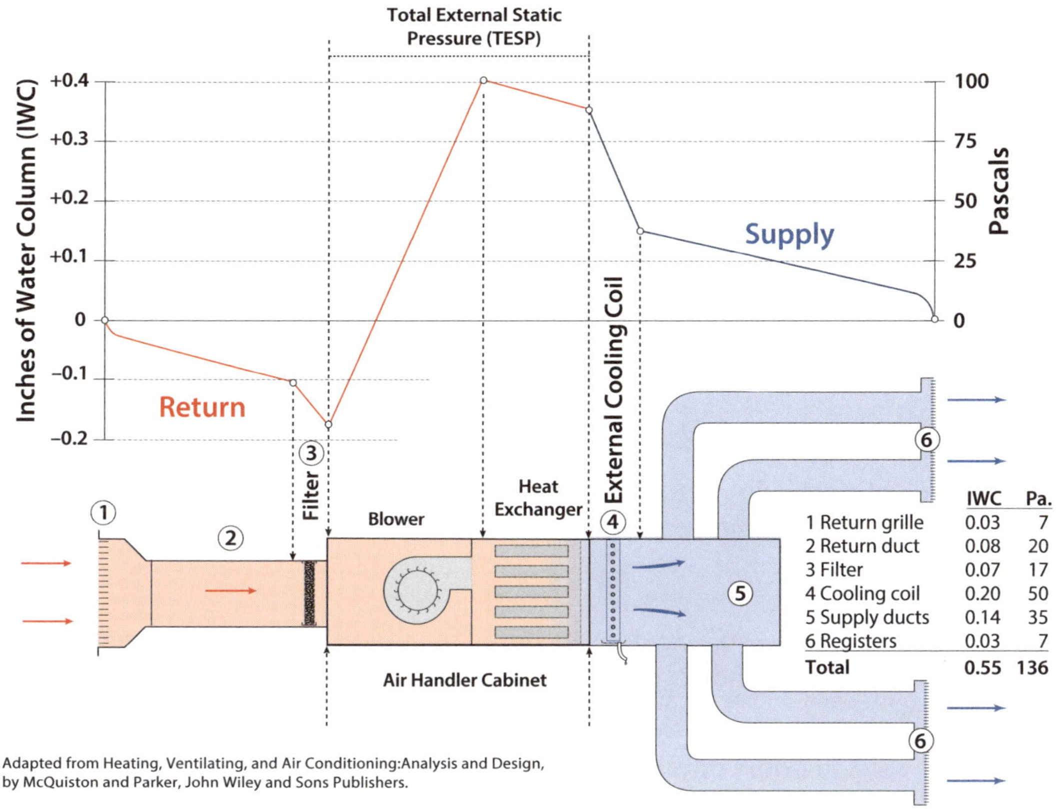

Adapted from Heating, Ventilating, and Air Conditioning:Analysis and Design, by McQuiston and Parker, John Wiley and Sons Publishers.

Duct Airflow Solutions

HVAC service technicians often find duct systems that are undersized for the heating or cooling output of the air conditioner. Under-sizing means that the ducts' cross-sectional area is too small. Also, the supply ducts and their fittings cause too much friction and air turbulence for adequate comfort. A too-high or too-low supply-air temperature or uncomfortable rooms suggests a duct-sizing or duct-restriction problem.

If a technician opens the door to a furnace's blower compartment during operation, this test can provide information about duct restriction. If opening the blower-compartment door changes supply temperature significantly, then increasing return air usually improves both comfort and efficiency. (The technician who tests needs to hold down a safety button to keep the blower on.)

Airflow Improvements at the Air Handler

Inadequate airflow is a common cause of residents' complaints. When the air handler is on, there should be a strong flow of air out of each supply register. Low register airflow may mean that a branch duct is partially blocked or disconnected. Or the return air from the room to the air handler isn't sufficient. When low airflow is a problem, consider specifying the following improvements as appropriate from your inspection.

- ✓ Clean air handler's blower.
- ✓ Clean the air-conditioning coil or heat-pump coil. If the blower is dirty, the coil is probably also dirty.
- ✓ On a condensing furnace, clean the secondary heat-exchanger coil.
- ✓ Lubricate the blower motor if it needs lubrication — many don't.
- ✓ Increase blower speed if necessary.

Airflow Improvements to Ducts

The methods of improving return ducts compared to improving supply ducts are somewhat different. If you use these methods, you will likely improve the airflow in any HVAC duct system.

Return-air solutions — Consider specifying the following return-air changes to improve system airflow and to reduce the imbalance between supply airflow and return airflow.

- ✓ Install vents in doors or walls to connect rooms without return grilles to rooms with returns.
- ✓ Install *transfer* or *jumper* ducts above ceilings to allow air to return from rooms that have no return register.
- ✓ Install a new return grille and duct, leading from a return-starved area directly to the air-handler cabinet.
- ✓ Consider filters in each return register, instead of a single one at the air handler.
- ✓ Repair or replace bent, damaged, or restricted return grilles. Change dirty filters in filter grilles.

Supply-air solutions — Consider specifying the following supply-duct changes to increase system airflow and to reduce the imbalance between supply and return airflow.

- ✓ Remove obstructions to registers and ducts such as rugs, furniture, and objects placed inside ducts (children's toys and water pans for humidification, for example).
- ✓ Clean registers, diffusers, blowers, heat exchangers, and air-conditioning coils.

Figure 6-33: *Dirty Indoor Coils Greatly Reduce Airflow and AC Efficiency*

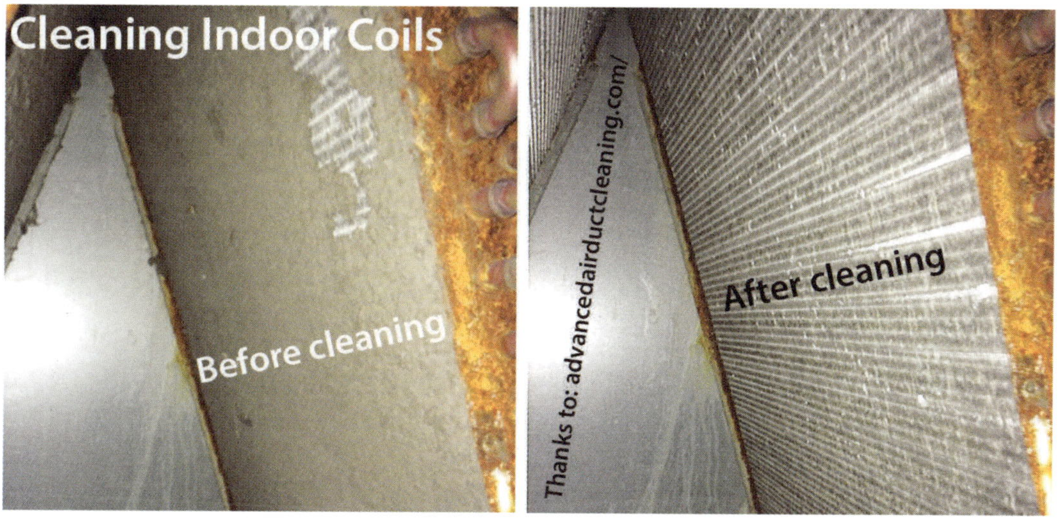

✓ Remove kinks from flex duct, and replace collapsed flex duct and fiberglass duct board.

✓ Remove excessive lengths of slacking flex duct, and stretch the duct to enhance airflow.

✓ Where flex ducts turn a 90-degree corner, install a smooth metal 90° fitting instead, with the flex duct stretched on both sides of that fitting.

✓ Install additional supply ducts as needed, to provide conditioned air throughout the building.

✓ Verify that balancing dampers to rooms, needing more airflow are wide open. Remove balancing dampers and register dampers to rooms with low airflow. (Balancing dampers are dampers in the takeoff near the plenum or main trunk.)

✓ Repair or replace bent, damaged, or restricted registers and diffusers. Low-resistance registers and diffusers are available, if your supplier has the information you need to specify them.

Rooms may be difficult to cool because of too small supply ducts or registers. Adding another supply duct and register is one option. Increasing the size of the existing supply duct and register is another option.

Figure 6-34: *Flex-duct Elbows — Bad and Good*

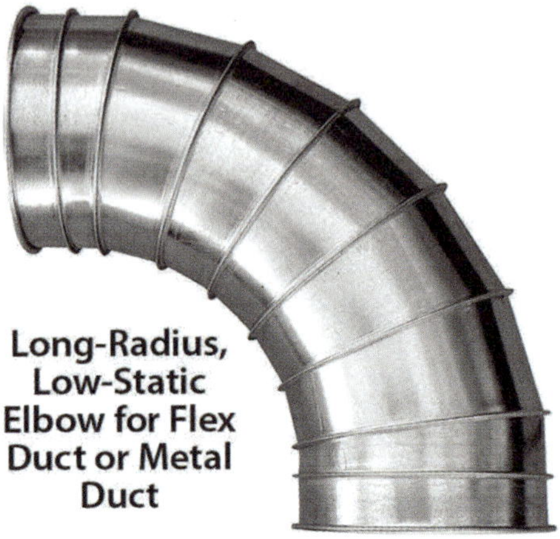

Other important system improvements — Consider the merits of the following suggestions.

- ✓ If supply and return ducts are outside the thermal enclosure or if condensation happens on ducts, insulate the supply ducts. With severe climates or high energy costs, insulate both the supply and return ducts.
- ✓ A contractor might need to repair or replace ducts and fixtures or install additional branch ducts to solve comfort or energy problems.
- ✓ Decide whether duct-sealing is important and if so, find and seal all the duct leaks you can access.
- ✓ If most of the HVAC system is outside the thermal enclosure, up to 25% of the system's energy may escape. You might reclaim this 25% by bringing the ducts inside the thermal enclosure by insulating and air sealing the enclosure, containing the ducts.
- ✓ The outdoor coil isn't part of the duct system. However, eliminate it as a possible AC-system problem because **it usually is a problem**. Clean the outdoor coils of air conditioners and heat pumps. No common maintenance improvement is more necessary than cleaning coils.

Grilles, registers, and diffusers — Here are definitions of the terminal devices that cover the openings and boots of the HVAC duct systems in rooms.

1. *Registers* supply conditioned air into an indoor environment. Registers can control the velocity and volume of air with their dampers.
2. *Diffusers* direct conditioned air into a space in a designed pattern of airflow distribution.

3. *Grilles* remove air from an indoor space and return it to the air handler. Grills don't have dampers or design patterns like diffusers.

These terminal devices are seldom serious airflow problems. However, if the device has these problems, consider replacement to improve airflow.

- Made of wood or decorative metal
- Damage to the vanes and the frame
- Made of low-quality materials

The standards and ratings for these three devices makes competing products difficult to compare. If you think that any of these three devices is a comfort problem or airflow problem, consult an engineer or product expert who is familiar with the ratings and standards of grills, registers, and diffusers.

Recommended Airflow for Air Handlers

The air handler's recommended airflow depends on its cooling capacity. Central air conditioners should deliver 325-to-475 cfm of airflow per ton (12,000 BTUs/hr) of cooling capacity or 161-to-237 cubic meters per hour per kilowatt of cooling capacity.

Unbalanced Airflow

Closing interior doors often separates supply registers from return registers in buildings with central returns. A bedroom with no return register and a closed door restricts the bedroom air from returning to the air handler. This restriction pressurizes bedrooms and depressurizes the central areas near return registers. These pressures drive air leakage through the building enclosure, waste energy, create moisture problems, and bring pollutants in from the crawl space, attic, or attached garage.

Unbalanced-airflow Test

The following test uses only the air handler and a digital manometer to evaluate whether the supply air can flow back through return registers to the air handler unobstructed. Activate the air handler, and close interior doors for this test.

1. Measure the pressure difference between the central zone and the bedroom as shown in "Measuring Pressures: Unbalanced Airflow between Supply and Return" on page 221.
2. Measure the pressure difference between the building's central living area and the outdoors with the digital manometer.
3. Disregard the positive or negative signs.

If the measured pressure difference is more than 3.0 pascals with the air handler operating, consider these two pressure-relief suggestions.

1. To estimate the surface area of pressure relief needed, slowly open the bedroom door until the pressure difference drops below 1 pascal.
2. Estimate the surface area of that door opening. This estimate is the area of the permanent opening required to provide pressure relief. Pressure relief may include undercutting the door, installing a transfer grille, or installing a jumper duct.

Figure 6-35: *Transfer Grills to Prevent Unbalanced Airflow*

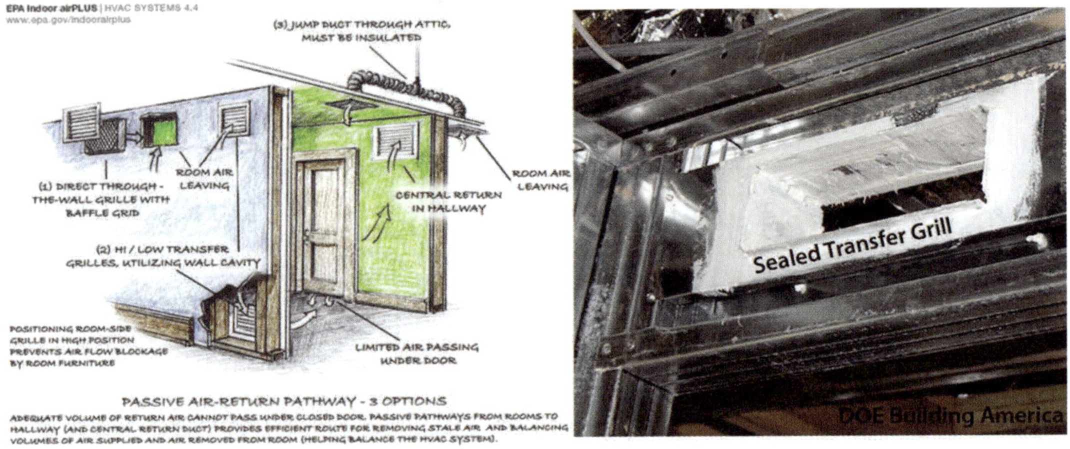

***Figure 6-36:** Measuring Pressures: Unbalanced Airflow between Supply and Return*

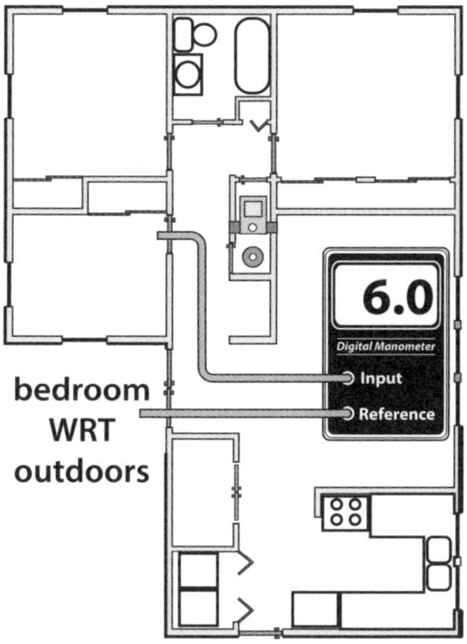

Bedrooms with supply registers but no return register are pressurized when the air handler is on and the doors are closed. Pressures this high can double or triple air leakage through the building enclosure.

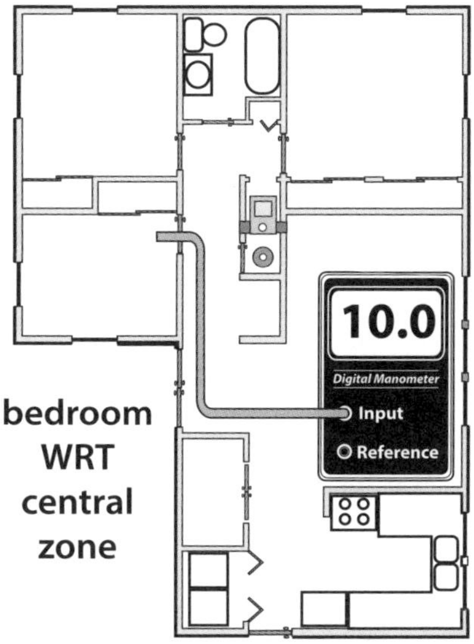

The air handler depressurizes the central zone and pressurizes the bedroom, when the bedroom doors are closed. This test measures the pressure difference between the pressurized bedroom and the depressurized central area.

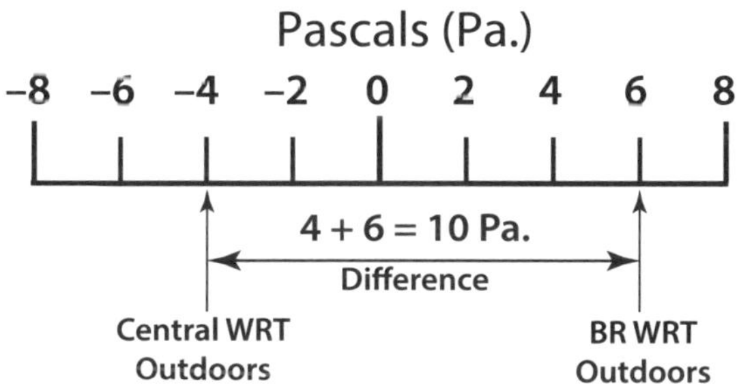

The distance on a number line represents the difference in pressure between the central zone and the bedroom.

***Dominant-duct-leakage test* —** The test for dominant duct leakage is easy. You use a manometer to measure *building with reference to outdoors* with the air handler operating. If the building is pressurized with all the interior doors open, the return air leakage is dominant. If the building is depressurized during the same conditions, the supply air leaks are dominant.

Figure 6-37: Panned Return Ducts

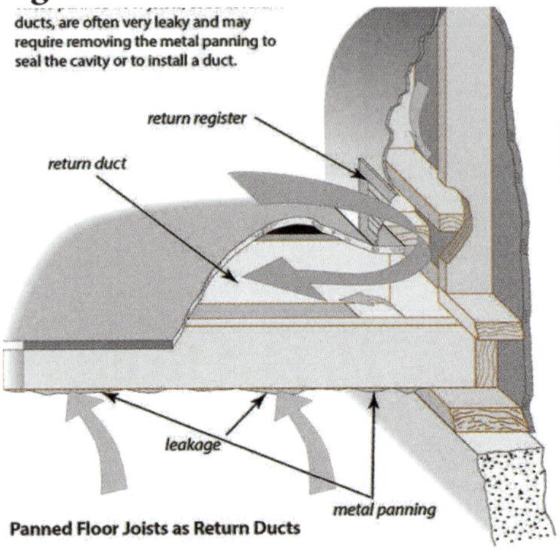

Installers sealed a panned cavity with foil-faced foam board, designed for lining such cavities, with duct mastic to provide an airtight return

Evaluating Duct Air Leakage

Duct leakage is a major energy-waster in buildings where the ducts are located outside the building's thermal enclosure — in a crawl space, attic, attached garage, or leaky unconditioned basement. When these intermediate zones remain outside the thermal enclosure, duct sealing is usually cost-effective in those zones.

Duct leakage inside the thermal enclosure is usually harmless, unless it creates uncomfortable temperature differences between rooms.

Solve Obvious Air-handler and Duct Problems

Most systems need essential maintenance, so don't expect to improve HVAC efficiency unless you find and solve these obvious problems. If you solve the obvious problems, the system may even perform adequately without more expensive changes.

Inspect the air handler and ducts — Ducts have a bad reputation, among energy specialists, as a cause of energy-and-comfort problems. However, be aware that an installer may not have adjusted the air conditioner or heat pump to the manufacturer's specifications during installation. Ducts are often a problem, but don't assume they are the main problem, until you prove it through testing.

Begin with a thorough inspection, and solve obvious problems with the air handler and duct system. Don't test yet. Instead start with these tasks.

- Look for disconnected ducts, restricted ducts, as well as dirty filters, coils, blowers, registers, and grilles. Someone should clean and repair these components before any testing.

- Look for closed registers or closed balancing dampers that could restrict airflow.
- If the indoor coil isn't easily visible, a dirty blower indicates that the evaporator coil is probably also dirty. Someone should clean the blower and coil before testing.
- If the blower and coil are dirty, the filter and its mounting allow dirt to bypass the filter.

Fix these problems before testing. Tests have little value until the you solve the obvious problems.

Troubleshooting Duct Leakage

Several simple procedures can help you find the locations of the duct leaks and evaluate their severity.

Finding duct leaks using touch and sight — One of the simplest ways of finding duct leaks is feeling with your hand for air leaking out of supply ducts, while the air handler's blower pressurizes the ducts. You can also locate duct leaks with a visual inspection. Use one of these 5 tests to locate air leaks.

Figure 6-38: *Finding Duct Air Leaks by Sight and Touch*

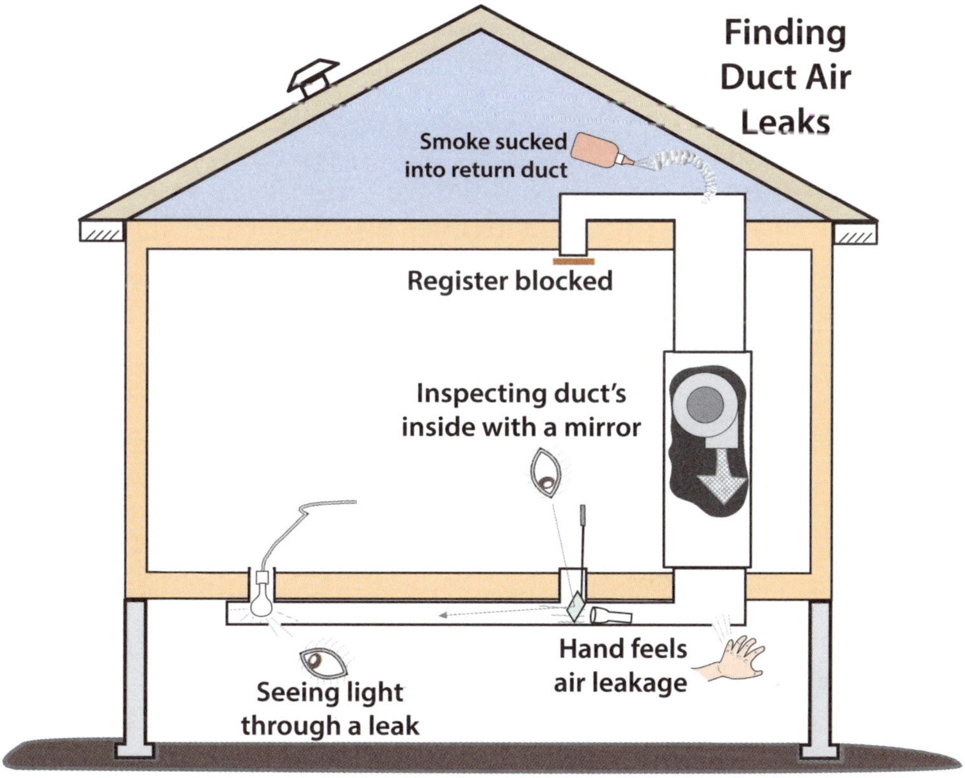

1. Use the air handler's blower to pressurize supply ducts. Close the dampers on supply registers temporarily or block the registers with pieces of carpet, magazines, or any object that won't be blown off by the register's airflow. The blockage increases the duct pressure and makes duct leaks easy to feel. Dampen your hand to make your hand more sensitive to airflow. This will help you to find duct air leaks.
2. Use chemical smoke to find return leaks by observing smoke entering return grilles through return-duct leaks. Smoke will blow back from supply-duct leaks.
3. Place a portable light inside the duct through a register or grille. Look for light shining out from exterior duct leaks.
4. Decide which duct joints were difficult to fasten and seal during installation. These joints are likely duct-leakage locations.
5. Use a light, and mirror, camera, or borescope to help you to visually examine duct interiors.

Feeling air leaks tells you their exact location. The air handler must pressurize the ducts in order to feel leaks. You can feel air leaking out of pressurized supply ducts, but you can't feel air leaking into depressurized return ducts. However, you can see smoke entering return ducts if you have a smoke generator.

Pressurizing the home with a blower door forces air from outdoor air leaks through duct leaks, where you can feel the leakage from both supply and return ducts.

Pressure pan testing — Pressure pan tests can identify leaky or disconnected ducts located in intermediate zones. With the house depressurized by the blower door to either –25 pascals or –50 pascals, record pressure pan readings at each supply and return register.

Figure 6-39: *A Pressure Pan*

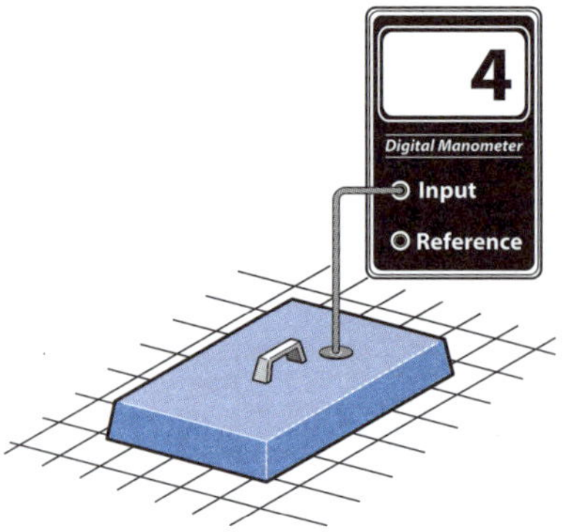

A pressure pan blocks a single register and measures the air pressure behind it, during a blower-door test. The magnitude of the measured pressure indicates the relative importance of the duct leaks near that register or grille.

Pressure-pan duct standards — If the ducts are perfectly sealed with no leakage to the outdoors, you won't measure any pressure (0.0 pascals) during a pressure-pan test with the blower door creating 50 pascals of house pressure. The higher the measured pressure reading, the more the ducts are connected to outdoors.

- If the pressure reading is 50 pascals, the register or grille connects directly to the outdoors.
- Following duct-sealing work, no more than three registers or grilles should have pressure-pan readings greater than 2 pascals. No single reading should be greater than 4 pascals.
- The best duct-sealing contractors use 1 pascal or less as a goal for all registers and grilles.

Figure 6-40: *Pressure-pan Testing*

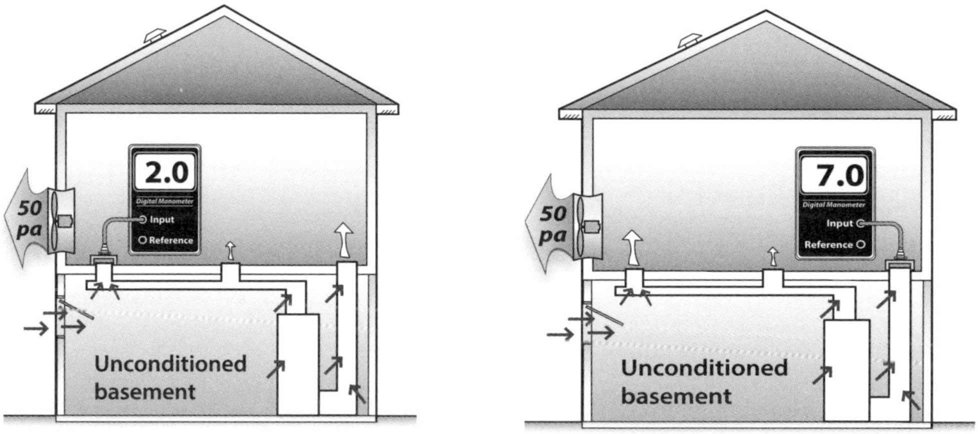

A pressure pan reading of 2 indicates moderate duct air leakage in the nearby supply ducts.

A pressure reading of 7 pascals indicates major air leakage near the tested return grille.

Measuring Duct Air Leakage with a Duct Blower

Pressurizing the ducts with a duct blower measures total duct leakage. The duct blower is the most accurate common measuring device to quantify duct air leakage. The duct blower includes a fan, a digital manometer, and a set of reducer plates for measuring different duct-leakage airflows. HVAC technicians use a blower door with a duct blower, to measure duct leakage to outdoors. This measurement is directly related to duct-sealing cost effectiveness.

Measuring total duct leakage — The total duct leakage test measures leakage to both indoors and outdoors. Begin by opening the building and its intermediate zones to the outdoors by opening windows, doors, or vents. Opening the intermediate zones to outdoors ensures that the duct blower is measuring only the ducts' airtightness — not the airtightness of ducts combined with other air barriers like roofs, foundation walls, or garages.

Technicians can test supply and return ducts separately, when they replace the air handler.

Figure 6-41: *Total Duct-air-leakage Test*

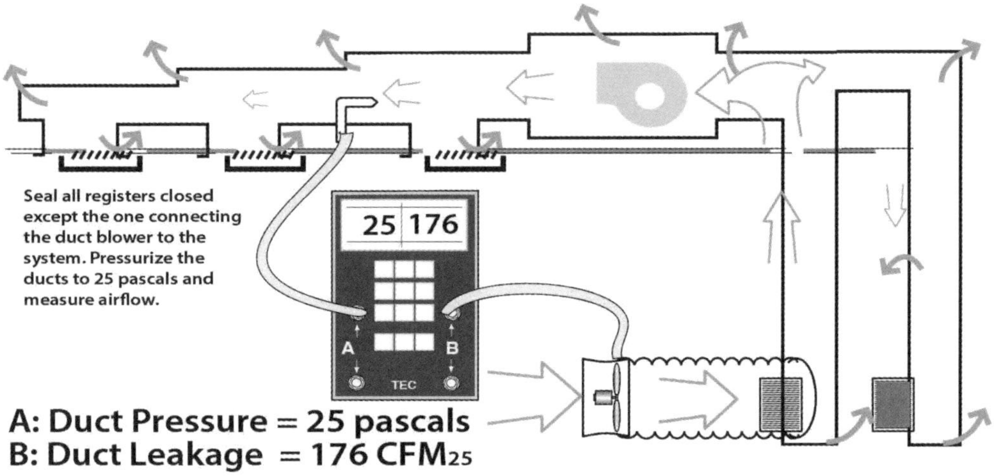

- ✓ Seal all registers, diffusers, and grilles.
- ✓ Remove the air filter(s) temporarily.
- ✓ Turn on the duct blower and pressurize the ducts to 25 or 50 pascals.
- ✓ Record duct-blower airflow.
- ✓ While the ducts are pressurized, start at the air handler and move outward feeling for leaks in the air handler, main ducts, and branches.
- ✓ After testing and associated air-sealing are complete, restore filter(s), remove seals from registers, and check air handler.

Sealing Duct Leaks

Seal all ducts located outside the thermal enclosure or in an intermediate zone such as a ventilated attic or crawl space. Leaks nearer to the air handler experience higher pressure and thus leak more air than leaks farther away. Lists of duct-leak locations follow in the upcoming subtopics.

Duct repair-and-sealing Methods

The first step in air sealing is to repair ducts and to make sure that the ducts are supported at regular intervals.

Duct repair, fastening, and support — Before you air seal ducts, make necessary repairs, using these guidelines.

- ✓ Attach flex duct to metal ducts or ductboard with a rigid metal or plastic take-offs, using two tensioned tie bands for each joint.
- ✓ Replace flex ducts' unreinforced 90-degree corners with smooth-flow metal corner fittings.
- ✓ Fasten round metal ducts to round ducts or to fittings with a minimum of three equally spaced sheet metal screws or rivets.
- ✓ Support flex ducts, duct-board ducts, and plenums with at least 1-$^{1}/_{2}$-inch (37 mm) wide material, installed every 4 feet (1.25 m) or less. Don't pinch flex ducts.
- ✓ Support metal ducts with $^{1}/_{2}$-inch-wide (12 mm) or greater metal straps, galvanized wire, or metal rods every 10 feet (3.0 m) or less.

Duct-sealing methods — Duct sealers install duct mastic and fiberglass mesh to seal duct leaks. When they need reinforcement or temporary closure for large leaks, the duct sealers use tape or sheet metal. Consider these standards.

- ✓ Remove any substance that would hinder sealant adhesion (tape, oil, dirt), using effective cleaning methods.
- ✓ Bridge seams, cracks, joints, and small holes, greater than ¼ inch (6 mm), with sheet metal or tape. Then cover the metal or tape completely with mastic reinforced by mesh.
- ✓ Seal seams, cracks, joints, and holes, less than ¼ inch (6 mm) using mastic with fiberglass-mesh reinforcement.
- ✓ Overlap the mastic and mesh at least one inch (25 mm) beyond the seams, repair tape, and metal reinforced areas of the ducts.

Sealing Return Ducts

Return leaks are important for combustion safety and efficiency, because these leaks can backdraft open-combustion furnaces with the pressure they create. Use the following techniques to seal return ducts.

- ✓ First, seal all return leaks within the combustion zone to prevent this leakage from depressurizing the combustion zone and causing combustion-appliance backdrafting.
- ✓ Seal all return ducts in crawl spaces, garages, and attics to prevent the ducts from sucking air pollution in from these spaces.

- ✓ Check filter slots for airtightness. Equip the filter compartment with a tight-fitting, durable, user-friendly filter-slot cover to allow easy filter changing.
- ✓ Seal the joint between the air handler and return plenum with a removable sealant such as aluminum-foil tape.

Figure 6-42: *Branch Ducts and Take-offs*

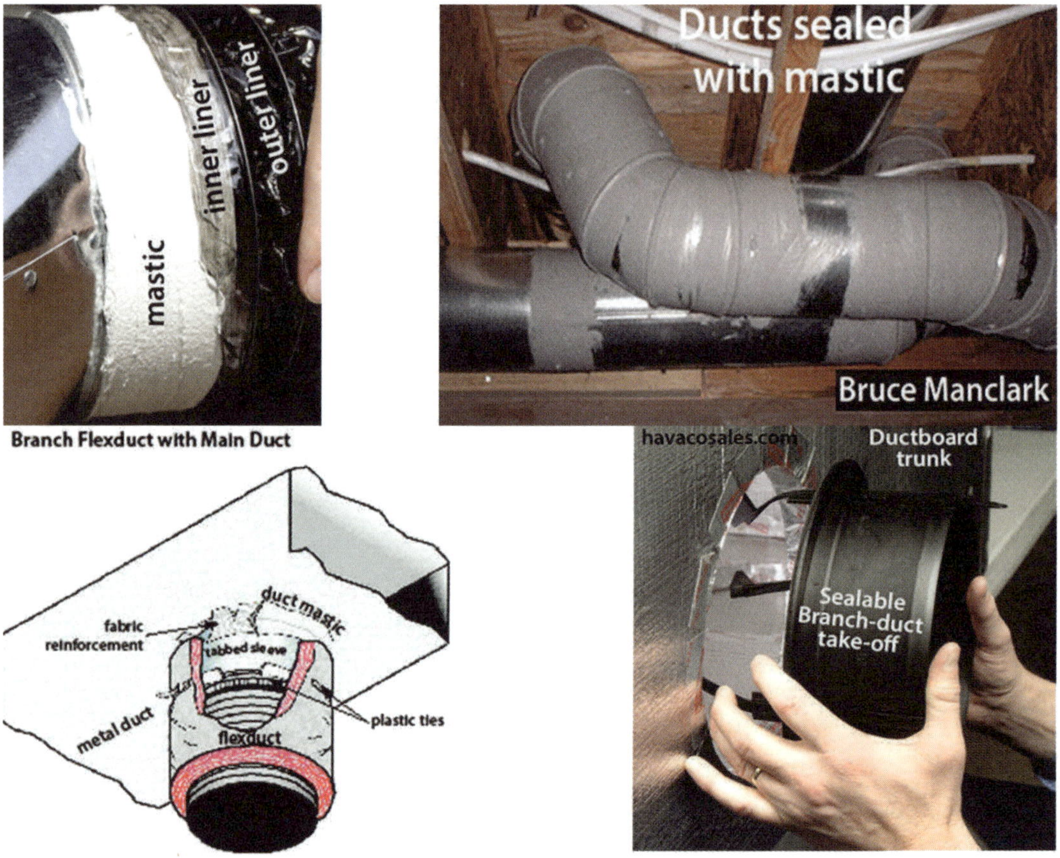

Panned-or-cavity return ducts **—** When installers use a floor, wall, or ceiling cavity for a return duct, that duct is usually very leaky. We call these ducts *panned returns* or *cavity returns*.

- ✓ Remove the panning to seal cavities containing joints in building materials.
- ✓ Use mastic to seal all cracks and gaps inside the return duct and register.
- ✓ Seal leaky joints between building materials that surround cavity return ducts, like panned floor cavities and air-handler return platforms.
- ✓ Carefully examine and seal leaks at transitions between panned floor joists and metal trunks that change the direction of the return air. You may need a mirror or camera to find the biggest return duct leaks in these tight areas.

Sealing Supply Ducts

Inspect these places in the supply ducts and seal them as needed.

- ✓ *Plenum joint at air handler:* Installers might have failed to seal these joints because of a lack of space. Seal these plenum connections thoroughly, even if you must cut an access hole in the plenum. Use silicone caulking or foil tape instead of mastic near the air handler for future access — air-handler replacement, for example.
- ✓ *Joints at branch takeoffs:* Seal these important joints with a thick layer of mastic. Fiberglas mesh tape should reinforce the mastic seal at gaps.
- ✓ *Tabbed sleeves:* Attach the sleeve to the main duct with 3-to-5 screws and apply mastic plentifully. Or better, remove the tabbed sleeve and replace it with a manufactured take-off that creates a tighter joint.

Figure 6-43: *Plenum Air Leakage and Air Sealing*

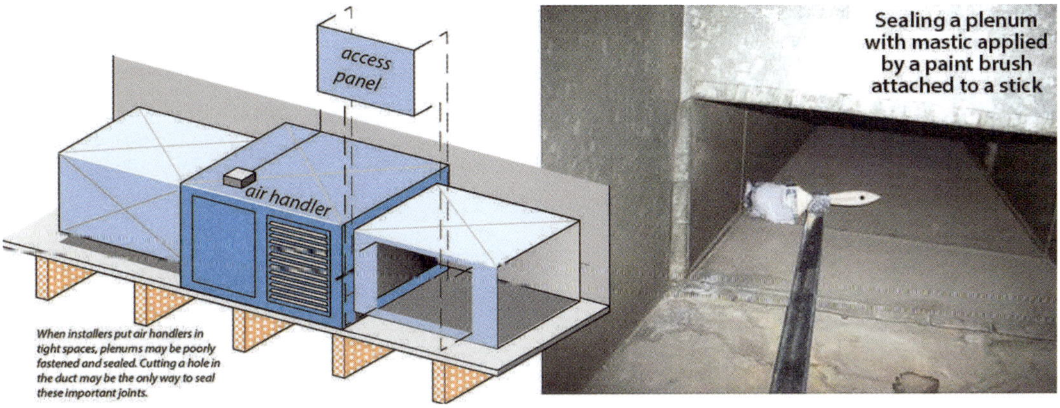

- ✓ *Flexduct-to-metal joints:* Apply a 2-inch band (50 mm) of mastic to the end of the metal connector. Attach the flexduct's inner liner with a tie band, and tighten it with a tie-band tensioner. Attach the insulation and outer liner with another tie band.
- ✓ *Damaged flex duct:* Replace flex duct when it is punctured, deteriorated, or otherwise damaged.
- ✓ *Deteriorating ductboard facing:* Replace ductboard, preferably with metal ducts, when the ductboard facing deteriorates because this deterioration leads to air leakage and dust pollution.

Figure 6-44: *Duct Sealing at Plenums and Trunks*

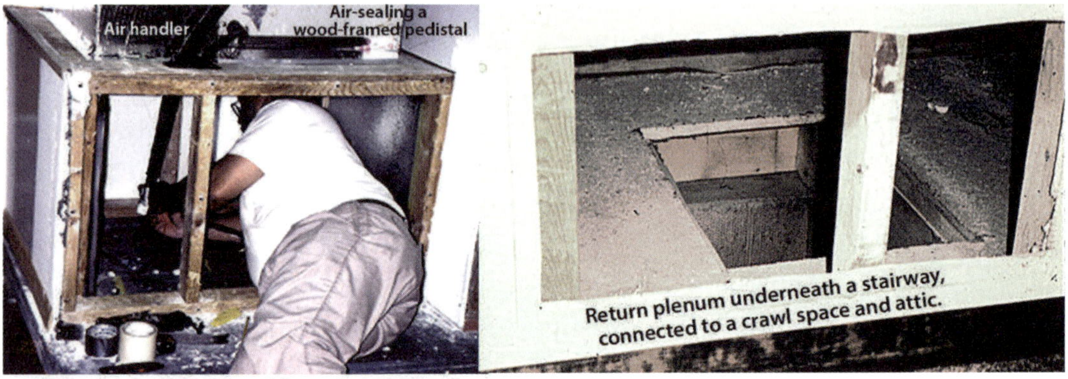

The installers built a wood-frame box to serve as a return plenum. That box was very leaky. This worker is sealing the leaks.

This return plenum returns outdoor air to the air handler from the attic and basement.

- ✓ Consider closing supply and return registers in unoccupied basements or crawl spaces.
- ✓ Seal penetrations made by wires or pipes traveling through ducts.
- ✓ Seal the joint between the boot and the ceiling, wall, or floor between conditioned and unconditioned areas.

Figure 6-45: *Sealing Register Boots*

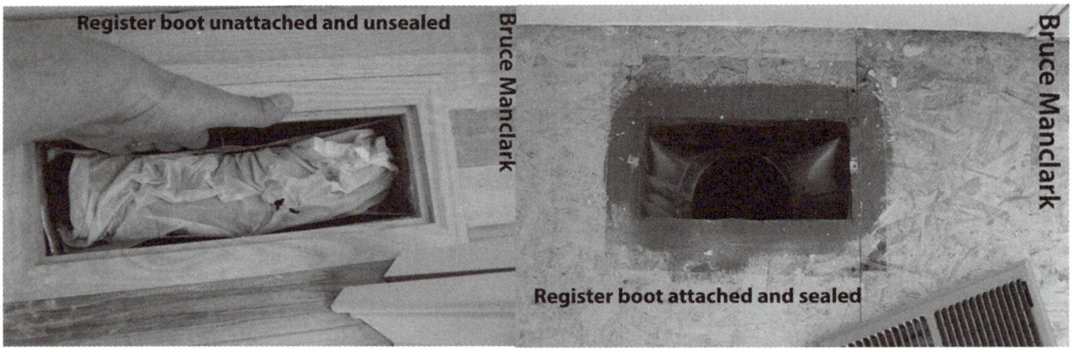

Materials for Duct Sealing

Duct mastic is the best duct-sealing material because of its superior adhesion and durability, compared to tape. Apply mastic at least $^{1}/_{16}$-inch thick (1.5 mm), and use reinforcing mesh for all joints wider than $^{1}/_{8}$-inch (3 mm) or joints that may move. Install screws to prevent joint movement or separation.

Cloth duct tape isn’t a good material for duct sealing because its adhesive often fails. Aluminum foil tape may adhere sufficiency to clean duct surfaces. Consider covering tape with mastic to prevent tape’s adhesive from drying out and failing.

Figure 6-46: *Insulated Ducts*

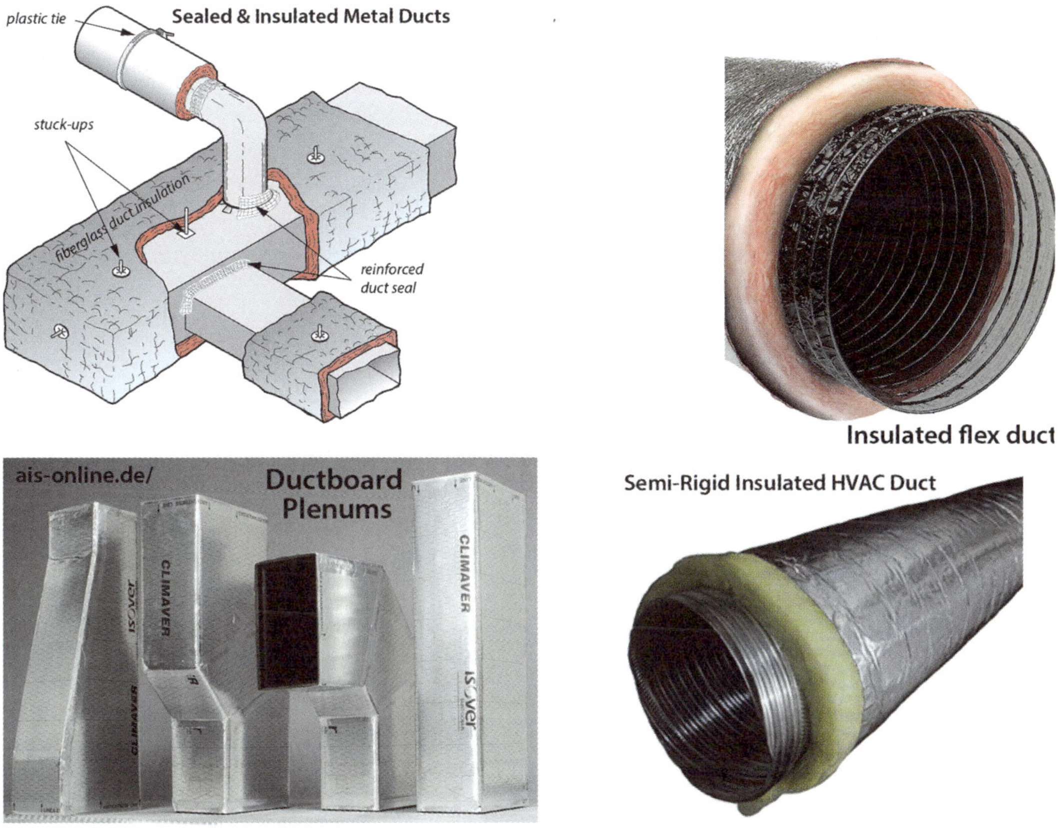

Duct Insulation

Insulate supply ducts, located in unconditioned areas outside the thermal enclosure, such as unconditioned basements, crawl spaces, attics, and attached garages. Use vinyl-faced or foil-faced duct insulation or closed-cell spray foam. For new ducts, consider factory-insulated ducts. Don't insulate ducts that run through conditioned spaces, unless they cause overheating in winter or condensation in summer.

Duct insulation materials and methods — Fiberglass duct and plastic bubble wrap insulation are common insulation materials for ducts.

Fiberglass duct board, insulated flexduct, and rigid or semi-rigid insulated ducts have built-in insulation. If you must replace ducts, these factory-insulated ducts save installation time, especially in tight spaces.

Installers sometimes insulate metal ducts on the inside for thermal resistance and sound insulation. This interior insulation is called *duct liner* or *liner board*. Flexduct, duct board, duct liner, and liner board may deteriorate with age and exposure to heat, moisture, and moving air. Deteriorating liner board or duct liner can reduce airflow and shed dust if it falls apart inside the duct.

Foil-faced or vinyl-faced fiberglass or plastic-bubble-wrap duct insulation fastens to sheet-metal ducts by various fasteners. To fasten or seal flexible duct insulation consider these suggestions.

- Apply twine or wire wrapped around the insulation and duct, then anchored to nearby structure.
- Install mechanical fasteners such as *stick pins* with self-locking washers, staples, or plastic tie bands.

Figure 6-47: *Insulation for Outdoor Ducts*

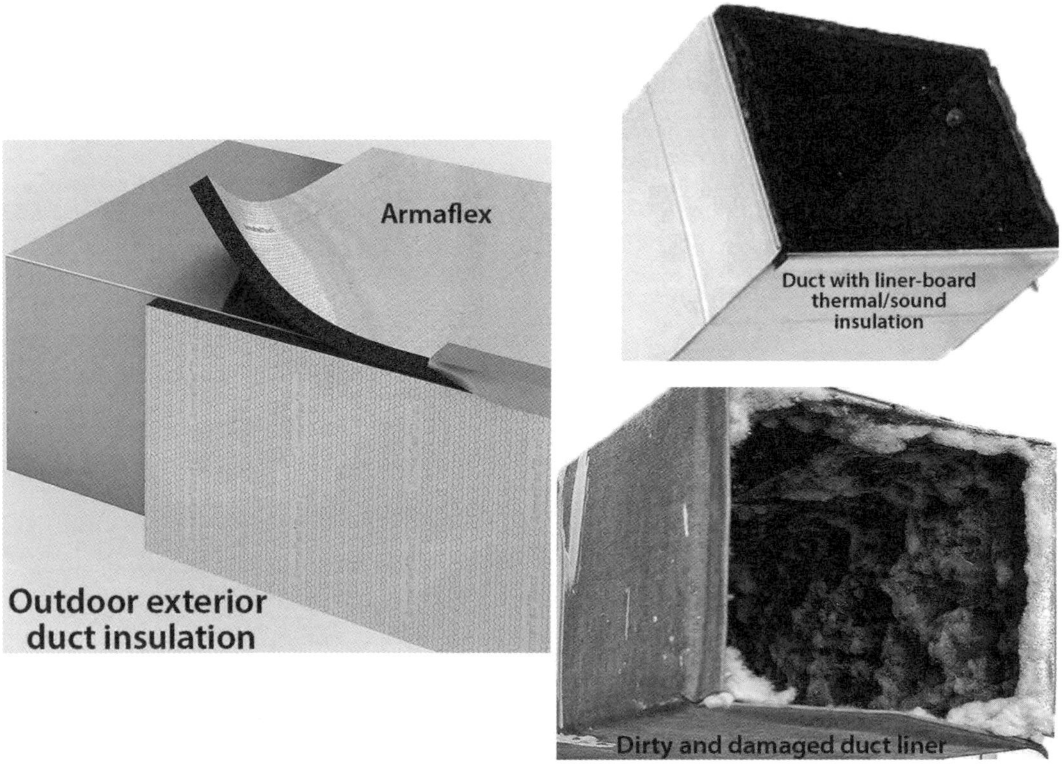

- Cover the insulation's joints with high-quality aluminum-foil tape or other recommended tape to seal gaps. Consider reinforcing the tape with duct mastic for extra durability.
- For outdoor external duct insulation, choose insulation with a damage-resistant covering. Or install a separate damage-resistant material as a covering to the insulation you install.

Duct-insulation caution — Tape can be effective for covering joints in the insulation to prevent air convection. However, tape fails if you expect it to resist the force of the insulation's compression or weight. Outward-clinch staples, plastic tie bands, plastic twine, and duct mastic help hold the insulation facing and tape together.

Burying ducts in attic insulation is common in some buildings and it reduces energy losses from those ducts. However, condensation on ducts in humid climates is common during the air-conditioning season, so don't allow blown cellulose to touch metal ducts to avoid corrosion from cellulose's Borate fire retardant.

Figure 6-48: *MERV-16 Air Filter with Manufactured, Airtight Filter Box*

Air Filtration

Manufacturers equip air handlers with air filters at the factory or recommend that installers install air filters in air handlers or return ducts. The purpose of these filters is to protect the blower and heat exchangers from fouling by dust, aerosols, and microbial slime. These contaminants can make the job of cleaning difficult or even impossible for air-conditioner coils. Contaminants also affect respiratory health, and filters remove respirable particles.

Protecting HVAC Equipment from Fouling

Dust, aerosols, and condensate combine to foul many evaporator coils within 5 to 10 years of operation from the time of installation. The causes of this fouling include the following.

1. Indoor concentrations of dust and aerosols
2. The amount of airflow leaking around the filter

3. The high MERV rating of the filter and the space around it may cause air to bypass the filter

Figure 6-49: *Filter Effectiveness = MERV Rating + Filter Airtightness*

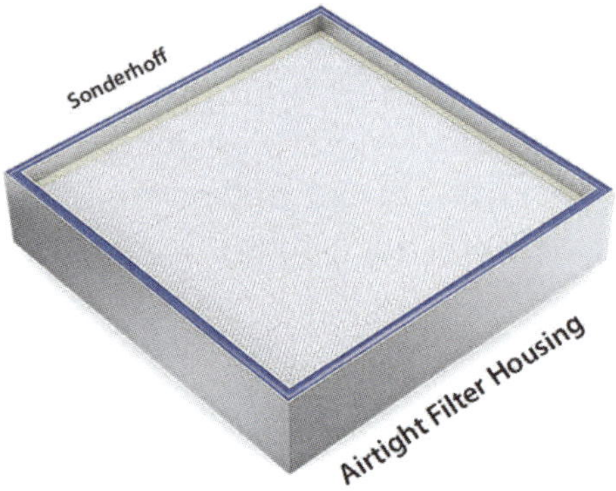

Protecting Lungs from Dust

Another function for air filters is to remove particles from indoor air for the health of residents. HVAC filters remove large-particle dust to protect heat exchangers from dust and aerosols. More efficient air filters reduce small-particle pollution in buildings, which promotes respiratory health. The best places for efficient filters are in forced-air HVAC heating-and-cooling systems, balanced ventilation systems, and ventilating dehumidifiers. Room air cleaners can remove particles effectively if air circulates among rooms.

This health benefit of good filters is greatest when the outdoor environment contains a lot of particles that leak indoors, such as in large cities, rural areas where wind and agriculture create airborne particulates, or regions with seasonal smoke during forest fires.

Figure 6-50: *Air-filter Leaks Cause Dirty Blowers and Dirty Coils*

Air-filter Ratings

The most common filter-rating method is the Minimum Efficiency Reporting Value (MERV). A high MERV rating means that the filter removes more particles and a larger fraction of smaller particles, compared to low MERV filters.

This MERV-rating system divides particles into three size categories: 0.3–1, 1–3, and 3–10 µm (microns). The two smallest dust categories are the most important for human health. Small respirable particles called PM2.5 (less than 2.5 microns) can more easily imbed themselves deep in lung tissue compared to larger particles.

The MERV ratings of available HVAC filters vary from MERV 3 to MERV 16. A MERV 3 filter captures large particles — clothing fibers, pollen, and dust mites, but few smaller respirable particles. A MERV16 filter captures more than 95% of all three particle sizes, including bacteria along with particles and aerosols from tobacco-smoke. The minimum MERV rating to remove 50% of the respirable 1–3 µm particle-size range is MERV 10.

The need for high MERV filters depends on the particle concentration in a building. Filtering air with a long-running central air handler or a portable air cleaner may or may not be effective, so filtration can waste energy because of unnecessary operation.

Table 6-3: *USA-ASHRAE MERV: Application and Static Pressure*

MERV	Particle Size and Dust-Type Removed	Application	Static-Pressure Drop
9–16	1.0 to 3.0 microns. Fine dust, bacteria, auto emissions, nebulizer drops, some tobacco smoke	Superior commercial and residential HVAC systems & sensitive people	63-125 pa 0.25-0.50 IWC
5–8	3.0 to 10.0 microns. Mold spores, pollen, hair spray, cement dust, oil smoke, powdered milk	Typical commercial and residential HVAC systems	38-63 pa 0.15-0.25 IWC
3–4	Larger than 10.0 microns. Most pollen, sanding dust, paint spray, textile fibers, animal hair	Low-dust environments	25-38 pa 0.10-0.15 IWC

Air-filter depth — Considering the same filter material and manufacturer, deeper filters are more effective than shallower ones.

- Deeper filters have more surface area to collect and hold particulates.
- They have less measured static-pressure drop across them, compared to shallower filters with the same length, width, and filter material.

- They collect a greater percentage of particulates.
- A 4-inch filter (100 mm) probably remains effective more than four times as long, compared to a 1 inch filter (25 mm).
- Deeper filters may provide better resistance to filter leakage compared to shallower fixtures because they have more contact area with the sides of their enclosure than shallower filters.

Air-filter pressure-and-airflow effects — Air filters affect static pressure, airflow, and energy consumption of forced-air HVAC systems, including balanced ventilation systems and ventilating dehumidifiers. Before choosing the type of air filter, consider its MERV rating and the building's need to remove particles.

HVAC designers designed air handlers to accommodate low-MERV filters with a small pressure drop. Changing to higher-MERV filters can increase filter pressure drop and thus decrease the system airflow.

Insufficient airflow may cause blowers to fail prematurely as they struggle to overcome system pressures that are higher than the blowers' design capability. The furnaces may overheat and cycle on their high-limit switches. Air conditioners may suffer insufficient airflow and poor efficiency. Insufficient airflow in cooling systems also risk evaporator-coil icing and premature compressor failure.

Figure 6-51: *Air Purifiers: Commercial and Do-it-yourself*

Permanent-split-capacitor motors (PSC) move air in proportion to the system's static pressure. Brushless-permanent-magnet (BPM) motors, also called electronically commutated motors (ECM), move air at a constant airflow rate with variable energy consumption.

The volumetric airflow rate of a blower is directly proportional to the energy consumed by the blower motor. Thus, as the HVAC system's static pressure increases, a PSC motor uses less energy because it moves less air. In contrast, an ECM motor uses more energy to maintain a constant airflow when static pressure increases.

Reduce the resistance of high-MERV filters — To reduce the resistance of an high-MERV air filter, consider these possible solutions.

- ✓ Evaluate and compare air filters: replace or re-specify deeper filters for less static-pressure drop.
- ✓ Install filters in multiple return registers instead of only one filter at the air handler.
- ✓ Install a slanted filter-bracket assembly to accommodate a larger filter with more surface area and less static-pressure drop, compared to the existing filter.
- ✓ Install an enlarged filter compartment to allow a larger filter or a deeper pleated filter (≥ 50 mm, 2 inches or 100 mm, 4 inches) with more filter surface area.
- ✓ If filter changing is currently difficult, modify the filter installation to allow easier filter changing.
- ✓ Verify that the filter fits tightly into its filter compartment.

Filter Recommendations

The need for high-MERV filters depends on four factors.

1. The type and concentration of airborne particles
2. The size range of those particles
3. The sensitivity of the residents to the particles
4. Tolerance of the air handler to high static pressure

In a geographical region with little air pollution or dust, a high MERV air filter probably isn't necessary. However a high-MERV filter may provide important IAQ benefits in a region with smog, pollen, road dust, or seasonal forest fires.

Remember that air handler must be maintain a low enough static pressure to move sufficient air, considering the static-pressure drop of the high-MERV filter.

Figure 6-52: Air-filter Locations

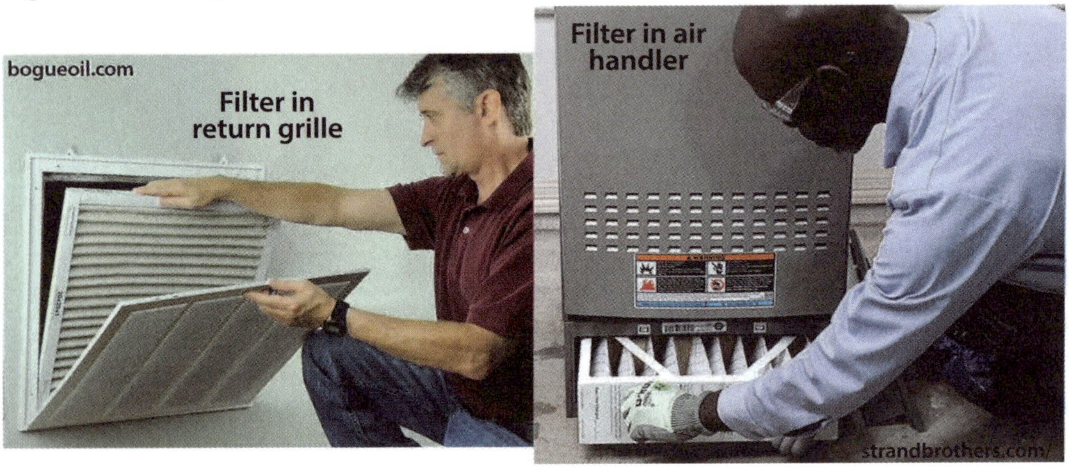

Changing air filters — Air filters prevent dirt from collecting on the blower and heat exchanger of forced-air heating, cooling, and ventilation systems. Dirt reduces HVAC airflow, especially if it builds up on the filter, the blower, and the heat exchangers.

The recommended frequency of filter changes depends on how much dirt is in indoor air and how long the air handler operates. Filters in some homes may last 6 months before they need replacement or cleaning. Other homes' filters may need cleaning or replacement every month.

Table 6-4: Forced-air-system Improvement: Sequence of Operations

Procedure	Evaluation/Measurement	Action
1. Find & fix obvious problems.	Observe system operation. Find enclosure problems, disconnected ducts, dirty components. Report possible HVAC settings problems.	Fix enclosure problems; fix obvious duct problems, then re-evaluate priorities & continue inspection and adjustments.
2. Improve airflow.	Measure supply and return static pressure. Measure filter and coil static if you suspect high resistance.	Decrease filter static; increase filter size; enlarge ducts. Specify cleaning, install new ducts, clean/repair registers.
3. Reduce duct-induced house pressures.	Measure house-pressure differences between rooms, central areas, & outdoors.	With excessive pressure, cut door bottoms, install door grilles & jumper ducts. Seal ducts to fix duct-leak imbalance.
4. Seal duct air leaks.	Use pressure pan or duct blower before and after duct sealing for feedback.	Seal as many duct air leaks as possible with mastic and webbing.
5. Install duct insulation.	Identify duct location, noise, and supply-versus-return priority.	Choose insulation's thermal resistance based on codes and climate.
6. Select a filter & eliminate bypass.	If blower or heat exchanger is dirty, the filter must be bypassing air.	Eliminate bypass and provide desired filter performance by choosing the best MERV.

Dehumidification

Air conditioners and dehumidifiers remove water vapor from indoor air. They refrigerate the air below its dew point to condense water vapor. The air conditioner (AC) should be the primary moisture remover whenever possible. However, when the AC doesn't activate because of a low outdoor temperature, the Rh may cause condensation and discomfort.

Many buildings have large sources of moisture, particularly from the ground. Plugged or damaged rain gutters and downspouts deposit water near the foundation. Air leaks bring humid air into the building. Residents shower and cook. Try to minimize these moisture sources to reduce latent cooling load, before purchasing a dehumidifier.

See "Controlling Water Vapor" on page 51.

In humid climates, the definition of comfort varies from person to person. Consider setting the thermostat for 78°F (26°C) or higher. Try for a relative humidity setting of 55% or less on the dehumidistat. The higher your thermostat setting and the higher your dehumidistat setting, the less electricity your dehumidifier consumes. However, for IAQ and building durability, keep the relative humidity below 60%.

See "Cooling Strategy for Humid Climates" on page 195.

Air Conditioners Dehumidify

If the AC dehumidifies enough, you don't need a dehumidifier. Air-conditioning systems may remove enough moisture to provide comfort and prevent condensation — or they may not.

Air conditioners remove water vapor in the process of cooling a building. However, AC efficiency and moisture-removal oppose one another. AC efficiency works best with warmer evaporators and higher airflow. AC moisture removal works best with colder evaporators and lower airflow.

Here are four conditions when air conditioners fail to dehumidify enough.

1. Oversized air conditioners don't run long enough to remove sufficient moisture.
2. Very efficient air conditioners operate with high airflows and warm evaporator coils. This mode of operation doesn't condense much water vapor.
3. In mild and humid weather, the thermostat doesn't activate the air conditioner because thermostats measure temperature, not Rh.
4. Airtight, well-insulated buildings with very small solar heat gains may need little or no AC. However persistently high indoor relative humidity remains a comfort and durability concern.

See "AC versus dehumidifier to control Rh" on page 199.

Energy-recovery ventilators (ERVs) and humidity — In relatively airtight buildings, ERVs help air conditioners maintain comfort by removing 25% to 50% of incoming water vapor, carried by the ventilation air. However, the ERV still brings humid air into the building. ERVs don't remove enough of the water vapor to act as dehumidifiers.

If you minimize ERV operation in hot humid weather, that helps the air conditioner or dehumidifier to maintain a comfortable Rh. If the ERV operates excessively, it can increase indoor relative humidity beyond the air conditioner's capacity to dry the indoor air for comfort and building durability.

See "Heat and Energy Recovery Ventilators" on page 130.

Dehumidification Strategies

A stand-alone dehumidifier condenses water vapor from the indoor air on its cold evaporator coil. The dehumidifier then heats the air after cooling and dehumidifying it. This *reheat* prevents the problem of over-cooling.

Reheat also reduces indoor relative humidity (Rh) by heating the cold dehumidified air. Heating the air further reduces its Rh. Heated air holds more water vapor than cooler air. Thus, the dehumidifier's reheated air is less saturated with water vapor, and its Rh is lower than after it flows through the cold evaporator coil.

See "Moisture's Behavior" on page 31.

In hot and very humid climates, this reheat may be an energy-and-comfort problem. For this reason, some dehumidifiers use outdoor condensers to reject the sensible heat produced by dehumidification.

Figure 6-53: *Dehumidifiers: Stand-alone / Portable*

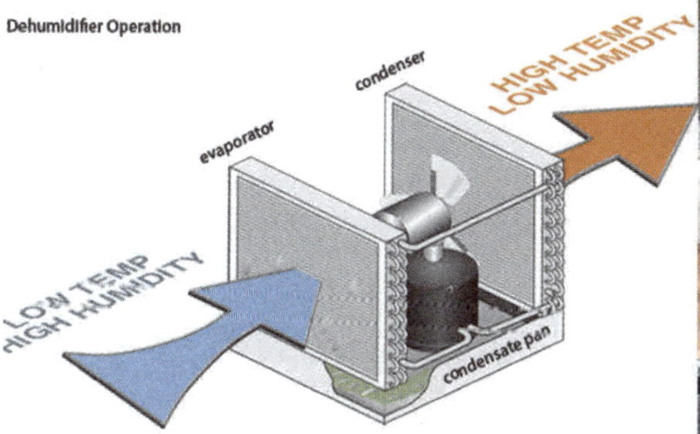

Stand-alone or Portable Dehumidifiers

The dehumidifier refrigerates air to remove moisture and then reheats that indoor air with the latent heat it collected by condensing the moisture. The energy factor (EF) expresses the dehumidifier's efficiency in liters per kWh (gallons per kWh). The range of EF values are between 0.5 and 2.5 liters per kWh or 0.13 and 0.66 gallons per kWh.

The dehumidifier functions as a heat-pump space heater with a COP of 1.5 to 2.0. If the building needs dehumidification and space heat at the same time, a stand-alone dehumidifier can save energy compared to resistance heating.

Whole-house Dehumidifiers

Whole-house dehumidifiers are specialty products that require careful research, installation, and commissioning. The three products we discuss here are for single-family buildings or small multifamily buildings.

1. Packaged, ducted whole-house dehumidifiers
2. Split-system ducted whole-house dehumidifiers
3. Ventilating ducted whole-house dehumidifiers

Whole-house dehumidifiers may connect to existing air handlers to distribute the dehumidified air, or they may use their own dedicated duct system.

These three types filter the incoming air to protect the heat exchangers and remove particles to improve IAQ.

Figure 6-54: *Wall-mount and Ceiling-mount Dehumidifiers*

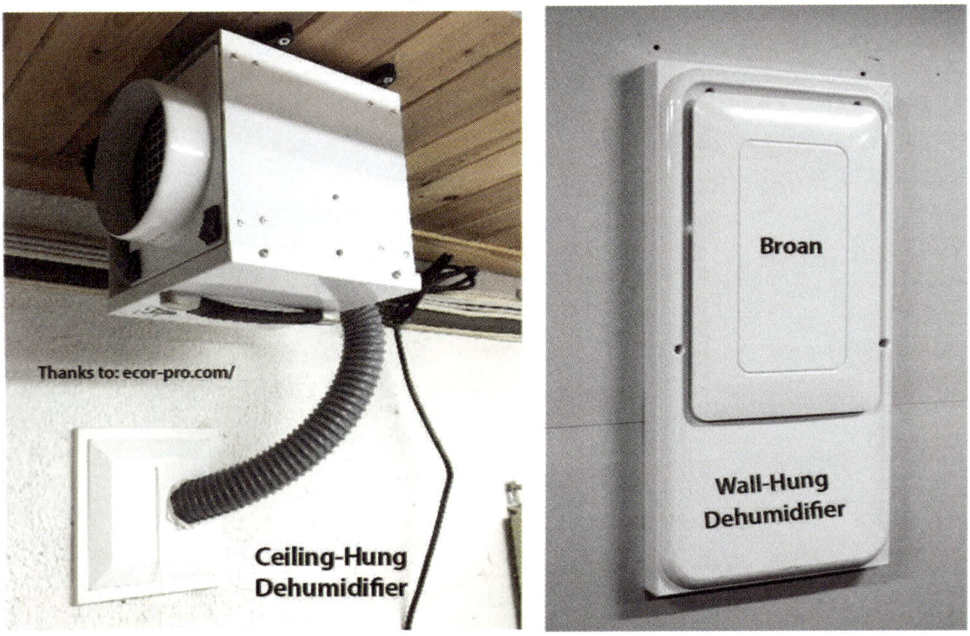

Dehumidification, reheat, and efficiency — Large HVAC systems in humid climates, like Hong Kong or Miami, refrigerate indoor air and then reheat that air to prevent over-cooling. The explanation of reheat is that the dew point (condensing temperature) is too cool to provide comfort (55°F or 16°C). Sometimes a chiller and a boiler run all summer side by side to chill a maintain a large building's comfort. The reheat air increases air temperature to provide both comfort and additional Rh reduction.

The AC and the dehumidifier oppose one another, because the dehumidifier adds sensible heat to the space and the AC removes the sensible heat. Avoid using dehumidifiers at the same time as air conditioners. However, dehumidifiers and air conditioners may cooperate to provide acceptable relative humidity and comfort in airtight buildings with small solar heat gains. The combination of the two systems may sometimes minimize energy use.

Figure 6-55: Ventilating-dehumidifier Design and Installation

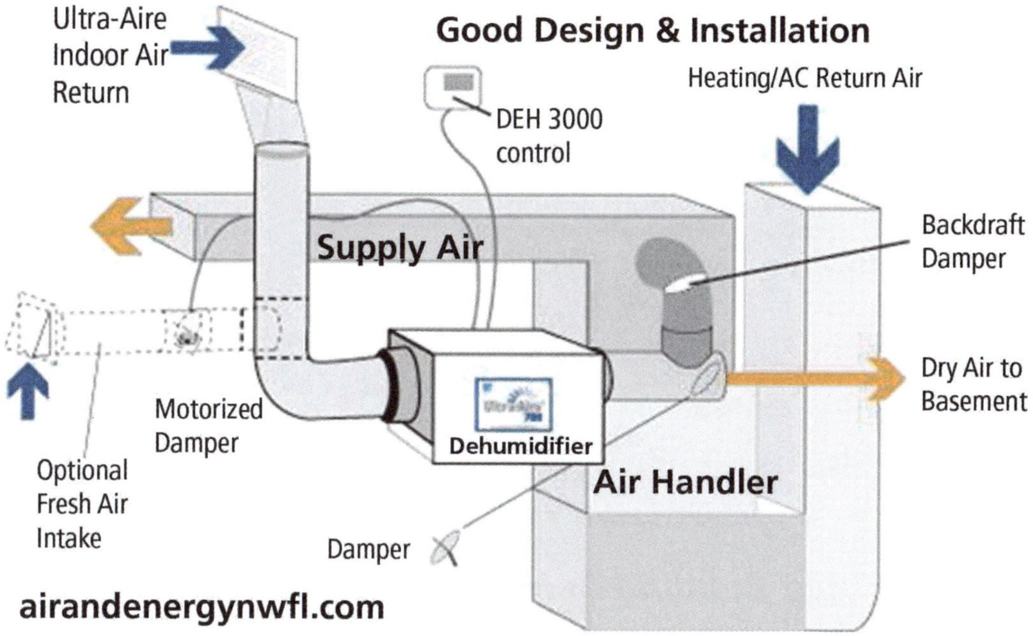

Disadvantages of dehumidifier supply connection to the AC return — If a contractor installs the dehumidifier supply duct into the AC return air, three disadvantages occur.

1. The warmer dryer air entering the AC evaporator denies that evaporator the opportunity to maximize its moisture removal.
2. The warmer air entering the AC evaporator increases the evaporator temperature so the AC creates less condensation.
3. The dehumidifier's warmer dryer, air entering the AC evaporator can re-vaporize the condensate on the evaporator and also condensate in the evaporator pan.

See "AC versus dehumidifier to control Rh" on page 199.

Figure 6-56: *Split-system-and-desiccant, whole-building Dehumidifiers*

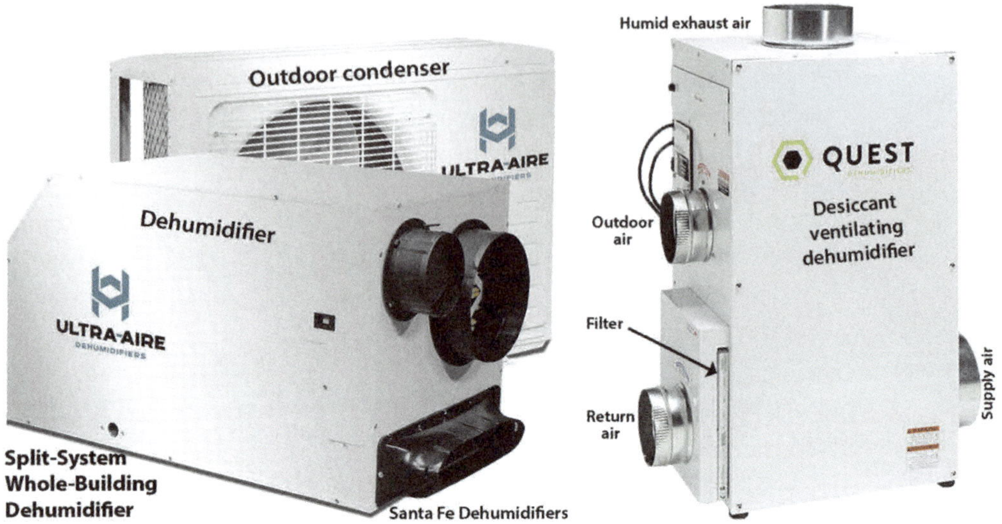

Types of Whole-Building Dehumidifiers

Before buying a whole-building dehumidifier, make sure that the existing AC equipment is optimized to remove both sensible and latent heat. As with any HVAC upgrade, improve the building enclosure with cost-effective retrofits. If this HVAC-and-enclosure optimization is successful, you may not need a dehumidifier.

See "Moisture Management" on page 47.

Here we discuss four whole-building dehumidifiers.

1. Ducted whole-building dehumidifiers
2. Split-system whole-building dehumidifiers
3. Ventilating whole-building dehumidifiers
4. Desiccant whole-building dehumidifiers

Ducted whole-building dehumidifiers — This type of dehumidifier works well in humid climates where the building needs more heating than cooling. In addition to dehumidifying, this most-common type of whole-house dehumidifier heats the air flowing through it. It's COP as a heater is between 1.5 and 2.0 and the heating capacity of available products is between 3000 and 8500 BTU/hr (1 to 2.5 kW).

Split-system whole-building dehumidifiers — This type of dehumidifier works well in a humid climate where a building needs more cooling than heating. The evaporator condenses water vapor on its cold surface. The split-system dehumidifier has an outdoor condenser to reject the heat collected by its evaporator. Thus, this dehumidifier functions as an air conditioner.

Its air-conditioning capacity is small, but may be enough for small airtight homes, and homes in mild but humid cooling climates. The air-conditioning capacity of available products is between 3000 and 5000 BTU/hr (0.9 to 1.5 kW).

Ventilating whole-building dehumidifiers — The ventilating dehumidifier combines an outdoor air inlet to one of the three whole-house dehumidifier types described above. This dehumidifier brings fresh air into a home like a supply only ventilation system. Make-up air enters the building through air leaks in the building enclosure.

Desiccant whole-building dehumidifiers — Desiccant dehumidifiers dehumidify air with a desiccant rather than an evaporator coil. A desiccant is a liquid that removes water vapor from the air passing through it, because it is so very dry. After the desiccant absorbs the water vapor, it needs heat to dry and to again be capable of removing more water vapor. A desiccant dehumidifier's efficiency is only about 25% compared to a refrigerated dehumidifier. However, desiccant dehumidifier's maintain their efficiency in extreme temperatures, while refrigerated dehumidifiers' efficiency drops quickly above 80°F or 26°C.

Commissioning Dehumidifiers

When a building owner hires a HVAC contractor or engineer to measure performance and make adjustments and repairs, this service is *commissioning*. If a dehumidifier becomes part of an existing HVAC system, re-commission that existing system when attaching a ventilator or dehumidifier.

A commissioning professional may do the following procedures to commission a whole-building ventilator or dehumidifier.

1. Examine ducts, filters, blower, and heat exchangers for dirt and other obstructions.
2. Verify correct control and system operation.
3. Find and seal duct leaks.
4. Verify adequate airflow by measurement.
5. Examine electric terminals — clean and tighten connections and apply non-conductive coating, if necessary.
6. Verify correct thermostat settings and dehumidistat operation.

See "Professional Service and Commissioning" on page 208.

Chapter Review

Table 6-5: *Decisions about improvements discussed in this chapter*

Procedure	Evaluation/Measurement	Action
1. Find & fix obvious problems with the building enclosure.	Inspect the building for air leakage, insufficient insulation, and inefficient windows.	Make improvements to the building enclosure before making changes in air conditioning or dehumidification systems.
2. Shade the building.	Evaluate the building's orientation with reference to the sun's travel through the sky.	Consider a cool roof, window treatments, and tree planting to minimize solar gain.
3.Evaluate alternatives to air conditioning.	Evaluate the potential to use fans and evaporative coolers to cool the building, instead of or in addition to air conditioners.	Research the type and size of these alternative cooling devices and seek professional guidance if necessary.
4. Evaluate existing air-conditioning equipment for opportunities to increase efficiency.	Hire an air conditioning expert technician to commission the existing air conditioning system.	Made necessary changes to airflow, duct leakage, and refrigerant charge. If appropriate, buy and install a new energy-efficient air conditioner.
5. Evaluate the building for moisture problems and excessive relative humidity.	Inspect for standing water near or under the building. Inspect roofs, gutters, downspouts, sprinkling systems, and ground drainage.	Make repairs and improvements as necessary to keep the ground near the building dry. Consider a dehumidifier to control indoor relative humidity.
6. Consider installing a solar photovoltaic system to provide electricity for cooling systems.	Evaluate your electricity bills and ask an engineer or PV expert to calculate savings and return on investment.	Install a PV system if the return on investment meets your standards. Do your research on installers before making a decision.

7 Photovoltaic Solar Systems

Photovoltaic (PV) devices contain semi-conducting materials that convert sunlight into electrical energy. A single PV device is a *cell*. Cells connect together in chains to form larger units, called *panels* or modules. We use the term: *panel* in this chapter. A group of PV panels is called an *array*.

Solar electricity from PV panels powers the lights, air conditioner, appliances, and electronics, which normally consume electricity from the grid through your electricity provider.

Homes, multifamily buildings, and commercial buildings have used photovoltaic (PV) systems for decades, but recently photovoltaic equipment has become much more affordable than in the past. So demand for PV is booming.

Figure 7-1: *Solar Cells, Panels, and Arrays*

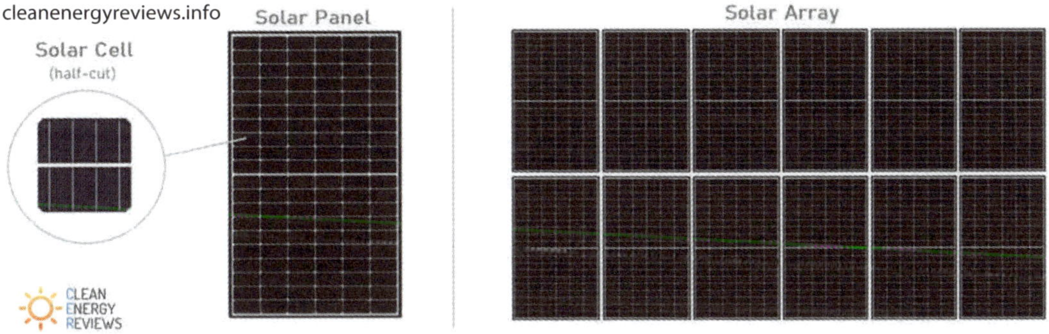

Why Solar PV?

If the building uses a lot of electricity for cooling even after building-enclosure improvements, solar photovoltaic systems can be a great investment. The hottest climates are often the sunniest ones. Electric cooling devices such as air conditioners, evaporative coolers, and fans can use solar electricity instead of expensive electricity from the grid. PV systems also shade roofs for an extra benefit.

How to Begin

PV solar systems are the most complex products, discussed in this book. To ensure your satisfaction, you must adopt one of these three approaches.

1. Work with the contractor that you know and trust already, based on experience or reviews from people you know.

2. Do the necessary research to understand what type of system you want, the system's benefits, what permits are required, how you will finance the system, and how you will choose an installer.
3. Higher an independent consultant to help you make decisions about a PV solar system. Verify that the independent consultant isn't also selling solar products.

For information about the intensity of soler radiation near you, see "Solar-Irradiation Map" on page 293.

Societal Benefits of PV

Photovoltaic systems provide two important societal benefits: renewable energy and distributed generation. Renewable energy reduces pollution and helps to minimize climate change.

Distributed generation systems provide electricity closer to the end user, and so incur fewer transmission losses than centralized power plants. Distributed systems posses less risk to system-wide outages, compared to central systems.

Figure 7-2: *A Simple Solar PV System*

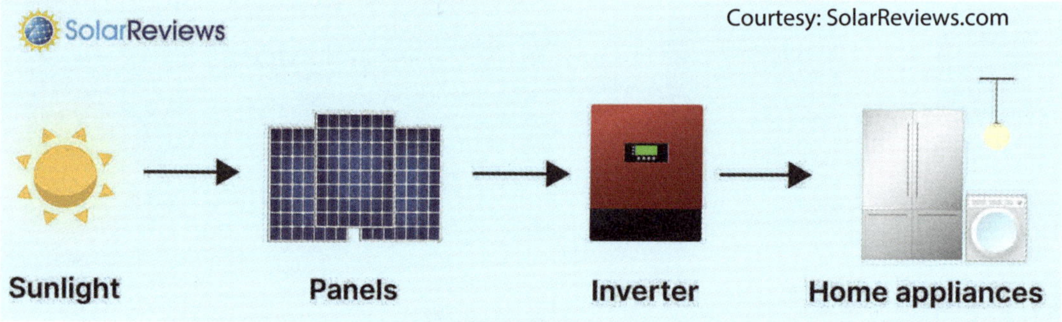

Types of PV Systems

PV systems collect solar radiation and convert it into direct-current (DC) electricity. The DC electricity flows through an inverter that converts DC to AC electricity for use in residential buildings. A majority of buildings throughout the world use AC powered electrical devices. Some remote buildings use electrical devices that use DC power, but this isn't common.

Most PV systems worldwide are either connected to the electrical grid for backup power, or they use batteries to store power for times when the sun isn't shining. In this chapter, we call the two systems: *grid-tied PV systems* and *off-grid PV systems.*

Photovoltaic Power Components

Solar photovoltaic systems include four major components.

1. Solar panels to collect sunlight and convert it to DC electricity
2. Inverters to convert DC electricity to AC electricity
3. Mounting racks to secure the solar panels
4. Backup systems to provide power when the sun isn't shining

PV Solar Panels

Solar cells are the smallest building blocks of photovoltaic systems. Cells connect to form panels. The silicon-based semiconductors in the cells and panels use the of sunlight's reaction with the semi-conductors to create an electrical current.

Groups of solar panels connect electrically and mount to racks. Polymer or glass sheet usually covers panels. These modular PV components are sold by manufacturers or distributors. Typical panels, produced for residential and light commercial installations, produce 100 to 400 watts of power.

Solar arrays are the groups of panels that you see on roofs of buildings or on ground-mounted racks. Some systems use PV roof shingles that don't need covers or racks.

Figure 7-3: Types of PV Panels

SolarReviews Courtesy: SolarReviews.com

Types of Solar Panels — Most of the PV solar panels on the market today for residential solar PV systems fit into three types: mono-crystalline, polycrystalline, and thin film solar panels.

- Mono-crystalline solar panels are the most efficient and popular type, but they are also the most expensive type.
- Polycrystalline solar panels are cheaper than mono-crystalline panels, however, they are less efficient.
- Thin film solar panels are the least expensive, but have the lowest efficiency rating and require much more space to produce similar power as the other two types. PV shingles use thin films.

Each type of solar panels has different characteristics. Some panels are better than others for particular PV installations.

DC-to-AC Inverters

Almost all the lights, appliances, and equipment in modern buildings use the alternating current (AC) that the electrical grid supplies.

- Inverters are electronic devices that convert DC power into AC power. If you intend to use your PV electricity to power your building's existing equipment, your system will require an inverter. An inverter is also necessary if you plan to connect your PV system to the utility grid. The majority of PV systems installed today utilize inverters.

- Inverters are specified by their capacity in kilowatts. This capacity should meet or exceed the peak output of the PV array, and also the output voltage of the array (usually 120–240 volts AC). Inverters convert input voltages of up to 600 volts from the PV array down to the 120–240 volts output needed by most equipment.
- Most inverters also include relays and circuit breakers that manage the task of safely connecting your solar array to your building's existing electrical system and to the utility grid. Inverters usually include a DC ground-fault circuit interrupter (GFCI) or arc-fault circuit interrupter (AFCI) that disconnects the PV array if a sensor detects dangerous electric currents.

Types of inverters — Failed inverters are the most common problem in most PV systems. Longevity and reliability are important issue with inverters, so read the reviews and specifications on the inverters you consider.

The two most common inverter systems are these.

1. The string inverter is a standalone unit that changes DC current from the panels to AC current to serve the building's loads. Electricians install them near the main service panel and electric meter.
2. Microinverters attached to each collector. They do the same DC-to-AC conversion as the string inverters, and they don't need a string inverter.

Figure 7-4: *PV System with String Inverter*

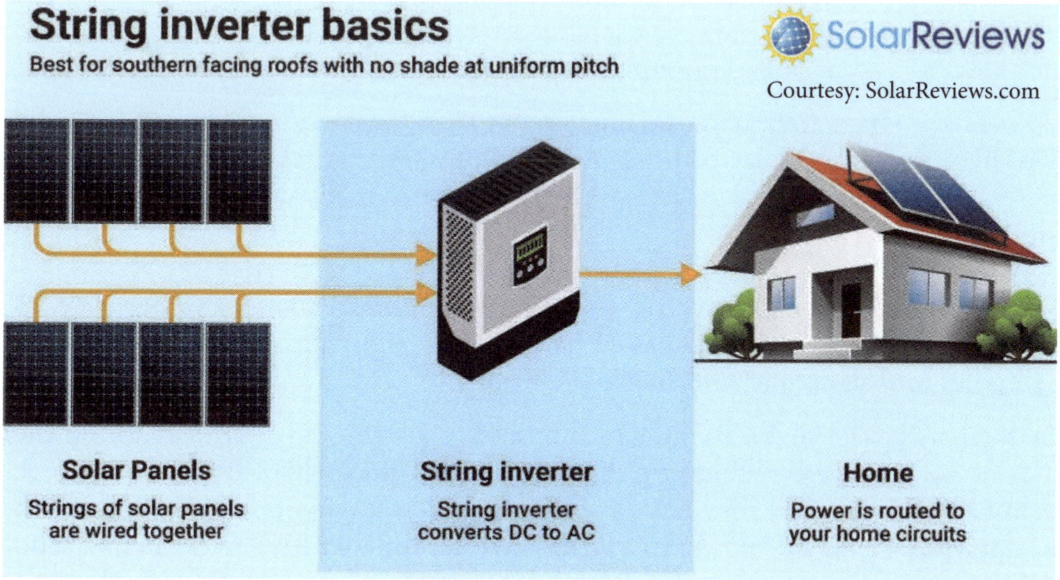

String inverters — The advantages of string inverters are there low cost, easy troubleshooting, and simple installation and replacement. Their disadvantages are low efficiency in partial shade, shorter lifespan, difficulty for expanding PV systems, and the lack of ability to monitor the panels.

Figure 7-5: *Solar PV Inverters*

solarrun.com.au/

Microinverters — The biggest advantages of microinverters are there superior efficiency in partially shaded solar arrays and other challenging installation conditions. Also microinverters shut down more rapidly than string inverters in emergency situations such as fires and other emergencies. Microinverters also have a longer lifespan than string inverters. They can provide collector-level monitoring, and they facilitate easy PV-system expansion by the ease of connecting new panels with microinverters.

The most important disadvantage of microinverters is that they attached to each PV panel. They may be numerous and difficult to identify as failed. Replacing a microinverter is usually more difficult than replacing a string inverter because microinverters are installed on roofs and string inverters are at ground level.

However, the string inverter costs much more than the micro inverter and may be more likely to fail. A solar technician must climb onto a roof to replace a failed microinverter. This may involve working with the racking system, which is time consuming.

DC-Optimizers for PV Systems — *DC-power optimizers* are another way to solve the problem of partially shaded panels. They allow for maximum electricity generation for solar panels on challenging roofs.

The total electricity that a PV system can send to a string inverter depends on the efficiency of the worst-performing collector. DC optimizers attached to each panel, like microinverters, and they have a similar panel-averaging effect. However each DC optimizer sends DC power to a string inverter unlike microinverters that send AC power directly to the home's electrical system. Optimizers are another option for getting the most out of panels that produce different amounts of electricity.

***Figure 7-6:** String Inverter / Microinverter — Comparison*

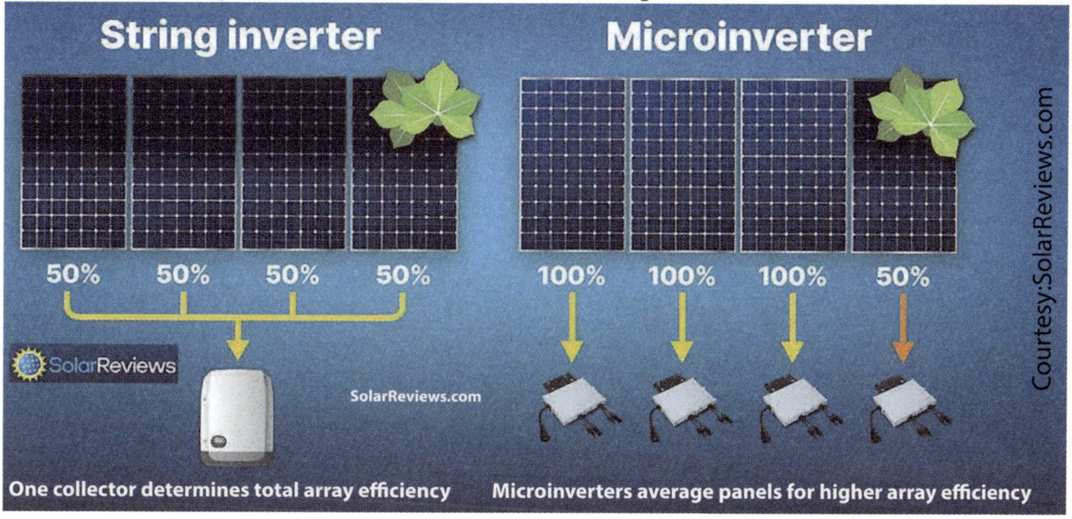

DC power optimizers are, as this book is published, new to the market so we don't have as much reliability data as we do for microinverters. Experts debate about which of the two solutions is better. Further discussion of DC optimizers is beyond the scope of this book.

***Figure 7-7:** DC Power Optimizers*

Mounting Racks

Your choice in mounting racks will determine how closely your array is aligned with the sun throughout the day and across the seasons. Your array will produce the most power if it faces the sun at a right angle at all times. This is difficult in practice since the sun crosses the sky from east to west each day, and the sun's path varies in its angle throughout days and years. **The angle of collectors is less important now than in the past because PV-panel performance has improved.**

Azimuth describes the sun's location on a horizontal plane that runs east and west. The sun's azimuth changes from morning to night as it rises in the east, crosses solar south, and sets in the west.

Figure 7-8: *Mounting Racks*

Altitude describes the sun's location in a vertical plane above the horizon. The sun's altitude above the horizon varies throughout the day as it rises from the horizon, passes high in the sky, and descends again to the horizon. Its altitude also varies throughout the year. In the northern hemisphere, you'll see it at its highest altitude on June 21, the summer solstice. Its lowest altitude will be on December 21, the winter solstice.

Your mounting rack also affects the outward appearance of the array. A fixed rack is the least obtrusive, especially if it's mounted parallel to the roof surface. Tracking arrays take up additional space, and are usually used in ground-mount installations.

- **Fixed Racks:** Fixed racks are the cheapest, strongest, and most common choice, especially for rooftop installations where space is at a premium. They hold the panels in an unchangeable orientation, usually facing an azimuth of 180 degrees, or due south. Their tilt, or altitude, can be set parallel to the roof surface on which they're mounted for the easiest installation. But on low-pitch roofs they are often raised above the roof angle to optimize output.
- **Manually Adjustable Racks:** Racks that are manually adjustable can be changed throughout the year to follow the sun's changing altitude. Their azimuth is usually fixed at due south. If adjusted faithfully—perhaps four times per year — they can increase the annual system output by 10 to 12 percent when compared to fixed-rack systems. This benefit must be weighed against the adjustable rack's increased cost and space requirements.
- **Single-Axis Trackers:** Single-axis trackers rotate east to west following the sun's daily path. Their altitude remains fixed throughout the year. Their motion is driven by either a small motor that is guided by a sun-seeking photocell, or by a passive gas-charge system in which a canister of refrigerant is heated by the sun and expands or contracts to move the array. Single-axis

tracking systems can increase annual output by up to 25 percent over fixed racks.

- **Dual-Axis Trackers:** Dual-axis trackers follow both the sun's altitude and azimuth throughout the day and year. These most complicated systems can produce up to 30 percent more energy than fixed-rack systems.

Figure 7-9: Racks Support Solar Arrays

Grid-tied PV Systems

Consumers living within an electric utility's service territory invest in PV power. In these common installation, the PV system is connected to the electric grid. These grid-tied systems feed excess PV power into the grid when the system's production is greater than residents' electricity consumption, such as during sunny weather. Residents draw power from the grid when the system's production is less than the consumption, such as at night and during cloudy weather.

Grid-tied systems are usually net-metered, using electric meters that record electricity going both into and out of a building. Many utility companies offer net metering agreements that compensate owners of qualified PV systems for power that is fed back into the electric grid.

Grid-tied systems don't require batteries, making them the most economical and popular type of PV system, though you can connect batteries to provide service if the electric grid fails.

Figure 7-10: *Net Metering*

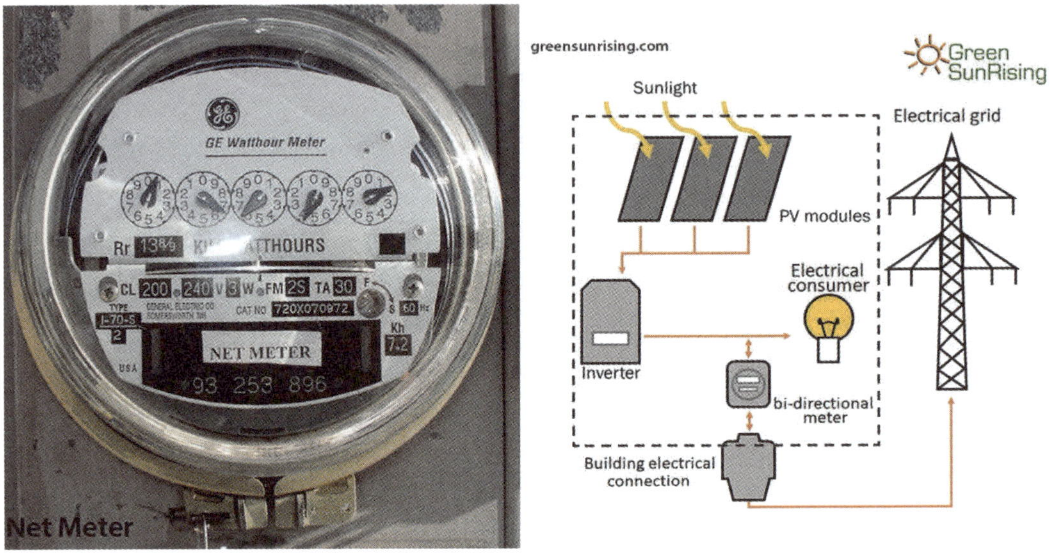

Net Metering Agreements — If you have a grid-tied PV system, you may be eligible for a net-metering agreement with your utility company. Net metering allows you to benefit from the excess power your system feeds into the electric grid. With net metering, you can effectively "bank" your power for later use without using batteries. Your utility company credits your account for the excess power you produce.

Off-grid PV Systems

Some building owners install photovoltaic (PV) systems because their buildings are located in remote areas beyond the reach of power lines. In these off-grid applications, the PV system includes batteries and often also a generator to provide electricity when the sun isn't shining. One advantage of these off-grid PV systems is that they allow property owners to avoid the cost of extending electric lines to their remote site.

Photovoltaic systems generate power only when the sun strikes the array. Backup systems provide an uninterrupted supply of solar electricity at night, during overcast weather, or during power failures.

1. The grid
2. Traditional battery storage
3. Smart battery storage
4. Generators

If you install a grid-tied system, you may not need a backup system, since you can usually rely on the grid to provide backup power when your PV system produces less power than you need.

Figure 7-11: *Inverter-charger and Batteries for Off-grid System*

Chris Dorsi

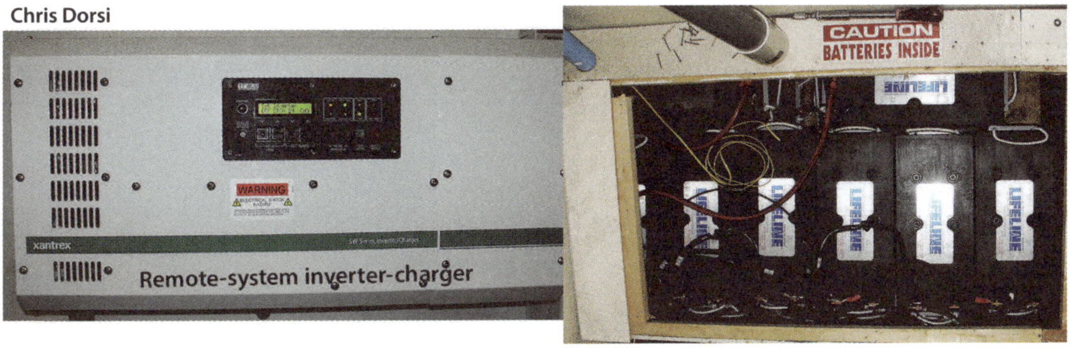

Battery systems — If you install a stand-alone or off-grid system, you'll probably want to install a backup system that includes batteries, a gas-powered generator, or both. You'll also need an inverter that can manage the charging of your batteries and can start your generator when it's needed.

Open-cell lead-acid batteries are most commonly used for PV backup systems because they provide the best ratio of storage capacity per dollar of cost. These are similar to the open-cell batteries used in automotive applications. But PV batteries are designed for deep-cycle applications, rather than the brief load experienced when starting a vehicle, and so aren't damaged by the deep cycling — from fully charged to completely drained — common to backup systems. Open-cell batteries do require periodic maintenance to replace water lost to evaporation. Sealed batteries, though more expensive, are better where low-maintenance equipment is needed, such as in remote locations.

The installation and maintenance of a battery bank adds a considerable expense to a PV system. If you need to power only a stock-tank pump or office lights, for example, you may be able to utilize the daytime-only power produced by a system with no batteries. But most PV systems include either a battery bank, a backup generator, or a connection to the utility grid.

Figure 7-12: The Market Choices for Battery Systems

Smart battery storage — New battery storage systems are coming to market. They include their own inverter, and energy management unit, and a smart phone app. The PV unit still contains its own inverter or inverters. The first of these on the market, at the time of this book's publication, claims to be able to start a 4 ton or 14 KW air conditioner. This battery system called the Franklin WH charges from PV, the grid, or a generator. This allows a range of options for off-grid homes and other buildings.

Generators — For off grid buildings, battery systems may need generators to start air conditioners because of their high peak-startup current draw. Many building owners install generators as backup electricity for emergencies. A variety of standard electrical generators work to back up a solar PV system, as well as the electrical grid. Install a generator in a permanent location **outdoors only**. Research the generator's noise level. In some residential areas, generators are forbidden because of noise and other reasons.

Does your utility company, local government, or federal government offer incentives for installing photovoltaic systems? Financial incentives, available in many regions can contribute up to half the cost of typical photovoltaic systems.

Evaluate Your Building's Solar Potential

Have you trimmed your annual electrical consumption to 1500 kilowatt-hours per person or less? Your investment in baseload improvements and building-enclosure improvements return your investment quickly and allow you to install a smaller photovoltaic system.

Does your building site receive sunlight for most of the day? Not every building site is appropriate for the installation of a photovoltaic system. To get the best production from photovoltaics, you should have mostly un-obstructed access to the sun from 9:00 a.m. to 3:00 p.m (09:00–15:00).

Figure 7-13: *AC-and-DC-Coupled Battery Systems*

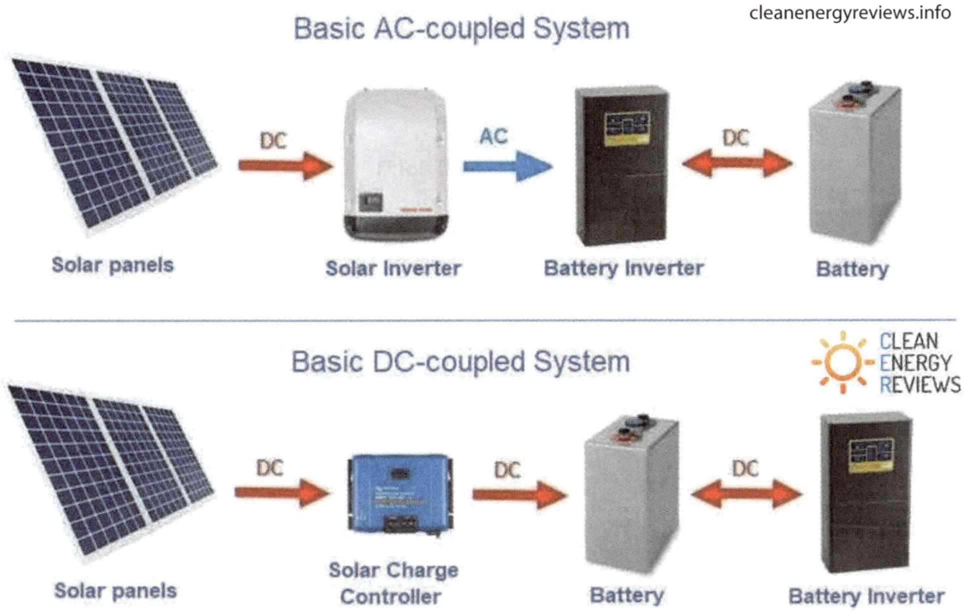

Solar Site Evaluation

Some sites are better than others for producing photo-voltaic power. The size and orientation of your building, the amount of shade, and possible zoning restrictions all affect the opportunity for PV systems for your site.

Photovoltaic systems are most commonly installed on the roofs of homes, garages, carports, and other structures. But they can also be installed vertically against a wall of your building, as part of an awning, or near the ground as a free-standing structure.

A preliminary site assessment helps you identify whether there are major shading obstacles. The information you gather in this process for deciding whether the PV system is cost-effective.

Access to the sun — Your PV system will produce the most power when exposed to the longest solar window (shade-free sun) available at your site. A perfect solar window with no shadows all day allows 100 percent of the potential solar energy to strike your array. Any shading, as from trees or an adjoining building, reduces your system's potential output compared to optimum exposure.

For the best performance your site should have an unobstructed solar exposure in the middle of the day. However, a little shade is acceptable, especially if it occurs early or late in the day. An hour of shade at 8:00 a.m. (08:00) reduces your system's output a lot less than an hour of shade at 1:00 p.m. (13:00)

Shade has a important effect on the PV output because of the series or parallel configuration in which solar cells are combined in the panels. A small strip of shade along the side of one module can reduce its output to zero since the shaded cells create resistance to the flow of current from the other cells in that array. In some configurations, an array that is shaded over 10 percent of its area could lose 80 percent of its output for as long as the shade persists.

Shading issues are most likely to occur in the winter when the sun's altitude is lowest and shadows are the longest. You may be able to tolerate the shadow of an adjoining building, for example, if it only passes over one corner of your array for a few months in winter.

Solutions to shade — If the site where you'll install your array receives some shade, you have three options to address its effects:

- Eliminate the shade by trimming the tree or removing obstacles.
- Move the array elsewhere, such as to a garage, patio cover, or barn.
- Increase the size of the array to make up for the impaired production.
- Install microinverters or DC optimizers instead of a simple string-inverter system.

Rooftop versus Ground Installation

Your analysis of the sun's path across your site gives you a better idea of the best place to locate your array of PV panels.

For many building owners, the roof is the preferred site for a PV array because it's out of the way, close to the existing electrical system, and above many objects that cast shade. Most rooftop systems mount to racks that stand above the roof. These racks can hold panels at either the same slope as the roof surface, or at a steeper angle that optimizes the exposure to incoming solar energy. Rooftop systems can also use solar "shingles," small thin-film panels that replace standard roofing shingles.

Compared to roof-mounted systems, ground-mounted systems may offer the advantages of safe maintenance access, more available room, and no risk of roof leaks.

Rooftop installation considerations — Your site assessment should address these issues if you plan to mount a PV array on your roof:

- *What is the condition of your current roof?* If your roof will be ready for replacement within the next five or ten years, consider replacing the roofing before installing the array.

- *What type of roofing material is installed on your building?* Wood and tile roofs can be difficult to seal, for example, where the array's support rack penetrates the roof.
- *Are there chimneys and other utility penetrations that break up the space?* If you'll install the panels around obstructions, consider if those objects will cast shade on the array.
- *Which face of your roof has the best exposure?* You'll get the best performance when your array faces solar south, though you may be able to use east or west faces if you can accept diminished output.
- *How much room will PV installers and other workers need to work around the array?* Be sure workers have a safe walkway at the edges of the roof.
- *Do you plan to install new equipment in the future on your roof, such as a satellite dish or air conditioner?* You should avoid filling your entire roof with a PV system.
- *Do local zoning or building regulations limit the height or appearance of buildings?* You can limit these impacts by installing your array close to the pitch of your roof, or by choosing a roof face that is out of view.
- *Will local authorities require permits, drawings, and specifications for electrical work, structural engineering, and other installation tasks?* Don't start the installation until you ensure that you and your contractor meet all requirements.

Figure 7-14: *Flat-roof-mounted 11 kW PV System*

Ground-mount considerations — If you mount a PV array on the ground, you should consider a different set of issues:

How far will the array be from the edge of your property? How high will it be? Local ordinances or zoning regulations often control these dimensions.

Will the array be subject to vandalism? This is a much greater issue for ground-mounted installations than for roof mounting.

Will you need to excavate for mounting racks or underground wiring? Be sure to locate any underground utility lines before digging.

Figure 7-15: *Ground-mounted PV Arrays*

Costs and Benefits of PV Systems

Your investment in photovoltaic power should depend upon the watt output of your system. Residential systems can vary from two to ten thousand watts (two to ten kilowatts), while commercial systems range up to thirty kilowatts or more.

Your Initial Investment

The typical installed cost for PV systems, including labor and materials, varies from $8 to $10 per watt. For a typical two-kilowatt (2 kw) system, the total cost is usually about $15,000 to $20,000. This base cost is for grid-tied systems with no battery storage or other backup system. Your cost will vary depending on the ease of access to your site, whether you choose to install a backup system, your choice of mounting racks, whether you install PV shingles or other architectural array options, and other factors.

Expected Power Production

The electrical power generated by a typical two-kilowatt system is usually about 2500 to 3000 kilo-watt-hours per year. The value of the energy produced by your system will vary depending on the cost of electricity in your area.

The average household in the United States uses about 11,000 kilowatt-hours of electricity per year. Thus this example two-kilowatt system can be expected to produce about 25 to 30 percent of a typical household's needs. Owners who have already improved the efficiency of their building find that a PV system can meet a larger portion of their electrical needs.

Equipment Lifespan

Modern PV systems have proven to be very long-lasting. The solar panels that make up about half the cost of most systems last 25 years or more, though their output tends to deteriorate about 1 percent per year. Inverters, which make up less than a quarter of the cost of most systems, aren't as reliable. They often fail prematurely. The remainder of the system — mounting racks, wiring, and disconnect equipment—should last longer than the panels.

Predicting PV Output

Given that probable lifespan, much of the burden of determining whether a PV system investment makes sense becomes an estimation of its annual output. In the USA, you can predict your system's annual output with an online calculator called PVWatts, hosted by the National Renewable Energy Lab (NREL). Throughout the world, you can use PVGIS, the Photovoltaic Geographic Information System, for PV calculations, thanks to the European Commission.

Either of these programs can produce an approximate estimate of your system's output with two factors: your location, and the peak output of your system's panels.

The rough calculation you'll make online can estimate your site's PV potential. But as you learn more about your system, you can fine-tune your online calculation by including other factors:

- The direction your panels will face in relation to solar south.
- The tilt of your panels in relation to your latitude.
- The losses inherent in your system's inverter and wiring.
- The cost of the grid-provided electricity your system will replace.

Working with PV Installers

Almost all photovoltaic systems are installed by electrical contractors. Your understanding of PV basics can help you choose a competent contractor and effectively supervise their work.

Finding a Contractor

Research before you hire one to install your PV system. Ask for written proposals from contractors who bid for your job. Be sure that each is bidding on the same size and type of system. If the bids don't specify similar details, you may need to go back and ask for adjustments so you can make an accurate comparison.

Choose a contractor who sells a variety of PV equipment, including string inverters, microinverters, DC optimizers, and panels. This choice reduces the tendency toward a single product that may not be optimal for your particular installation. A versatile contractor can give you a choice of a number of options that you can discuss with a consultant or compare with other contractors.

Before asking the contractor for an estimate, do your research on the average cost per watt for solar PV systems in your region. The cost that you get from contractors will depend on the amount of competition. However, the cost per watt should be within a fairly narrow range. The correct contractor choice doesn't depend entirely on cost per watt. A more experienced and reliable contractor who sells high-quality equipment may charge more simply because their installed PV system is worth more in the long run. A contractor's customer service may be better and they may offer better warranties. Consider all the benefits and risks, and don't decide only on cost.

Figure 7-16: *PV Installers*

Don't pick a contractor on price alone. Evaluate your contractor based on her entire proposal, and be sure to ask specific questions:

- How long has the contractor been in business?

- Has the contractor installed PV systems before? Can you contact previous customers?
- Does the contractor have a valid electrical contractor's license? Does he have any industry PV certifications?
- Has the contractor provided a rough estimate of the power the system will produce? How does that estimate compare to competing contractors?
- Will the contractor apply for the required permits, including electrical permits and building permits?
- Does the contractor know about local financial incentives. What do these incentives require of the contractor?
- Will the contractor provide an installation warranty, beyond the equipment manufacturer's warranty?

Contracts and Financing

Years ago when PV systems were much more expensive, many people signed up for power-purchasing agreements (PPAs) because the initial cost was less than purchasing a PV system. A PPA is a form of lease agreement where the customer has no equity in the system. After bank financing became widely available, PPA's became less popular. A PPA isn't an investment, and the customer doesn't benefit from tax credits. The leasing contracts often have inflexible terms including prepayment penalties and escalating monthly payments.

Today, most people recognize that PV is a good investment and thus PPA isn't a wise choice. Expect the range of between 5% and 10% return on investment (ROI) or a 10 to 20 year payback. Maybe local or national incentives increase the return on investment or decrease the payback in years.

Many new companies have entered the PV business recently. Many PV companies provide optional financing. Many banks also offer financing for PV systems. The way that PV companies compete with banks is to offer lower interest rates through a finance company. However, they raise the initial cost of the system to finance the lower interest rates offered to the customers.

Customers won't get the best financing deal unless they do calculations by hand or online. If your future might involve the sale of your home within a few years, you might prefer the lower product cost and higher interest rate. If you're sure that you will keep your home, the lower interest rate may be a better choice. The only way to tell is to calculate.

PV Summary

Solar energy can be an appealing solution to your electrical needs. But if your building is built to typical standards, you'll see a much more favorable payback for improved insulation, air-sealing, shading, and efficient appliances. After you make these improvements, a photovoltaic system is the next important next step toward a building that produces as much energy as it consumes.

To-do List

- Focus your first conservation dollars on reducing your building's consumption for heating, cooling, and water heating.
- Minimize your baseload electrical consumption to use less than 1500 kilowatt-hours annually for each resident.
- Evaluate your site's potential for solar power by considering climate, shade, and a good place to mount solar panels.
- Contact your utility company and local energy office to learn if you can benefit from financial incentives for photovoltaic systems.
- Work with an experienced PV contractor to design and install a high-quality system that is productive, reliable and trouble-free.

What to Avoid

- Power-Purchase Agreements (PPAs) to lease a Solar PV system.
- Contractor with the lowest price or one type of PV system for sale.
- Assumption that the contractor has the necessary permits (electrical, building).
- Don't fail to analyze financing to find the best plan, especially if you may sell your home or building in the future.
- Install a PV system in an unsuitable site.

Cooling Glossary

Absolute humidity - Air moisture content expressed in grains (or pounds) of water vapor per pound of dry air.

Absorbance - The ratio of a solar energy absorbed to incident solar energy. Also called absorptivity.

Absorptance - The ability of a surface to absorb solar energy. Similar to absorbance but not the same.

Absorption - A material's ability to draw in and hold particles of a liquid or gas.

Adsorption - A material's ability to transmit and store particles of a liquid or gas.

Air barrier - A building material or building assembly that provides high resistance to air leakage. Also called: *air control layer.*

Air-control layer - A building material or building assembly that provides resistance to air leakage that includes the terms: *air-control material* and *air-control system.*

Air changes at 50 pascals (ACH_{50}) - The number of times that the complete volume of a building is exchanged for outside air when a blower door depressurizes the home to 50 pascals.

Air conditioning - Cooling buildings with a refrigeration system. More generally means both heating and cooling.

Air exchange - The total building air exchanged with the outdoors through air leakage and ventilation.

Air handler - A steel cabinet containing a blower with cooling and/or heating coils connected to ducts.

Air-impermeable insulation - An insulation having an air permanence equal to or less than 0.02 L/s-m^2 at 75 Pa pressure differential tested according to ASTM E 2178 or E 283.

Ambient - Of the surrounding area or environment.

Ambient air - Air in the habitable space. Also the air around a human observer.

Arc-fault interrupter - A circuit breaker that interrupts the circuit when it detects an electric arc.

Array - A group of PV panels or collectors.

Ampere - A unit of measurement of electrical current flow. A coulomb per second.

Annual return - The yearly savings divided by the initial cost needed to achieve the savings, expressed as a percent.

Atmospheric pressure - The weight of air and its contained water vapor on the surface of the earth. At sea level this pressure is 101,325 pascals or 14.7 pounds per square inch.

Awning - An external window-shading device that may be stationary or adjustable, large or small.

Backer rod - Polyethylene foam rope used as a backer for caulking.

Badgir - A tower with a wind catcher used to cool buildings.

Band joist - See *Rim joist.*

Batt - A narrow blanket of fiberglass insulation that insulates a building cavity.

Battens - A wood or metal non-structural element for fastening or creating a space, similar to a furring strip.

Block frame - A non-finned window frame for new or retrofit installation in a rough opening.

Blower - The squirrel-cage fan in an air handler.

Blower door - A device that consists of a fan, a removable panel, and gauges used to measure and locate air leaks.

Boot - A duct fitting that connects a duct to a register.

British thermal unit (BTU) - The quantity of heat required to raise the temperature of one pound of water one degree Fahrenheit.

BTUs/hr - British thermal units per hour.

Building cavities - The spaces inside walls, floors, and ceilings between the interior and exterior sheeting.

Building enclosure - The building assemblies that separate the indoors from the outdoors, which should contain insulation and an air barrier.

Building science - Branch of science dealing with construction, maintenance, safety, and energy efficiency of buildings.

Canopy - A large piece of fabric, used to shade and outdoor area.

Capillary action - The ability of water to move through materials, even upward against gravity, through small tubes or spaces. Also called "capillarity".

Capillary barrier - A material or air space designed to stop capillary action from carrying water into a building.

Cathedral ceiling - A sloped ceiling insulated between the roof deck and the finish ceiling material.

Cathedralized attic - An attic that is insulated at the underside of the roof deck rather than at the ceiling.

Caulking - A mastic compound for filling joints and cracks.

Celsius - A temperature scale on which water freezes at 0°C and boils at 100°C.

Cell - A single PV device.

Cellulose insulation - Insulation, packaged in bags for blowing, made from newspaper or wood waste and treated with a fire retardant.

Center-of-wall R or RSI - R-value or RSI of the most insulated part of a wall.

Centigrade - See Celsius

CFM_{50} - The number of cubic feet per minute of air flowing through the fan housing of a blower door when the house pressure is 50 pascals (0.2 inches of water). This figure is the most common and accurate way of comparing the airtightness of buildings that are tested using a blower door.

CFM_n - The number of cubic feet of air flowing through a house from indoors to outdoors during typical, natural conditions. This figure can be roughly estimated using a blower door.

Cladding - The exterior covering or coating on a structure, such as wood siding, stucco, or brick veneer. Synonym for "siding".

Clean and tune (C&T) - A procedure performed on a cooling system by a qualified technician to optimize its efficiency.

Clear-wall R or RSI - The area-weighted R or RSI of a wall, including studs and plates.

Coefficient of performance (COP) - A heat pump or air conditioner's output in watt-hours of heat moved divided by watt-hours of electrical input.

Coil - A snakelike piece of copper tubing surrounded by rows of aluminum fins that clamp tightly to the tubing in order to aid in heat transfer.

Commissioning - The process of testing and adjusting building mechanical systems after building construction or as a retrofit measure.

Compressor - A motorized vapor pump that compresses the gaseous refrigerant and sends it to the condenser where collected heat is released.

Condensate - Liquid formed by condensing vapor.

Condense - When a gas turns into a liquid as it cools, we say it condenses. Condensation is the opposite of evaporation.

Condenser - The coil in an air conditioning system where the refrigerant condenses and releases heat.

Conditioned - Heated or cooled areas of a building are said to be conditioned, either intentionally or unintentionally.

Conductance - The ability of a material to conduct some energy form like heat or electricity.

Conduction - Heat flow from molecule to molecule in a solid substance.

Control circuit - A circuit whose work is switching a power circuit or opening an automatic valve.

Control layer - A building material or building assembly that controls the flow of water, air, water vapor, or heat.

Convection - Heat transfer caused by the movement of a fluid like water or air. Warmer fluids rise and cooler fluids fall causing convection.

Cool roof - A reflective roof or a roof that, through other strategies, minimizes solar gain entering the building.

Cooling load - The maximum rate of heat removal required of an air conditioner when the outdoor temperature and humidity are at the highest expected level.

Cost-effective - Having an acceptable payback, return-on-investment, or savings-to-investment ratio.

Cubic foot per minute (cfm) - A measurement of air movement past a certain point or through a certain structure.

DC-power optimizer - A control that optimizes electrical production in an PV array that is partially shaded.

Decking - The wood material installed under roofing material to support the roofing.

Dehumidifier - A device that removes water vapor from the air.

Dehumidification - The removal of water vapor from the indoor air through mechanical means.

Dehumidistat - A control that senses humidity and turns a dehumidifier on and off to control humidity.

Degree-days - A measure of the temperature element of climate produced by multiplying temperature difference by time.

Demand - The peak need for energy. Some utilities levy a monthly charge for demand.

Dense packing - Blowing insulation with sufficient force to create a high density to reduce settling and minimize air leakage and air convection.

Density - The weight of a material divided by its volume, usually measured in pounds per cubic foot or kilograms per cubic meter.

Deposition - Describes water vapor turning from vapor to ice, without turning to water first.

Depressurize - Cause to have a lower pressure or vacuum with respect to a reference of a higher pressure. A fan's intake creates the suction to depressurize a space.

Desiccant - A liquid or solid material used to adsorb water vapor.

Design temperature - A high outdoor temperature used to design cooling systems.

Desorption - Particals, formally adsorbed, leaving a substance.

Dew point - The warmest temperature of an object in an environment where water condensation from the surrounding air would form on that object.

Diffusion - Movement of water vapor through a material, driven by the *vapor pressure* across a material and the enabled by the vapor permeability of that material. See also: *vapor permeable*

Diffuser - A supply-air grille that distributes air in specific directions into a room.

Distribution system - A system of pipes or ducts used to distribute cooling energy.

Drainage plane - A *water-resistant barrier,* combined with a space, that allows water drainage.

Draft - A pressure caused by a difference in temperature, in a vertical shaft or chimney.

Drywall - Gypsum wallboard used to produce a smooth and level interior wall surface and to resist fire.

Dry-bulb temperature - Normal ambient air temperature measured by a thermometer.

Duct blower - A blower-door-like device used for testing duct leakiness and airflow.

Earth tube - An earth-to-air heat exchanger, used for cooling and/or heating.

Eave - The part of a roof that projects beyond its supporting walls (See - *Soffit*).

Eave chute - Device that maintains air space between the insulation blanket and the roof sheathing and prevents insulation from clogging eave vents.

Eave vent - Vent opening located in the soffit under the eaves of a house to allow the passage of air through the attic and out the roof vents.

Economizer - A subsystem in an HVAC system that saves energy by circulating outdoor air if the temperature and humidity are favorable to condition building air, in a process called *free cooling*.

Efficiency - The ratio of output divided by input.

Efflorescence - White, brown, yellow, or green salt deposits masonry materials carried by water from a within a masonry assembly to its surface.

Elastomeric - A characteristic of a material that is flexible and permits joint movement.

Elastomeric coating - Polymeric material, such as acrylic, that is used to coat roof leaks and to reduce solar heat gain.

Emittance - The ability of a material to emit radiant energy from its surface. Also called emissivity.

Energy - A quantity of heat or work.

Energy audit - The process of identifying energy conservation opportunities in buildings.

Energy conservation - Actions to reduce the amount of energy that a building or community uses.

Energy consumption - The conversion or transformation of potential energy into kinetic energy for heat, light, electricity, etc.

Energy efficiency - Term describing how efficiently an device, system, or building uses energy.

Energy-efficiency ratio (EER) - A measurement of energy efficiency for air conditioners. The EER is computed by dividing cooling capacity, measured in British Thermal Units per hour (BTUs/hr), by the watts of power.

Energy-recovery ventilator - A ventilator that recovers latent and sensible energy from the exhaust airflow and imparts it to the incoming airflow.

Envelope - The *building enclosure*. The exterior walls, floor, and roof assembly of a building. Also sometimes describes a building cavity.

Equivalent length - The length of straight pipe that has equivalent resistance to a duct fitting. Used for duct design.

Evaporation - An event where a liquid becomes a gas. Evaporation is the key process in the operation of air conditioners and evaporative coolers.

Evaporative cooler - A device for cooling homes in dry climates that reduces the temperature of incoming air by humidifying it.

Evaporator - The heat-transfer coil of an air conditioner or heat pump that cools the surrounding air as the refrigerant inside the coil evaporates and absorbs heat.

Evapotranspiration - Plants evaporate their absorbed water like humans evaporate sweat, which cools the outdoor environment around the plants.

Exfiltration - Air flowing out of a residence from its conditioned space through the enclosure.

Expansion valve - A valve that meters refrigerant into the evaporator. Also called: "expansion device".

Fahrenheit - A temperature scale used in the United States and a few other countries. On the Fahrenheit scale water boils at 212°F and freezes at 32°F.

Fenestration - Window and door openings in a building enclosure.

Fiberglass - A fibrous material made by spinning molten glass.

Fill tube - A plastic or metal tube used for its stiffness to blow insulation inside a building cavity.

Fin - Attachments to HVAC coils to help heat or cool air passing through. Or, an part of a window frame to provide attachment to a building.

Fin comb - A comb-like tool used to straighten bent fins in air-conditioning coils.

Finned tube - A length or coil of pipe with heat transfer fins attached to enhance heat transfer.

Fire barrier - A tested building assembly, designed to contain a fire for a particular time period: typically 1-to-4 hours.

Fire blocking - Building materials installed to resist the free passage of flames and smoke, to adjacent areas of the building.

Fire resistance - The property of building materials or assemblies that prevents or retards the passage of heat, hot gases, or flames during a fire.

Fire resistance rating - The period of time a building element, component or assembly maintains the ability to confine a fire, continues to perform a given structural function, or both.

Fire stop - Framing member designed to stop the spread of fire within a wall cavity.

Fire tape - Tape and drywall-finishing compound applied to create a fire-resistant assembly.

Firewall - A fire-resistance-rated smoke-tight wall with protected openings that restricts the spread of fire. A firewall extends continuously from the foundation to or through the roof. The firewall is designed to remain standing even if the assemblies on either side collapse during a fire.

Flame spread - A fire rating for materials in a fire test that compares the spread of flame to red oak, which has a flame spread of 100.

Flammable - The rating for building materials that will burn readily when exposed to a flame.

Flashing - Waterproof material used to prevent water leakage through intersections between building assemblies or around penetrations.

Floor joists - Structural beams that support the floor.

Foamboard - Rigid plastic foam insulation.

Fouling - The deposit of particles and fluids on a heat exchanger or other HVAC component.

Free cooling - Nighttime cooling provided by an economizer.

Furring - Wood or metal strips that create a space for insulation or ventilation.

Gasket - Elastic strip that seals a joint between two materials.

Glazing - Glass installation. Pertaining to glass assemblies or windows.

Glass load factor - A number combining glass' solar heat transmission and its heat conduction. Used for cooling load calculations.

Grade - The level of the ground around a building.

Grade - The level of the ground around a building.

Grid-tied - A PV system tied to the local electrical grid.

Ground fault circuit interrupter (GFI or GFCI) - An electrical connection device that breaks a circuit if current flows in a grounding wire.

Heat capacity - The quantity of heat required to raise the temperature of 1 volumetric unit of a material 1°F or 1°C.

Heat gains - Heat that accumulates in buildings; this is desirable during the heating season and undesirable during the cooling season.

Heat loss - The amount of heat escaping through the building enclosure during some period of time like a month or a year.

Heat pump - A reversible air conditioner that provides both heating and cooling.

Heat-recovery ventilator - A central ventilator that transfers heat from exhaust to intake air.

Heat transmission - Heat flow through the walls, floor, and ceiling of a building. Does not include air leakage.

Heat-transfer coefficient - See U-factor.

Heating degree day - Each degree that the average daily temperature is below the base temperature (usually 65°F) constitutes one heating degree day.

House pressure - The difference in pressure between the indoors and outdoors measured by a manometer.

Humidistat - An automatic control that switches a fan, humidifier, or dehumidifier on and off to control relative humidity.

Humidity ratio - Same as absolute humidity. The absolute amount of air's humidity measured in pounds of water vapor per pound of dry air or grams per gram in the SI system.

HVAC - Short for heating, ventilation, an air conditioning.

Hydronic - A heating system using hot water or steam as the heat-transfer fluid.

Ignition barrier - A material installed to prevent another material, often plastic foam, from catching fire.

Imperial measurement system (IP) - A system of measurement created by Great Britain and now used mainly by Americans.

Indoor air quality (IAQ) - The quality of indoor air relative to its acceptability for healthful human habitation.

Inch of water column (IWC) - Small air pressure differences are measured in inches of water column (IWC) in the *Imperial measurement system*. Equal to 249 pascals.

Infiltration - The inflow of outdoor air into the indoors, which is accompanied by an equal outflow of air from indoors to the outdoors.

Infrared - Describes heat radiation emitted by the sun or by warm objects on earth.

Input rating - The rate at which an energy-using device consumes electricity or fossil fuel.

Insolation - The amount of solar radiation striking a surface.

Insulated glass unit (IGU) - A glass assembly with two or more glass panes, spaced apart and sealed in a factory.

Insulation restrainer - A flexible material, such as netting or fabric, used to hold blown fibrous insulation in place.

Insulation shield - A fire-barrier erected around a heat producing device to prevent insulation from covering or contacting hot surfaces.

Insulation - Material with relatively high thermal resistance.

Internal gains - The heat generated by bathing, cooking, and operating appliances, that must be removed during the summer to promote comfort.

Intermediate zone - A zone located between the building's conditioned spaces and outdoors, like a crawl space or attic.

Inverter - A device that changes DC electricity to AC electricity.

IWC - Inches of water column (IWC): IP measurement system.

Jacket - The cabinet of an air handler.

Jacketing - Installing a reinforcing layer of masonry mortar or stucco to existing masonry to strengthen and air seal the masonry assembly.

Jamb - The side or top piece of a window or door frame.

Jamb clips or plates - Structural devices used to fasten a block-frame window to its opening.

Joist - A horizontal wood framing member that supports a floor or ceiling.

k-value - Here used to describe the measurement of *thermal conductivity.*

Kilowatt (kW) - A unit of electric power equal to 1000 joules per second or 3412 BTUs per hour.

Kilowatt-hour (kWh) - A unit of electric energy equal to 3600 kilojoules or 3412 BTUs.

Kinetic energy - Energy in transition or motion.

Latent heat - The heat absorbed or released by a substance when it changes state. For example, from a liquid to a gas.

Landscape - The features of the outdoor environment, including trees and other plants that may shade buildings and cool the environment through *evapotranspiration.*

Liner board - A type of duct insulation, installed inside a duct.

Low-e - Short for low emissivity, which means the characteristic of a metallic glass coating to resist the flow of radiant heat.

Masonry - Construction that uses: stone, sand, lime, clay, gypsum, and Portland cement.

Manifold - A section of pipe with multiple openings.

Manometer - Device that measures fluid pressures.

Mastic - A thick creamy substance used to seal seams and cracks in building materials.

Metabolic process - Chemical and physiological activities in the human body.

Metering device - In refrigeration, an orifice or capillary tube that meters refrigerant into an evaporator

Microclimate - A very localized climatic area, usually of a small site or habitat.

Mortar - A mixture of sand, water, and cement used to bond bricks, stones, or blocks together. Also called, *render*

Natural ventilation - Ventilation using only natural air movement, without fans.

Net free area - The area of a vent adjusted for insect screens, louvers, and weather coverings. The net free area is always less than the actual area.

Off-cycle losses - The energy losses in an HVAC cooling system because of heating of the evaporator coil by surrounding air after the cooling system shuts off.

Off-grid - A PV system that isn't tied to a local electrical grid and instead uses batteries to store electricity for nights and cloudy weather.

Output - The useful energy that a device produces after accounting for waste involved in the energy transfer.

Packaged air conditioner - An HVAC system that contains the compressor, evaporator, and condenser in a single cabinet.

Panel - A group of PV cells that form a collector.

Pascal - An SI unit that measures air pressure. Equal to one newton per square meter.

Payback period - The number of years that an investment in energy conservation requires to repay its cost in energy savings.

Perlite - A heat-expanded mineral used for insulation.

Perm - An Imperial and SI measurement unit of how much water vapor a material transmits per unit of time.

Permeability - A measurement of a solid material's ability to transmit fluids.

Permeance - The ability of a material to transmit another material.

Plenum - A large duct that connects the air handler to smaller main supply ducts.

Polyethylene - Polymer plastic used for vapor barriers, air barriers, and foam backer rod.

Polyisocyanurate (PIR) - A plastic foam insulation sold in sheets, similar in composition to polyurethane.

Polystyrene - A rigid plastic foam insulation, usually white or blue in color.

Polyurethane - A versatile plastic foam insulation, usually yellow in color.

Potential energy - Energy in a stored or packaged form.

Power - Energy divided by time. The capacity to do work or to provide or remove heat, measured in BTU/hr or watts.

Pressure - A force driving flow because of a difference in some condition between two areas.

Pressure boundary - An air barrier or plane where the largest blower-door-induced pressure is measured.

Pressure diagnostics - The practice of measuring building air pressures and airflows in order to find air leaks and sources of unwanted pressures.

Pressure pan - A device used to block a duct register, in order to measure the static pressure behind it, during a blower door test.

Psychrometer - A device that measures relative humidity.

Psychrometrics - The science of the relationship between air, water vapor, and heat and the effect of these factors on comfort, health, and durability.

R-value or RSI-value - An Imperial or SI measurement of *thermal resistance.*

Radiant barrier - A foil sheet or coating designed to reflect heat radiation or retard the emission of heat radiation.

Radiant temperature - The average temperature of objects in a building like walls, ceiling, floor, furniture, and other objects.

Radiation - Heat energy that originates on a hot body like the sun and travels from place to place through the air or through outer space.

Rain screen - The combination of a *water-resistive barrier* and a space, used to drain and ventilate wall assemblies.

Reflectance - The ratio of radiant heat reflected from a given surface to the total radiation falling on the surface. Also called *reflectivity.*

Refrigerant - A fluid used in air conditioners and heat pumps that heats when it condenses and cools when it evaporates.

Register - A grille covering a supply duct that may have a control damper.

Reheat - Heating refrigerated air above its dew point to provide comfortable supply air to a building, during air conditioning or dehumidification.

Relative humidity - The percent of moisture absorbed in the air compared to the maximum amount possible. Air that is saturated has 100% relative humidity.

Render - Another term for *mortar* or *stucco.*

Resistance - The property of a material to resist the flow of electricity, water vapor, heat, or air.

Retrofit - An energy conservation measure that is applied to an existing building. Also means the action of improving the thermal performance or maintenance of a building.

Return air - Air circulating back to the air conditioner from the house, to be re-cooled by the AC and supplied to the rooms.

Room air conditioner - An packaged air conditioner, installed through a wall or window, which cools the room by removing heat from the room and releasing it outdoors.

Rough opening - The framed opening in a wall into which a door or window is installed.

Sash - See *Window sash*

Saturation - Describes the mixture of vapor and liquid at the phase-change point in a refrigeration coil.

Savings-to-investment ratio (SIR) - Measures how many times an energy retrofit pays for itself during its lifetime.

Scale - Dissolved minerals that precipitate inside evaporators or drain pans.

Seasonal energy efficiency ratio (SEER) - An Imperial measurement of energy efficiency for central air conditioners. The SEER equals cooling capacity, measured in BTUs/hr, divided by the watts of electrical power.

Sensible heat - The heat absorbed by a substance which raises its temperature.

Shade - 1. Blockage of direct solar heat. 2. A soft movable shading device that opens and closes on a roll, installed on the interior or exterior of a window.

Shading coefficient (SC) - A decimal describing how much solar energy is transmitted through a window opening, compared to clear single glass, which has an SC of 1.0.

Sheathing - A structural *sheeting*, attached on top of a framed wall or roof and underneath the building's siding and roofing.

Sheeting - Any building material used for covering a building surface.

Shutter - A solid and stable window covering with a frame and stiles or slats, installed on the interior or exterior of a window.

Sill - The bottom of a window or door frame.

Sill pan - A flashing device that sits on a rough-framed window sill to prevent water penetration.

SIR - See *Savings-to-investment ratio.*

Sling psychrometer - A device holding two thermometers that is slung through the air to measure dry-bulb temperature, wet-bulb temperature, and relative humidity.

Slope - The roof section of an attic with the roof and ceiling surfaces attached to the rafters.

Solar absorption - The ratio of absorbed solar radiation to incident solar radiation.

Solar chimney - A solar collecting chimney that creates draft by heating the air inside it and causing the air to rise.

Solar gain - Radiant solar heat that is absorbed by a building and contributes to the need for cooling.

Solar heat - Radiant energy from the sun with wavelengths between 0.7 and 1 micrometers.

Solar heat-gain coefficient (SHGC) - The ratio of solar heat gain through a window to incident solar heat. Includes both transmitted heat and absorbed and re-radiated heat.

Solar reflectance - The ratio of reflected solar radiation to incident solar radiation. See also albedo.

Solar screen - A stationary external window shading device that intercepts and absorbs solar heat before in enters the building.

Solar transmittance - The percent of total solar energy transmitted by a material.

Solenoid - A magnetic device that moves a switch or valve stem.

Sorption - Particles of a fluid moving in or out of a solid.

Spalling - Surface degradation of masonry materials because of moisture movement.

Split-system air conditioner - An air conditioner that has the condenser and compressor outdoors and the evaporator indoors.

Stack effect - The draft established in a building from air infiltrating low and exfiltrating high.

State point - Air at a particular temperature and humidity occupies a single point on the *psychrometric* chart called a state point.

Stick pins - Duct-insulation fasteners that glue sticks to a duct, equipped with a cap to hold the insulation.

Stucco - Cement and sand plaster applied to the building's exterior walls. Also called *render*.

Subcooling - The number of degrees Fahrenheit that a condenser and nearby piping cools the liquid refrigerant below its saturation temperature.

Sublimation - Describes when ice turns to water vapor without turning to water first.

Superheat - The number of degrees Fahrenheit that an evaporator and nearby piping heats gaseous refrigerant above its saturation temperature.

Supply air - Air that has been heated or cooled and is then moved through the ducts and out the supply registers of a home.

Temperature - A measure of *kinetic energy* that indicates the amount or heat stored in a unit of material.

Tensioner - A tool to tighten tie bands.

Thermal barrier - A material that protects materials behind it from reaching 250° F or 121°C during a fire. Drywall (1/2 inch or 25 mm) is a 15-minute thermal barrier.

Thermal boundary - A line or plane where insulation and *air barrier*(s) exist or should exist in order to resist *thermal transmission* and air leakage through or within a *building enclosure.*

Thermal break - A piece of relatively slow-conducting material between two fast-conducting materials.

Thermal bridging - Rapid heat conduction resulting from direct contact between very thermally conductive materials like metal and glass.

Thermal bypass - A large air leak that allows air to flow around insulation.

Thermal conductivity - General term applied to k-value, meaning conduction heat-flow rate through a homogeneous material.

Thermal enclosure/envelope - The insulated and air-sealed boundaries of a dwelling, which surrounds the conditioned space.

Thermal resistance - Same as R-value or RSI, expressing ability to resist heat flow.

Thermal transmittance - Expressed as *U-factor* or *U-value*, thermal transmittance is heat flow by *conduction*, *convection*, and *radiation* through a non-uniform layered building component like a wall.

Thermostat - A device used to control a heating or cooling system to maintain a *setpoint* temperature.

Tie band - A fastener to attach a round flex duct to a round metal duct that is tightened by a tensioner.

Ton of refrigeration - The capacity to remove 12,000 BTUs per hour of heat from a building.

U-factor - The amount of heat that flows through a square foot of building cross-section experiencing a temperature difference of 1° F. In general U-factor and U-value are synonymous.

U-value - The amount of heat that flows through a meter of building cross-section experiencing a temperature difference of 1° C.

Vapor barrier - A material that resists the flow of water vapor to less than 0.1 Imperial perms or 0,065 metric perms.

Vacuum - Low air pressure compared to a zone with higher air pressure.

Vapor diffusion - The flow of water vapor through a solid material.

Vanes - Flat metal slats that direct air through a grille, register, or diffuser.

Vapor permeable - A material with a water vapor *permeance* of more than 10 perms.

Vapor pressure - The pressure exerted by a vapor, which increases with temperature.

Vapor retarder - A vapor barrier or vapor control layer with a permeability rating of 0.1 to 10 Imperial perms or 0.065 to 6.5 metric perms. (Also called: *vapor control layer*)

Vaporize - Change from a liquid to a gas.

Veneer - The outer layer of a building component that protects or beautifies the component.

Ventilation - The movement of air through an area for the purpose of removing moisture, air pollution, or unwanted heat.

Vermiculite - A heat-expanded mineral used for insulation. Sometimes contains asbestos.

Visible transmittance - The percent of visible light transmitted by a glass assembly.

Water-resistive barrier - A water-resistant material used to prevent water from wetting a building's structural sheathing and other vulnerable components. Also called: *water-control layer* or *weather-resistant barrier.*

Watt - A unit of electrical power equivalent to one joule per second or 3.4 BTUs/hr.

Watt-hour - A unit of electrical energy equivalent to 3600 joules or 3.4 BTUs.

Weather-resistant barrier - An term for a material that prevents water from wetting the wall assembly. See instead: *water-resistive barrier.*

Weatherstripping - Flexible gaskets, often mounted in rigid metal strips, for limiting air leakage.

Webbing - A reinforcing fabric used with mastics and coatings to prevent patches from cracking.

Weep holes - Holes drilled for the purpose of allowing water to drain out of an area in a building where it has collected.

Weep screed - Openings at the bottom of a stucco wall to drain water.

Wet Bulb Depression (WBD): - The difference between the dry-bulb and wet-bulb temperatures.

Wet-bulb temperature - The temperature of the dampened thermometer of a *sling psychrometer*.

Wet spray - Fibrous insulation mixed with water and sometimes also a binder during installation.

Whole-house fan - A fan that draws fresh outside air into the living space, flushes hot air up the attic and exhausts it to the outside.

Whole-wall R or RSI - The area-weighted average R or RSI of a framed building including studs, plates, corners, headers and all other framing members, but not windows or doors.

Wind catcher - An opening in a wind tower that catches wind to cool a building.

Wind tower - A tower, built to rise above the ground, where wind velocity is higher than near the ground.

Window blinds - Hard window coverings, made with slats or vanes that appear the you lower them. The vanes or slats are adjustable to block solar heat and light or let it in. Installed on the interior or exterior of a window.

Window film - Plastic films, coated with a metalized reflective surface, that are adhered to window glass to reflect heat rays from the sun.

Window frame - The sides, top, and sill of the window that forms a box around window sashes and other components.

Window low-e coatings - Metalized coatings applied to glass in a factory to increase the reflectance and decrease the emittance of the glass.

Window sash - A movable or stationary part of a window that frames a piece of glass.

Window shade - A soft movable shading device that opens and closes on a roll, installed on the interior or exterior of a window.

Window shutter - A solid and stable window covering with a frame and stiles or slats, installed on the interior or exterior of a window.

With reference to (WRT) - Compared to another measurement. In weatherization, a way to evaluate pressure differences between *zones* of the building.

Zone - A room or portion of a building separated from other rooms by an air barrier — not usually an effective air barrier.

Appendices

A-1 Online Energy-Information Sources

Search your browser for the organization's name, or enter the link in your browser's URL address bar.

Organization	URL / Link
Building Enclosure: Codes, Standards, and Authoritative Information	
International Code Council (ICC)	iccsafe.org/
2021 International Existing Building Code (IEBC)	codes.iccsafe.org/content/IEBC2021P1
Building Performance Institute of Europe (BPIE)	bpie.eu/knowledge-hub/
Air Barrier Association or America (ABAA)	airbarrier.org/technical-information/
Building Science Corporation	buildingscience.com/
International Assoc. of Certified Home Inspectors	nachi.org/sop.htm
Building Enclosure (BE)	buildingenclosureonline.com/publications/
Building Performance Institute (BPI)	bpi.org/
Heating, Ventilation, and Air Conditioning Codes, Standards, and Authoritative Information	
Air Conditioning Contractors of America (ACCA)	acca.org/standards/technical-manuals
American Society of Heating, Refrigeration, and Air Conditioning Engineers (ASHRAE)	ashrae.org/technical-resources/standards-and-guidelines
Federation of European Heating, Ventilation, and Air Conditioning Associates	rehva.eu/eshop
HVAC School	hvacrschool.com/
International Energy Agency (IEA)	https://www.iea.org/
Building Intelligence Center (HVAC Manufacturers' Product Information)	building-center.org/hvac-table-of-contents-2/
Authoritative General Information on Energy Efficiency for Buildings	
U.S. Department of Energy (DOE)	energy.gov/
European Commission: Energy	energy.ec.europa.eu/publications_en
American Counsel for an Energy-Efficient Economy	aceee.org/
European Counsel for an Energy-Efficient Economy	eceee.org/
Rocky Mountain Institute (RMI)	rmi.org/
Alliance to Save Energy (ASE)	ase.org/resources
U.S. Energy Information Administration (EIA)	eia.gov/
Direct Energy	www.directenergy.com/

A-2 R & RSI-Values of Materials & Assemblies

Table 9-1: *R & RSI-Values of Materials & Assemblies*

Material / Description	R-value hr•sf•°F/ BTU	RSI-value m^2K/W
Air Films		
Exterior air film	0.17	0.030
Interior air film	0.68	0.12
Interior ceiling air film	0.61	0.11
Building enclosure materials / inch or 25 mm		
Wall stud / inch or 25 mm	1.25	0.22
Asbestos-cement board	0.25	0.044
Gypsum or plaster board	0.9	0.16
Plywood or wood panels	1.2–1.4	0.22
Softwoods	1.0–1.4	0.21–0.25
Hardwoods	0.60–0.80	0.11–0.14
Fiberboard sheathing	2.4–2.7	0.42–0.48
Particleboard (high to low density)	0.85–1.4	0.15–0.25
Flooring Materials R or RSI of given thickness		
Carpet 0.25 inch thick	1.3	0.23
Carpet 0.50 inch thick	2.0	0.35
Cork tile, 0.125 in	0.25	0.044
Tile-asphalt, linoleum, vinyl, rubber	0.05	0.0088
Wood, hardwood finish, 0.75 in	0.68	0.12
Insulation / inch or 25 mm		
Blanket and Batt / inch or 25 mm		
Fiberglass batts and blankets, 0.4–2.0 lb/ft^3 [1]	2.6–4.2	0.46
Low-density batts and blankets / inch or 25 mm	2.6–2.8	0.74
Standard density / inch or 25 mm	3.1	0.55
Medium density / inch or 25 mm	3.7	0.65
High density / inch or 25 mm	4.3	0.76
Cotton batts	3.7	0.65
Board and Slabs / inch or 25 mm		
Glass fiber, organic bonded, 3.8-4.2 lb/ft^3	4.0–4.4	0.70–0.77
Mineral fiber 4.0-5.0 lb/ft^3	4.0–4.4	0.70–0.77
Extruded polystyrene, 1.6-2.0 lb/ft^3	5.0	0.88
Expanded polystyrene, molded bead, 1.4-2.0 lb/ft^3	3.6-4.4	0.63–0.77
Cellular polyisocyanurate (faced with foil or glass fiber)	5.6–7.6	0.99–1.3

Table 9-1: *R & RSI-Values of Materials & Assemblies*

Material / Description	R-value hr•sf•°F/ BTU	RSI-value m^2K/W
Loose Fill / inch or 25 mm		
Cellulose insulation, 1.2-2.0 lb/ft^3 horizontal application, open blow	3.2–3.6	0.56–0.63
Cellulose insulation, 3.0-4.0 lb/ft^3 vertical wall application	3.0–3.4	0.53–0.60
Mineral wool, 0.6–1.2 $lb/ft^3$0.6-1.2 lb/ft^3, horizontal application, open blow	2.6–3.6	0.46–0.63
Fiberglass: chopped batt waste, 0.4-0.8 lb/ft^3, horizontal application	2.4–3.0	0.42–0.53
Fiberglass virgin fiber, 0.4-0.8 lb/ft^3, horizontal application, open blow	2.6–3.0	0.46–0.53
Fiberglass virgin fiber, 1.0-2.5 lb/ft^3, vertical wall application	3.6–4.4	0.63–0.77
Vermiculite, exfoliated	2.1	0.37
Perlite, expanded, 2.0-3.0 lb/ft^3	2.4	0.42
Sprayed or Injected Insulation / inch or 25 mm		
High-density polyurethane foam	6.0–7.0	1.1–1.2
Low-density polyurethane foam	3.5–3.8	0.62–0.67
Spray-on cellulose, included wetted cavity fill	3.0–3.6	0.53–0.63
Spray-on fiberglass, included wetted cavity fill	3.2–3.8	0.56–0.67
Air-Krete cementitious foam 2.1 lb/ft^3	3.9	0.69
Masonry Materials[2] / inch or 25 mm or as listed		
Concrete, stucco, mortar / inch or 25 mm	0.10–0.20	0.017–0.035
Aerated cellular concrete / inch or 25 mm	1.0–2.0	0.18–0.36
Adobe (clay/mud): / inch or 25 mm	0.20–0.35	0.035–0.062
Insulating clay block: / inch or 25 mm	1.2–2.2	0.21–0.39
Brick, common / inch or 25 mm	0.15–0.30	0.026–0.053
8-inch (200 mm) concrete block	1.1	0.19
with vermiculite fill in cores	2.1	0.37
12-inch (300 mm) concrete block	1.2	0.21
with vermiculite fill in cores	2.9	0.51
Roofing for standard thickness or as listed		
Asbestos-cement shingles	0.21	0.037
Asphalt roll roofing	0.15	0.026
Asphalt shingles	0.44	0.077
Built-up roofing, 0.375 in	0.33	0.058
Slate, 0.5 in	0.05	0.0088
Wood shingles	0.94	0.17

Table 9-1: R & RSI-Values of Materials & Assemblies

Material / Description	R-value hr•sf•°F/ BTU	RSI-value m^2K/W
Siding for standard thickness or as listed		
Asbestos-cement	0.20	0.035
Asphalt roll siding	0.15	0.026
Asphalt insulating siding (0.5 in bed.)	1.5	0.26
Wood siding over sheathing thickness as listed		
Hardboard siding, 0.4375 in	0.7	0.12
Wood, drop, 1 • 8 in	0.8	0.14
Wood, bevel, 0.5 • 8 in, lapped	0.8	0.14
Wood, bevel, 0.75 • 10 in, lapped	1.0	0.18
Wood, plywood, 0.375 in, lapped	0.6	0.11
Aluminum or Steel, over sheathing		
Hollow-backed	0.6	0.11
Insulating-board backed nominal 0.375 in	1.8	0.32
Insulating-board backed nominal 0.375 in, foil backed	2.9	0.51
Wall assemblies tested on a Guarded Hot Box Whole-wall R/RSI-values including window framing, corners, etc.		
Standard wood 2x4, R-11 Fiberglass-batt insulated	9.7	1.71
Standard wood 2x6, R-19 Fiberglass batts, installed perfectly	12.8	2.25
Standard wood 2x6, R-19 Fiberglass batts, installed typically	11.7	2.06
Standard wood 2x6, R-19 Fiberglass batts, installed poorly	11.0	1.93
Steel frame wall C-stud, Fiberglass batts, installed typically	5.6	0.99
Steel frame wall, Fiberglass batts, 1-inch EPS sheathing	10.5	1.85
Steel frame wall Fiberglass batts, 0.5 inch EPS sheathing	8.0	1.41
Steel frame wall with cavity-sprayed polyurethane and fiberglass batts	8.2	1.44
Structural insulated panel, 6 inches of EPS foam	21.6	3.80
Concrete block 12-inch, insulated with EPS inserts into cores	3.9	0.69
Light-weight EPS-bead concrete block, with EPS inserts	14.7	2.5
Insulating concrete form	11.1	1.95
Autoclaved concrete block	8.6	1.51

1. This wide variability of R-values per inch or cm is due to density and installation
2. Effective R-value may be 1.2–2.6 times typical due to the thermal mass factor.

A-3 Household Moisture Sources

Moisture Source	Estimated Moisture Amount (pints or 1/2 liters)
Bathing: tub, excludes towels and spillage	0.12/standard size bath
shower, excludes towels and spillage	0.52/5-minute shower
Clothes washing: automatic, lid closed	0+/load (usually nil)
Clothes drying: vented outdoors	0+/load (usually nil)
not vented outdoors, or indoor line drying	4-6/load (more if gas dryer)
Combustion: unvented kerosene space heater	0.5 - 2.0/hour, or 7.6/per gallon
Cooking: three meals (four people)	2.10 (plus 2.84 if cooking over gas)
simmer at 203° F, 10 min., 6 in. pan	0.01 covered, 0.13 uncovered (more if gas)
boil 10 min., 6 in. pan	0.48 covered, 0.57 uncovered (more if gas)
Dish washing: three meals (four people)	1.05/day
Firewood storage indoors (1 cord of green wood)	400-800/6 months
Gas range pilot light (each)	0.37 or less/day
House plants (5 to 7 average plants)	0.86 to 0.96/day
Respiration and perspiration (four people)	0.44/hour average
Refrigerator defrost (per day)	1.03/day average
Saunas, steam baths, and whirlpools	0 to 2.7+/day
Evaporation (desorption) from materials:	6.3 to 16.9/day average
seasonal: new construction	10+/day average
Ground moisture	0 to 105/day (highly variable)

Adapted from: *Moisture Sources Associated with Potential Damage in Cold-Climate Housing*, University of Minnesota, 1988.

A-4 Heat Illnesses: Symptoms & Solutions

HEAT-RELATED ILLNESSES

WHAT TO LOOK FOR	WHAT TO DO
HEAT STROKE	
• High body temperature (103°F or higher) • Hot, red, dry, or damp skin • Fast, strong pulse • Headache • Dizziness • Nausea • Confusion • Losing consciousness (passing out)	• Call 911 right away-heat stroke is a medical emergency • Move the person to a cooler place • Help lower the person's temperature with cool cloths or a cool bath • Do not give the person anything to drink
HEAT EXHAUSTION	
• Heavy sweating • Cold, pale, and clammy skin • Fast, weak pulse • Nausea or vomiting • Muscle cramps • Tiredness or weakness • Dizziness • Headache • Fainting (passing out)	• Move to a cool place • Loosen your clothes • Put cool, wet cloths on your body or take a cool bath • Sip water **Get medical help right away if:** • You are throwing up • Your symptoms get worse • Your symptoms last longer than 1 hour
HEAT CRAMPS	
• Heavy sweating during intense exercise • Muscle pain or spasms	• Stop physical activity and move to a cool place • Drink water or a sports drink • Wait for cramps to go away before you do any more physical activity **Get medical help right away if:** • Cramps last longer than 1 hour • You're on a low-sodium diet • You have heart problems
SUNBURN	
• Painful, red, and warm skin • Blisters on the skin	• Stay out of the sun until your sunburn heals • Put cool cloths on sunburned areas or take a cool bath • Put moisturizing lotion on sunburned areas • Do not break blisters
HEAT RASH	
• Red clusters of small blisters that look like pimples on the skin (usually on the neck, chest, groin, or in elbow creases)	• Stay in a cool, dry place • Keep the rash dry • Use powder (like baby powder) to soothe the rash

CDC

A-5 Annual-Average-Temperature Map

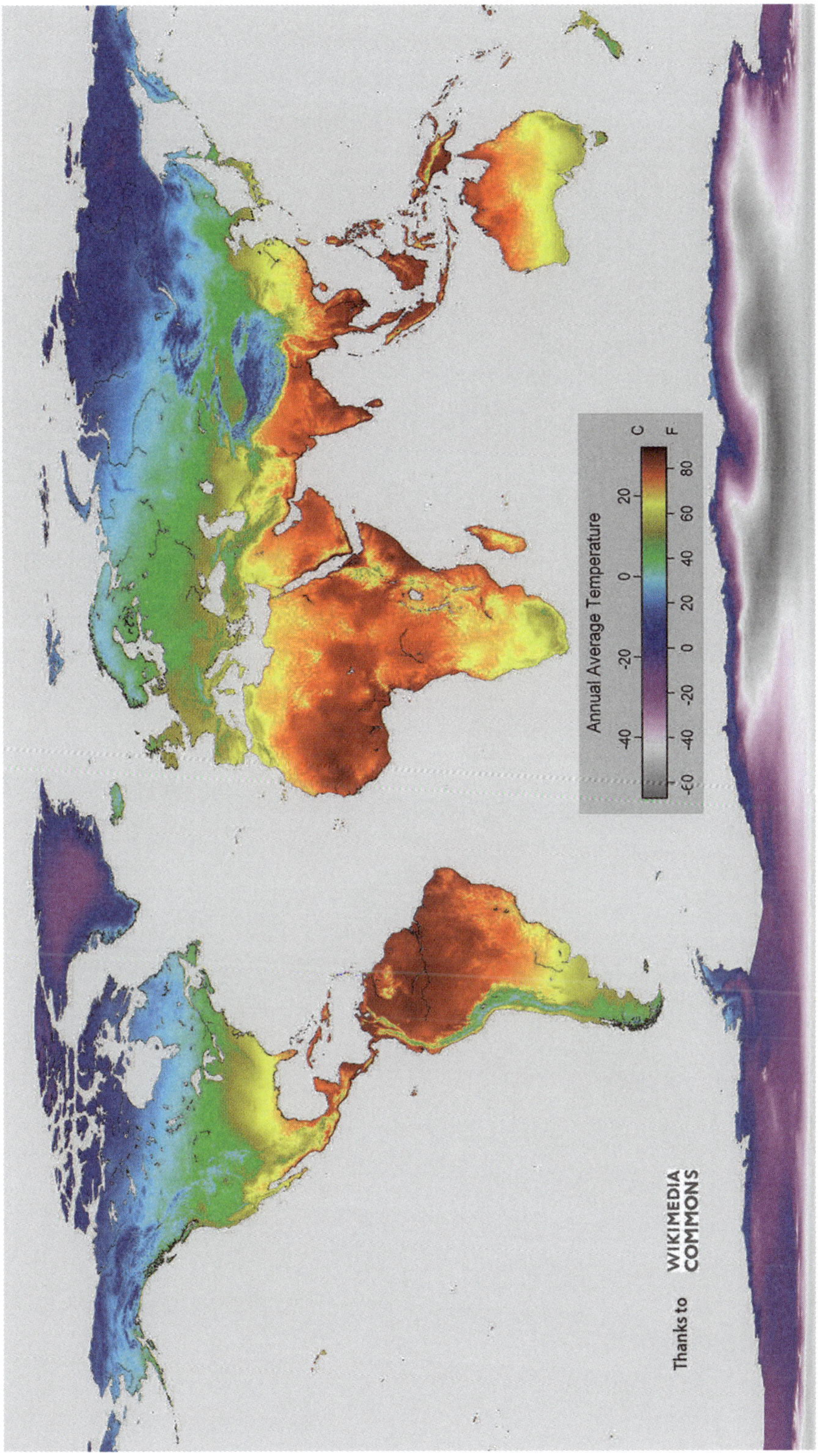

A-6 Yearly-Average-Relative-Humidity Map

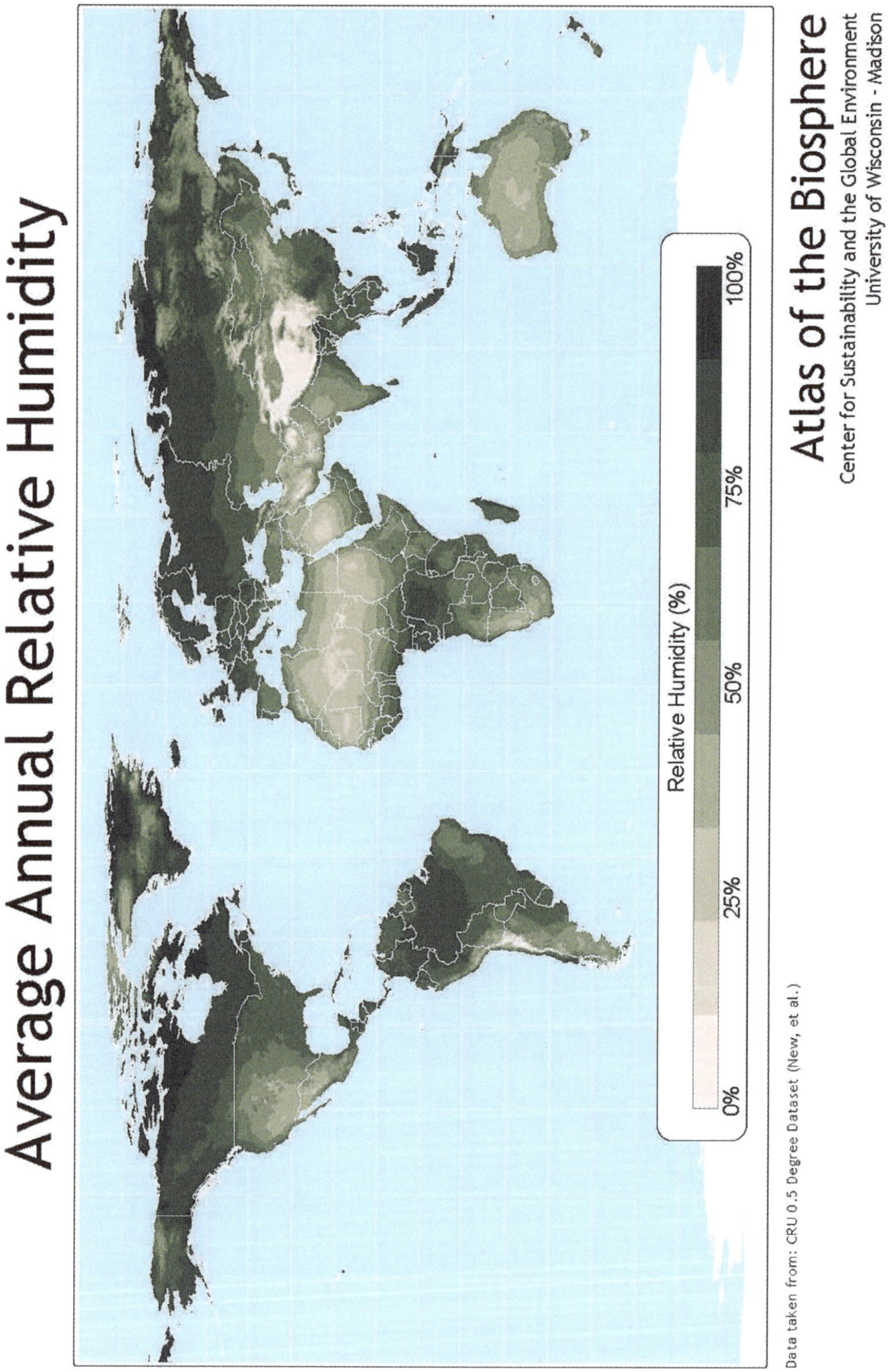

A-7 Solar-Irradiation Map

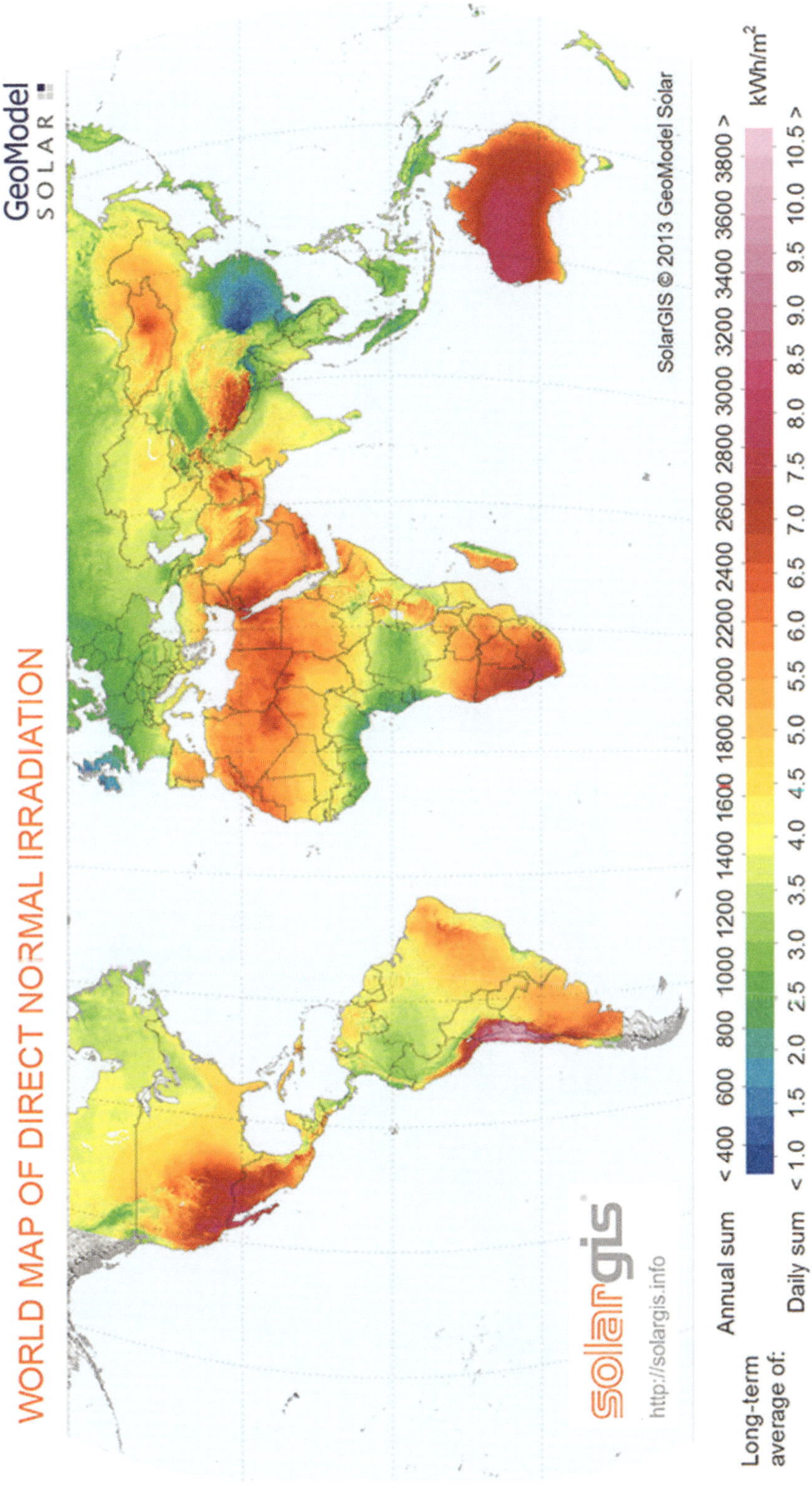

A-8 World Wet-Bulb-Globe-Temperature Map

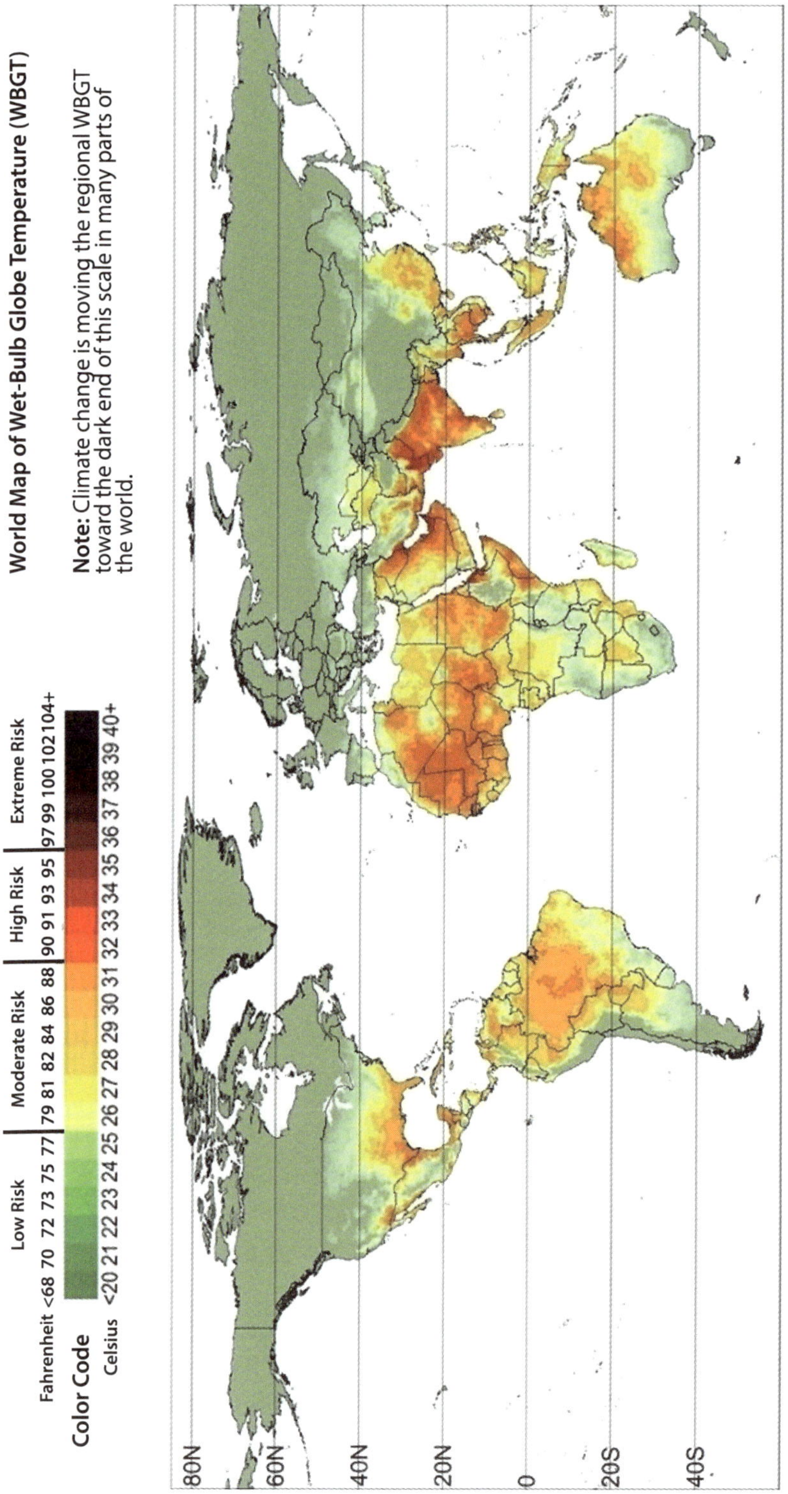

Cartographer Unknown

A-9 Estimating Wet-Bulb Globe Temperature

Wet Bulb Globe Temperature (WBGT) from Temperature and Relative Humidity

Temperature in Degrees Celsius

Relative Humidity (%)	20	22	24	26	28	30	32	34	36	38	40	42	44	46	48	50
0	14.8	16.1	18.0	18.6	19.8	21.1	22.3	23.5	24.7	25.8	27.0	28.1	29.3	30.3	31.4	32.5
5	15.3	16.7	18.7	19.4	20.7	22.0	23.3	24.6	25.9	27.2	28.4	29.6	30.9	32.2	33.4	34.6
10	16.0	17.4	19.4	20.2	21.6	23.0	24.3	25.7	27.1	28.4	29.7	31.1	32.4	33.8	35.1	36.4
15	16.5	18.0	20.1	20.9	22.4	23.8	25.2	26.7	28.1	29.6	31.0	32.4	33.8	35.2	36.7	38.1
20	17.1	18.7	20.8	21.6	23.1	24.6	26.2	27.7	29.2	30.6	32.1	33.6	35.1	36.6	38.2	39.7
25	17.6	19.3	21.4	22.3	24.0	25.5	27.0	28.6	30.1	31.7	33.2	34.8	36.3	37.9	39.5	
30	18.2	19.8	22.0	23.0	24.6	26.2	27.8	29.4	31.0	32.7	34.2	35.9	37.4	39.1		
35	18.7	20.3	22.6	23.6	25.3	26.9	28.6	30.2	31.9	33.5	35.2	36.8	38.5			
40	19.3	20.9	23.2	24.3	26.0	27.6	29.4	31.0	32.7	34.4	36.1	37.8	39.5			
45	19.7	21.5	23.8	24.9	26.6	28.3	30.1	31.8	33.5	35.2	37.0	38.7				
50	20.2	22.0	24.3	25.5	27.3	29.0	30.8	32.5	34.3	36.1	37.9	39.6				
55	20.7	22.4	24.8	26.0	27.8	29.6	31.4	33.3	35.0	36.8	38.6					
60	21.1	22.9	25.4	26.6	28.4	30.2	32.1	34.0	35.7	37.5	39.4					
65	21.6	23.2	25.9	27.1	29.0	30.9	32.7	34.5	36.4	38.2						
70	22.1	23.9	26.4	27.6	29.4	31.4	33.3	35.1	37.0	38.9						
75	22.5	24.4	26.9	28.2	30.1	32.0	33.8	35.8	37.7	39.5						
80	22.9	24.8	27.4	28.7	30.6	32.5	34.4	36.3	38.2							
85	23.3	25.2	27.8	29.2	31.1	33.0	35.0	36.9	38.9							
90	23.7	25.7	28.3	29.6	31.6	33.5	35.5	37.5	39.5							
95	24.2	26.1	28.7	30.1	32.0	34.0	36.0	38.0	40.0							
100	24.5	26.5	29.1	30.5	32.5	34.5	36.5	38.5								

NOTE: This chart is calculated using temperature and humidity, assuming a very clear sky (maximal solar load), and atmospheric pressure of 1ATA (760 mmHg). Chart A was developed by Professor Yoram Epstein to be used in Ariel's Checklist for hikers in Israel.

Wet Bulb Globe Temperature (WBGT) from Temperature and Relative Humidity

Temperature in Degrees Fahrenheit

Relative Humidity (%)	68.0	71.6	75.2	78.8	82.4	86.0	89.6	93.2	96.8	100.4	104.0	107.6	111.2	114.8	118.4	122.0
0	58.6	60.9	64.3	65.5	67.7	69.9	72.1	74.3	76.4	78.5	80.6	82.6	84.7	86.6	88.6	90.5
5	59.6	62.1	65.6	67.0	69.3	71.7	74.0	76.4	78.6	80.9	83.1	85.3	87.5	89.9	92.1	94.2
10	60.7	63.3	66.9	68.4	70.8	73.3	75.8	78.2	80.7	83.0	85.5	88.0	90.3	92.8	95.1	97.6
15	61.7	64.5	68.1	69.6	72.2	74.8	77.4	80.0	82.6	85.2	87.8	90.2	92.8	95.4	98.0	
20	62.7	65.6	69.4	70.9	73.6	76.3	79.2	81.8	84.5	87.1	89.8	92.5	95.2	97.8		
25	63.8	66.7	70.5	72.2	75.1	77.8	80.6	83.4	86.2	89.0	91.8	94.6	97.4			
30	64.8	67.6	71.7	73.4	76.3	79.2	82.1	84.9	87.8	90.8	93.6	96.6	99.4			
35	65.6	68.6	72.7	74.6	77.5	80.5	83.5	86.4	89.4	92.4	95.3	98.3				
40	66.7	69.6	73.8	75.7	78.8	81.8	84.8	87.8	90.9	94.0	97.0					
45	67.5	70.6	74.8	76.8	79.9	83.0	86.1	89.2	92.3	95.4	98.6					
50	68.4	71.5	75.8	77.8	81.1	84.1	87.4	90.5	93.7	96.9						
55	69.3	72.4	76.7	78.8	82.1	85.3	88.5	91.9	95.1	98.3						
60	70.1	73.3	77.7	79.8	83.2	86.4	89.8	93.1	96.3	99.6						
65	70.9	73.8	78.6	80.9	84.2	87.5	90.8	94.1	97.5							
70	71.7	75.0	79.5	81.7	84.9	88.6	91.9	95.3	98.6							
75	72.4	75.9	80.3	82.7	86.1	89.6	92.9	96.4								
80	73.2	76.7	81.2	83.6	87.1	90.4	93.9	97.4								
85	74.0	77.4	82.0	84.5	88.0	91.5	94.9	98.5								
90	74.7	78.2	82.9	85.3	88.9	92.3	95.9	99.4								
95	75.5	78.9	83.6	86.1	89.6	93.2	96.8									
100	76.1	79.7	84.4	86.9	90.5	94.1	97.7									

Many thanks to Professor Yoram Epstein & Ariel's List for this valuable information
Visit: arielschecklist.com/

NOTE: This chart is calculated using temperature and humidity, assuming a very clear sky (maximal solar load), and atmospheric pressure of 1ATA (760 mmHg). Chart A was developed by Professor Yoram Epstein to be used in Ariel's Checklist for hikers in Israel.

A-10 Advice for Coping with WBGT Conditions

Heat Stress Category (WBGT)	Moderate Work		Hard Work	
	Work/Rest Cycle	Water Intake Per Hour	Work/Rest Cycle	Water Intake Per Hour
White ≤76.9°F (≤24.9°C)	60/15 MINUTES	300 ml (1/3 qt)	40/20 MINUTES	500 ml (1/2 qt)
Green 77-81.9°F (25-27.7°C)	60/15 MINUTES	750 ml (3/4 qt)	40/20 MINUTES	1000 ml (1 qt)
Yellow 82-84.9°F (27.8-29.4°C)	40/20 MINUTES	1000 ml (1 qt)	30/30 MINUTES	1000 ml (1 qt)
Red 85-88.9°F (29.5-31.6°C)	30/30 MINUTES	1000 ml (1 qt)	**Exercise is forbidden. Very high risk for heat casualties.**	
Black ≥89°F (≥31.7°C)	**Exercise is forbidden. Very high risk for heat casualties.** Developed by Professor Yoram Epstein for Ariel's Checklist for hikers in Israel.			

Climate Categories

Kestrel Instruments

Cat 3	Cat 2	Cat 1	Activity Guidelines
< 82.0°F <27.8°C	< 79.7°F <26.5°C	< 76.1°F <24.5°C	Normal Activities – Provide at least three separate rest breaks each hour with a minimum duration of 3 min each during the workout.
82.2 - 86.9°F 27.9-30.5°C	79.9 - 84.6°F 26.6-29.2°C	76.3 - 81.0°F 24.6-27.2°C	Use discretion for intense or prolonged exercise; Provide at least three separate rest breaks each hour with a minimum duration of 4 min each.
87.1 - 90.0°F 30.6-32.2°C	84.7 - 87.6°F 29.3-30.9°C	81.1 - 84.0°F 27.3-28.9°C	Maximum practice time is 2 h. For Football: players are restricted to helmet, shoulder pads, and shorts during practice. If the WBGT rises to this level during practice, players may continue to work out wearing football pants without changing to shorts. For All Sports: Provide at least four separate rest breaks each hour with a minimum duration of 4 min each.
90.1 - 91.9°F 32.2-33.3°C	87.8 - 89.6°F 31.0-32.0°C	84.2 - 86.0°F 29.0-30.0°C	Maximum practice time is 1 h. For Football: No protective equipment may be worn during practice, and there may be no conditioning activities. For All Sports: There must be 20 min of rest breaks distributed throughout the hour of practice.
≥ 92.1°F ≥ 33.4°C	≥ 89.8°F ≥32.1°C	≥ 86.2°F ≥30.1°C	No outdoor workouts. Delay practice until a cooler WBGT is reached. Thanks to: Kestrel Instruments

A-11 Solar Radiative Properties of Materials

Description/Composition	a *	e *	t *
Aluminum			
polished	0.09	0.03	
anodized	0.14	0.84	
foil	0.15	0.05	
Asbestos, felt	0.25	0.50	
Asphalt pavement		0.85-0.93	
Brick, red (Purdue)	0.63	0.93	
Concrete	0.60	0.88	
Concrete, rough	0.60	0.97	
Galvanized sheet metal			
clean, new	0.65	0.13	
Oxidized, weathered	0.80	0.28	
Glass, 1/8 inch thick			
float or tempered			0.79
low iron oxide type			0.88
Gypsum or plaster board		0.90-0.92	
Ice, with sparse snow cover	0.31		
Marble, polished	0.5-0.6	0.9	
Mylar, 3/64 inch thick			0.87
Paints			
black (Parsons)	0.98	0.98	
green, oil base	0.50	0.90	
grey	0.75	0.95	
red, oil base	0.74	0.90	
white, acrylic	0.26	0.90	
white, on aluminum	0.20	0.91	
white, zinc oxide	0.16	0.93	
Plexiglas, 1/8 inch thick			0.90
Snow, fine particles, fresh	0.13	0.82	
Snow, ice granules	0.33	0.89	
Water		0.94	0.95-0.96
Wood		0.82-0.92	
White plaster	0.07	0.91	

* Alpha (a) -absorptance; epsilon (e) -emittance; tau (t) -transmittance

A-12 Preventing Fire and Flood Damage

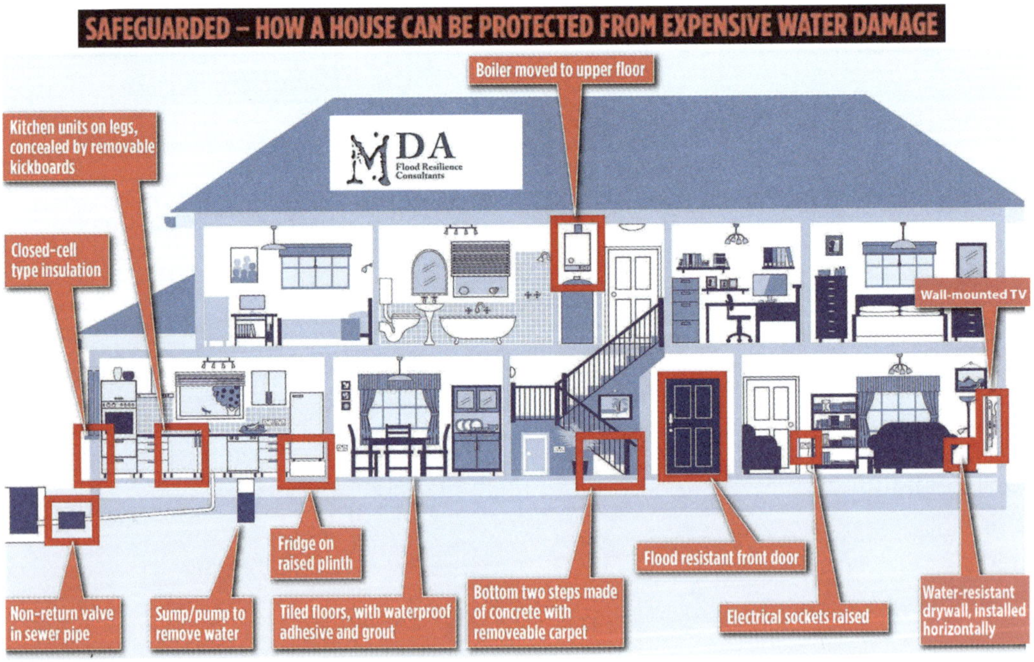

A-13 Evaporative Cooler Air Temperatures

IP Supply-Air Temperature from Outdoor RH% and Outdoor Dry-Bulb Temperature

Outdoor Dry-Bulb Temp. °F \ Outdoor Relative Humidity %	2	5	10	15	20	25	30	35	40	45	50	55	60	65	70	75	80
75	54	55	57	58	59	61	62	63	64	65	66	67	68	69	70	71	72
80	57	58	60	62	63	64	66	67	68	69	71	72	73	74	75	76	77
85	61	62	63	65	67	68	70	71	72	73	74	75	76	77	79	61	
90	64	65	67	69	70	72	74	76	77	78	79	81	82	83	84	86	
95	67	68	70	72	74	76	78	79	81	82	84	85	87				
100	69	71	73	76	78	80	82	83	85	87	88						
105	72	74	77	79	81	84	86	88	89								
110	75	77	80	83	85	87	90	92									
115	78	80	83	86	89	91	94										
120	81	83	86	90	93	95											
125	83	86	90	93	96												

Supply Air Temperature

Metric Supply-Air Temperature from Outdoor RH% and Outdoor Dry-Bulb Temperature

Outdoor Temp. °C \ Outdoor Relative Humidity %	5	10	15	20	25	30	35	40	45	50	55	60	65	70	75	80	85
25	12.7	13.6	14.4	15.2	15.9	16.6	17.3	18	18.7	19.4	20	20.6	21.2	21.8	22.4	22.9	23.4
27	14	14.9	15.8	16.6	17.4	18.2	18.9	19.7	20.4	21.1	21.7	22.4	23	23.6	24.2	24.8	25.4
29	15.2	16.2	17.1	18	18.9	19.7	20.5	21.3	22.1	22.8	23.5	24.2	24.9	25.5	26.1	26.7	27.3
31	16.4	17.5	18.5	19.5	20.4	21.2	22.1	22.9	23.7	24.5	25.3	26	26.7	27.3	28	28.6	29.2
33	17.6	18.8	19.8	20.9	21.8	22.8	23.7	24.6	25.4	26.2	27	27.8	28.5	29.2	29.9	30.5	31.2
35	18.8	20	21.2	22.3	23.3	24.3	25.3	26.2	27.1	28	28.8	29.6	30.3	31.1	31.8	32.5	33.1
37	19.9	21.3	22.5	23.8	24.8	25.9	26.9	27.9	28.8	29.7	30.5	31.4	32.2	32.9	33.7	34.4	35
39	21.1	22.5	23.9	25.2	26.3	27.4	28.5	29.5	30.5	31.4	32.3	33.2	34	34.8	35.6	36.3	37
41	22.2	23.8	25.2	26.5	27.7	28.9	30.1	31.2	32.2	33.2	34.1	35	35.8	36.7	37.4	38.2	38.9

Evaporative Cooler Approximate Air Changes per Hour (ACH) by Climate

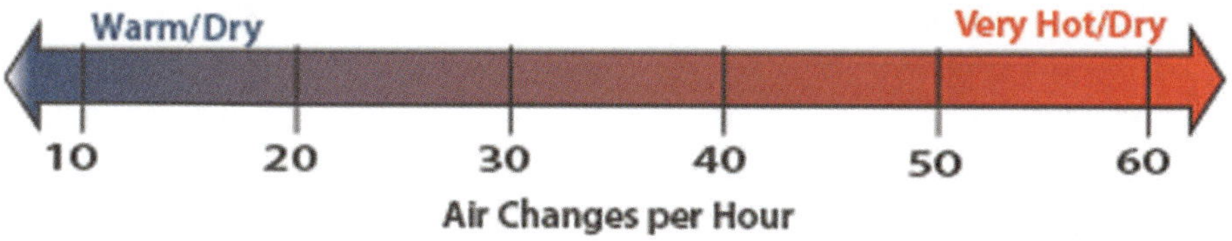

A-14 Inch-Pound — Metric Conversions

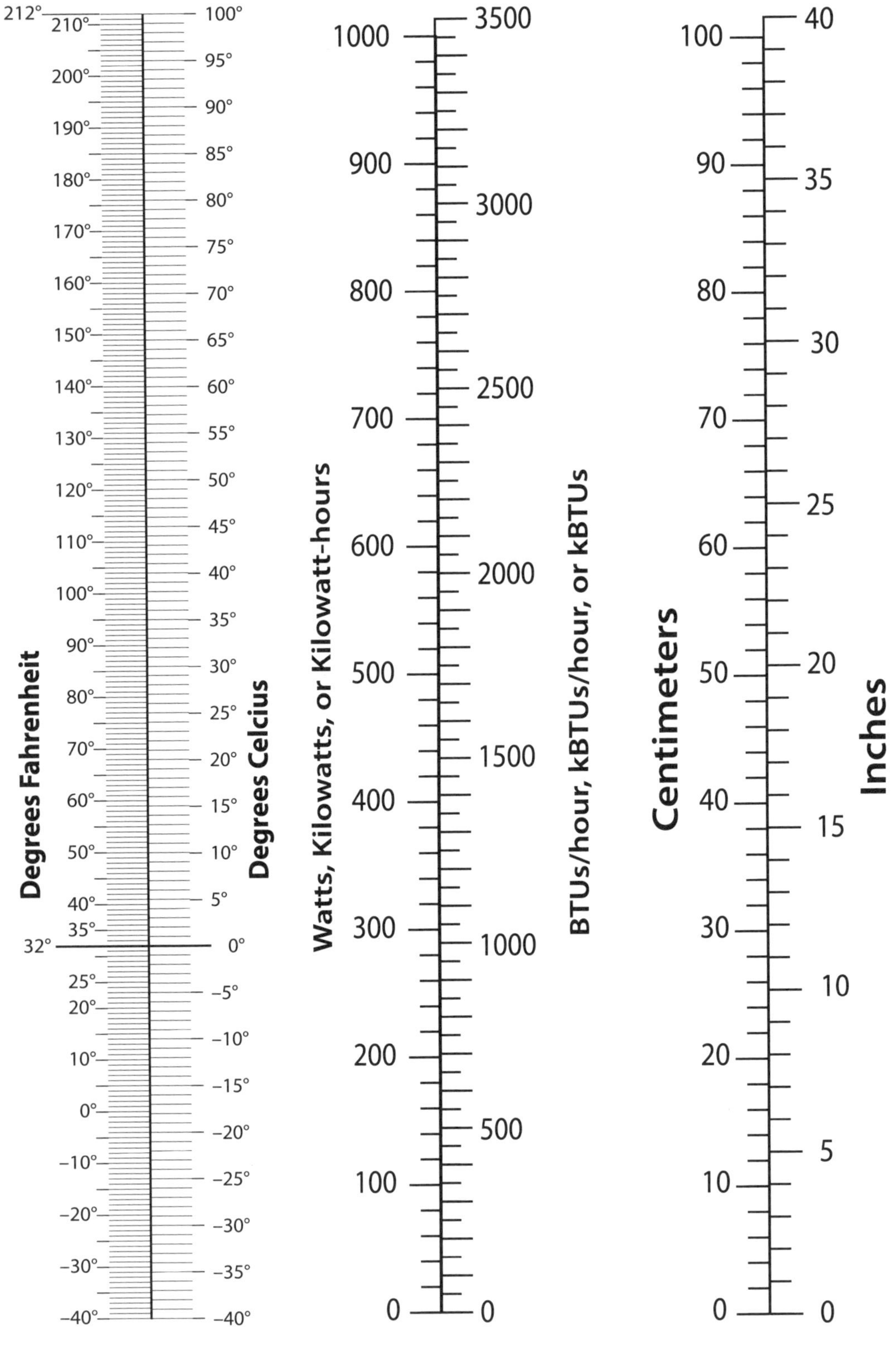

A

B

C

D

E

F

G

H

I

J

K

L

M

N

O

P

R

S

W

Z